Sustainable Development:
Science, Ethics, and Public Policy

Environmental Science and Technology Library

VOLUME 3

The titles published in this series are listed at the end of this volume.

Sustainable Development: Science, Ethics, and Public Policy

Edited by

John Lemons

Department of Life Sciences,
University of New England,
Biddeford, ME, U.S.A.

and

Donald A. Brown

Bureau of Hazardous Sites and Superfund Enforcement,
Pennsylvania Department of Environmental Resources,
Harrisburg, PA, U.S.A.

KLUWER ACADEMIC PUBLISHERS
DORDRECHT / BOSTON / LONDON

Library of Congress Cataloging-in-Publication Data

Sustainable development : science, ethics, and public policy / edited by John Lemons and Donald A. Brown.
p. cm. -- (Environmental science and technology library)
Includes bibliographical references and index.

1. Sustainable development. 2. Environmental protection--Decision making. I. Lemons, John. II. Brown, Donald A. III. Series: Environmental science and technology (Dordrecht, Netherlands)
HC79.E5S868 1995
338.9--dc20 95-10769

ISBN 978-90-481-4559-1

Published by Kluwer Academic Publishers,
P.O. Box 17, 3300 AA Dordrecht, The Netherlands.

Kluwer Academic Publishers incorporates
the publishing programmes of
D. Reidel, Martinus Nijhoff, Dr W. Junk and MTP Press.

Sold and distributed in the U.S.A. and Canada
by Kluwer Academic Publishers,
101 Philip Drive, Norwell, MA 02061, U.S.A.

In all other countries, sold and distributed
by Kluwer Academic Publishers Group,
P.O. Box 322, 3300 AH Dordrecht, The Netherlands.

Printed in the Netherlands

To Linda for her continued support - D.A.B.

To Sage for the hope of the future,
and to John and Doris for the inspiration of the past - J.L.

ACKNOWLEDGMENTS

This book evolved out of a conference held at the United Nations in New York in January of 1994. The conference was entitled "The Ethical Dimensions of the United Nations Program on Environment and Development, Agenda 21." The conference was organized by the Earth Ethics Research Group- Northeast Chapter with the cooperation of the United Nations Environmental Programme. The book and conference were supported by grants from the World Bank, the United States Environmental Protection Agency, the Commonwealth of Pennsylvania Department of Environmental Resources, the National Association of Environmental Professionals, and the Hastings Center.

We also extend our thanks to Dr. Noel Brown of the United Nations Environment Programme, George Bortnyk, President of Earth Ethics Research Group, Inc., Mr. Brown's staff at the Pennsylvania Department of Environmental Resources, Brenda Smith of the University of New England for general typing and other assistance, and Kathy Sammis for copyediting.

Finally, we extend our thanks to Cheryl Miller for her editorial and technical assistance in preparing the chapter manuscripts, and for her handling of all of the numerous details that exceeded our abilities. We also appreciate her for performing her work with good humor and understanding despite our constant demands.

ABOUT THE EDITORS

Donald A. Brown is director of the Bureau of Hazardous Sites and Superfund Enforcement in the Office of Chief Counsel for the Pennsylvania Department of Environmental Resources. He is interested in and has written and lectured extensively on the interface between environmental science, law, economics, and environmental ethics.

Mr. Brown represented Pennsylvania at the Earth Summit and was recently director of a conference held at the United Nations as a follow up to the Earth Summit on the ethical dimensions of the United Nations program on environment and development. He formerly served as Chief of the Central Office of the Bureau of Litigation and Assistant Attorney General with the Pennsylvania Department of Environmental Resources. Before that he served as a lawyer with the New Jersey Department of Environmental Resources where his last position was director of the Office of Regulation and Enforcement.

John Lemons is a professor of biology and environmental science in the Department of Life Sciences at the University of New England, Biddeford, Maine. He also is a former Editor-in-Chief of *The Environmental Professional*, the official journal of the National Association of Environmental Professionals. Dr. Lemons has published extensively on problems of nuclear waste, biodiversity, national park management, climate change, and environmental ethics.

CONTENTS

Chapter 1
INTRODUCTION

Donald A. Brown[1]
John Lemons[2]

1. The International Acceptance of the Concept of Sustainable Development

This book examines the role of science, economics, law, and ethics in implementing sustainable development programs. In a relatively short time, many international leaders throughout the world have accepted the concept of sustainable development as a way of reconciling potential conflicts between environmental protection and human development goals. What is meant by "sustainable development," and what factors explain this rapid acceptance of the idea of sustainable development in international affairs?

As early as the United Nations Conference on the Human Environment held in Stockholm in 1972, some members of the international community believed that progress toward protection of the environment was linked to progress in elimination of poverty throughout the world. In 1980 the International Union for the Conservation of Nature adopted a World Conservation Strategy that called for sustainable use of species and ecosystems. Although the concept of sustainable development had been used in some international circles for at least 15 years, a report prepared for the United Nations by the World Commission on Environment and Development (WCED 1987) in 1987 pushed the concept of sustainable development to center stage in international affairs. This report, entitled *Our Common Future,* received international attention because it concluded that rapid deterioration of the global environment was threatening life on earth and that decisive political action was needed to ensure human survival. *Our Common Future* identified several environmental trends that threaten to radically alter the planet, and many species upon it, including the human species. Environmental deterioration identified in the report included: (1)rapid loss of productive dryland that was being transformed into desert, (2)rapid loss of forests, (3)global warming caused by increases in greenhouse gases, (4)loss of the atmosphere's protective ozone shield due to industrial gases, and (5)the pollution of surface water and groundwater.

The scientific evidence of growing environmental degradation relied upon in *Our Common Future* was of even greater concern because the earth's environment was exhibiting stresses at a current population of approximately 5.5 billion people. These visible signs of deterioration became even more ominous when one considered the rapid growth in population expected for our planet in the 21st century. Because population may grow to 10

[1]Bureau of Hazardous Sites and Superfund Enforcement, Pennsylvania Department of Environmental Resources, Commonwealth of Pennsylvania, 400 Market St., Harrisburg, PA 17101-2301, U.S.A.; [2]Department of Life Sciences, University of New England, Biddeford, ME 041005, U.S.A.

J. Lemons and D. A. Brown (eds.), Sustainable Development: Science, Ethics, and Public Policy, 1–10.

billion by 2050 and between 12 and 14 billion people by the end of the next century, *Our Common Future* concluded that urgent and decisive political action was necessary to prevent widespread environmental destruction. For the first time in history, humanity must face the risk of unintentionally destroying life on earth.

Until very recently, the problems of environmental degradation and poverty were viewed as unrelated. Of equal historical significance as its environmental conclusions, *Our Common Future* also focused world attention on the futility of separating economic development problems from environmental issues. The report explained how some forms of development eroded the environmental resources upon which they must be based, and how environmental degradation undermines economic development. For instance, development that can't afford to pay for treatment of sewage creates water pollution, and polluted water limits future development options. In addition, in many developing countries, in the absence of help from the developed world, rapid depletion of natural resources is the only hope of eradicating poverty. Thus, the report concluded that "poverty is a major cause and effect of global environmental problems." That is, there is no hope of solving the global environmental problems unless the international community works rapidly to resolve problems of human development throughout the world. Thus, *Our Common Future* forced the international community for the first time in human history to see problems of poverty, population growth, industrial and social development, depletion of natural resources, and destruction of the environment as closely interrelated.

To solve the twin problems of environmental degradation and development, *Our Common Future* called for a political transformation that supported sustainable development throughout the world. Sustainable development was defined as development that meets the needs of the present without compromising the ability of future generations to meet their needs. *Our Common Future,* because of its identification of the interrelationship between environmental destruction and poverty, put sustainable development on the front burner throughout the world.

In December 1989, the General Assembly of the United Nations, in reaction to the problems identified by *Our Common Future,* called for an unprecedented international meeting of all the nations of the earth. The United Nations Conference on Environment and Development, generally known as the Earth Summit, was held in Rio de Janeiro in June of 1992 in response to *Our Common Future*. The Earth Summit was the largest and most ambitious international conference in history as measured by the number of issues under consideration and the size and number of international delegations. About 110 heads of state assembled at the Earth Summit, more than at any other previous international conference.

Five documents were signed in Rio de Janeiro that will be implemented in the years ahead and that will keep sustainable development in the center of international affairs (see, e.g., Johnson 1993). They were: (1)the treaty on climate change, (2)the treaty on biodiversity, (3)the convention on forest principles, (4)the Rio Declaration, and (5)Agenda 21.

2. The Rio de Janeiro Documents

2.1. THE CLIMATE CONVENTION

The United Nations Framework Convention on Climate Change requires that signatory states reduce greenhouse gases to "earlier levels" by the year 2000. Although the convention does not require states to hold greenhouse gases to a specified level, it requires states to issue reports detailing their actions to mitigate climate change for review by the Conference of Parties created by the convention. The target of reducing carbon dioxide emissions to 1990 levels by the end of the decade is stated as a goal rather than as a binding provision.

2.2. THE BIODIVERSITY CONVENTION

The Convention on Biological Diversity requires that states develop national strategies for the conservation and sustainable use of biological diversity and for inventories of species to be preserved. Access to the genetic resources of a state is subject to the prior consent of the country providing the resources and on mutually agreed terms.

2.3. THE FOREST PRINCIPLES

The Authoritative Statement of Forest Principles is a nonbinding agreement that urges states to develop forests according to their socioeconomic needs on the basis of national policies for sustainable development. The principles also encourage states to make efforts to promote reforestation and forest conservation. Because of strong opposition from developing countries, the principles contain few provisions that actually limit sovereign rights to exploit forests. Nevertheless, the signatory nations agreed to keep the principles under assessment for their adequacy with regard to further international cooperation on forest issues

2.4. THE RIO DECLARATION

The Rio Declaration on Environment and Development is a nonbinding set of 27 principles on sustainable development. Although the Rio Declaration is not binding on signatory nations, the principles are understood to be a description of norms that should guide national behavior in the future. These principles could have far-reaching political consequences in the years ahead because they include agreement on sustainable development concepts that are without historical precedent. Some of the more significant principles include the following norms: (1)nations should not cause damage to the environment of other states and areas beyond their borders; (2)eradicating poverty and reducing disparities in worldwide standards of living are indispensable requirements for sustainable development; (3)the polluter in principle should pay the cost of pollution; (4)states should discourage or prevent transboundary movements of activities and substances that endanger health or the environment; and (5)scientific uncertainty should not be a reason for postponing urgent measures to prevent environmental degradation. (This principle is generally referred to as the precautionary principle, a doctrine that will be referred to several times throughout this book.)

2.5. AGENDA 21

Although it did not receive as much publicity in the United States and some parts of the world as the treaties on climate change and biodiversity, Agenda 21 may prove to be the most significant of all the Earth Summit agreements. This document is a blueprint for international action in the 21st century. It contains 40 chapters focused on solving the twin problems of environmental protection and sustainable development. Each of the 40 chapters includes a statement of objectives, an outline of required activities, guidelines for developing a framework of action, necessary institutional changes, and identification of the needs of implementation, including indications of necessary research and a financial and cost analysis.

Agenda 21 is the international community's response to the issues raised by *Our Common Future*. It calls for the governments not only to adopt new environmental programs but also to commit to significant economic, social, and international institutional reforms. Agenda 21 is arguably the most important of the Rio documents from the standpoint of its potential to change unsustainable behavior, because it is the first international agreement that creates expectations for nations to integrate environmental, economic, and social planning.

The purpose of Agenda 21 is to transform human life on earth so as to make it harmonious with nonhuman life and environmental constraints. It is premised on the notion that sustainable development is not an option but is an urgent requirement. The preamble to Agenda 21 demonstrates the strong sense of urgency that motivated its authors:

> Humanity stands at a defining moment in history. We are confronted with a perpetuation of disparities between and within nations, worsening poverty, hunger, ill health and illiteracy, and the continuing deterioration of ecosystems on which we depend for our well-being.

Underlying Agenda 21 is the notion that the human community can either: (1)continue present policies, which both increase poverty and disparities between rich and poor and destroy ecosystems; or (2)change course. To change course, the governments of the world must integrate environmental, economic, and social programs in a new historically unprecedented global partnership between the developed and developing worlds. Because strong and systematic national and community support will be needed to move the world toward sustainable development, the world community, according to Agenda 21, is urgently challenged to develop an ethic that will recognize the duties that people have to care for not only other humans but also future generations and other forms of life with which we share this planet.

The international order already may have been transformed by Earth Summit developments, because Agenda 21 has added to the list of recognized universal rights two new ones: (1)the right to an equitable international order and (2)the right to an environment with health and dignity. Many commentators agree that Agenda 21 is one of the historically most important international agreements because it is the first international agreement that attacks, in an integrated manner, the twin problems of environment and development. Yet many also assert that Agenda 21 is seriously flawed. Common criticisms of Agenda 21 include: (1)failure of Agenda 21 to provide necessary financial commitments of the developed to the developing world, (2)failure to suggest major changes in worldwide energy-use patterns, (3)failure to deal with unsustainable forest practices, and (4)failure to make population control a central feature of sustainable development.

A common criticism of Agenda 21's philosophical underpinnings is a failure to make preservation of the natural environment valuable in and of itself, that is, for reasons that transcend human purposes or goals. Rather than exhibiting a new and transformed respect for nature, Agenda 21 is a compromise among the rich and poor nations, industrial workers, private individuals, and indigenous people. As a result, Agenda 21 follows a narrow anthropocentric approach to the twin problems of environment and development so that only human beings are of ultimate concern. That is, there is little evidence in Agenda 21 that humans owe moral duties to the natural environment, to animals, plants, and ecosystems, and that these things may possess a value of their own independent of their usefulness to humans.

Agenda 21 also can be criticized because of its failure to give a coherent definition to the concept of sustainable development. One general meaning might be "the continued satisfaction of basic human needs such as food, water, and shelter as well as higher-level social and cultural necessities such as security, freedom, education, employment, and recreation." Another might be "the continued productivity and functioning of ecosystems." *Our Common Future* defined sustainable development as that which "meets the needs of the present without compromising the ability of future generations to meet their own needs." Gowdy (1994) discusses three meanings of sustainability: (1)sustaining intergenerational economic welfare, (2)maximizing the time of existence of the human species, and (3)sustaining nature and its diversity. Several organizations—such as the World Conservation Strategy, the World Resources Institute, the International Institute for Environment and

Development, the Ecological Society of America, and the World Bank—regard the term "sustainability" as acceptable but ill-defined (Shearman 1990). Regardless of the precise meaning, it is clear that the term has implications for ecological, social, and economic systems.

Although the meaning of the concept of sustainable development can be generally understood to be consistent with its definition in *Our Common Future,* this definition is so vague that it is not helpful in providing a specific rule of action when there are conflicts among environmental, economic, or social goals. Does, for instance, the concept of sustainable development require humans to preserve all species of animals and plants even if an animal or plant is without a known use value to present or future generations of people? What does the concept of sustainable development require of nations in regard to conservation of nonrenewable resources? What are we trying to sustain under the concept of sustainable development—animals, plants, species, ecosystems, people, jobs, cultures, communities, ways of life? A prescriptive rule about these and many other issues cannot be derived from the concept of sustainable development or its definition adopted by *Our Common Future*.

However, the concept of sustainable development can be understood to have been interpreted by provisions of Agenda 21 and the principles contained in the Rio Declaration. Under the assumption that the Rio Declaration and Agenda 21 are interpretations of sustainable development, some definitional substance to the concept of sustainable development is contained in specific sections of Agenda 21 and the Rio Declaration. For instance, the precautionary principle in the Rio Declaration and obligations in Agenda 21 that require nations to move away from polluting transportation systems are examples of specific obligations that flow from the concept of sustainable development that have been identified in the Rio documents. However, Agenda 21 and the Rio Declaration leave most questions about the meaning of sustainable development unresolved. For these unresolved issues, the concept of sustainable development's most important practical use is not in its ability to prescribe specific action but as an invitation to citizens and governments to begin a process that will fill in the missing details in specific programs.

Despite the many unresolved issues raised by Agenda 21, most analysts view it as an ambitious and significant attempt to develop principles to guide future action (see, e.g., Brown 1994). Unlike the treaties on climate change and biodiversity, Agenda 21 is not binding on signatory nations but is to serve as a set of normative principles that will guide nations in developing specific laws and programs. Agenda 21 is generally referred to as soft law, meaning that it is not enforceable in an international forum. Soft law operates as a set of international expectations about future actions. Even though the specifics of Agenda 21 are not legally binding, nations in signing it agreed to adopt their own strategies toward sustainable development and report to United Nations on a periodic basis progress made in implementing sustainable development programs. Agenda 21 also urges governments at the subnational level—that is, at the state, provincial, county, and municipal levels—to develop local Agenda 21 implementation plans by 1996.

2.6. OTHER INTERNATIONAL AGREEMENTS CONCERNED WITH SUSTAINABLE DEVELOPMENT

In addition to the Rio documents, many other regional and international agreements will have an effect on the rate and degree to which the international community moves toward sustainable development. Some of the most important international agreements include the following:

1. The Convention on the Prevention of Marine Pollution by Dumping of Waste and Other Matter, generally known as the 1972 London Dumping Convention. This

treaty deals with wastes from ships, aircraft, barges, and other non-land-based sources of pollution.

2. The U.N. Convention on the Law of the Sea. This is a comprehensive 1982 recodification of international law of the sea.
3. The 1985 Vienna Convention for the Protection of the Ozone Layer. This agreement originally only required scientific studies, but when the "hole" in the ozone layer was reported in 1986, the signatory nations adopted in 1987 the Montreal Protocol requiring a 50 percent reduction in ozone-depleting gases. As the seriousness of the threat to disruption of the ozone layer became more apparent, the government parties in 1990 meeting in London agreed to a near total ban on ozone-attacking chemicals.
4. The 1989 Basel Convention on the Control of Transboundary Movements of Hazardous Wastes and Their Disposal. This agreement sets up a manifest system to control the transboundary migration of hazardous waste.
5. The Program of Action on World Population agreed to at the United Nations Conference on Population and Environment at Cairo in September of 1994. This document contains 16 chapters that represent a strategy for stabilizing the world's population. The major objective of this plan is to achieve and maintain a harmonious balance among population, resources, food supplies, the environment, and development.
6. Agreements reached at the United Nations World Summit for Social Development in Copenhagen in March of 1995. The intent of this conference was to find new ways to integrate social development goals into more traditional economic development planning. Since Agenda 21 calls for the integration of economic, environmental, and social planning, and because little attention has been paid to social planning, the Copenhagen Summit is expected to give significant content to the notion of social planning in the context of sustainable development.

2.7. THE U.N. COMMISSION ON SUSTAINABLE DEVELOPMENT

In December 1992, the United Nations created the United Nations Commission on Sustainable Development (CSD) to monitor worldwide progress of implementation of Agenda 21 and other Rio documents. The CSD is comprised of the representatives of 53 United Nations member nations elected by the Economic and Social Council of the United Nations for three-year terms; 13 members are elected from Africa, 11 from Asia, 10 from Latin America and the Caribbean, 6 from Eastern Europe, and 13 from Western Europe and other nations.

Among other duties, the CSD is charged with the responsibility of reviewing national reports submitted to the United Nations regarding each nation's activities undertaken to implement Agenda 21. Governments are asked to submit voluntary, annual national reports to the CSD for its review. In some countries, nongovernment organizations (NGOs) have prepared these reports for the national governments. CSD is also expressly charged with reviewing national commitments on financial targets set by Agenda 21 and in particular the Agenda 21 goal that each developed nation should provide 0.7 percent of gross national product for development assistance to the developing world.

The CSD meets once a year for a period of two to three weeks at the United Nations headquarters. The CSD met for the first time in June 1993 and established a program of work that gave guidance to member nations as to how they should develop progress reports on Agenda 21 implementation. At this meeting, CSD divided the 40 Agenda 21 chapters into thematic clusters and scheduled the submission of national reports to coincide with cluster arrangements. The thematic clusters to be reviewed by CSD are as follows: (1)critical elements of sustainability (Agenda 21 Chapters 2-5); (2)financial resources and mechanisms

(Chapter 33); (3)education, science, and technology (Chapters 16 and 34-36); (4)decisionmaking structures (Chapters 8 and 37-40); (5)roles of major groups (Chapters 23-32); (6)health and human settlements (Chapters 6, 7, and 21); (7)land, forests, and biodiversity (Chapters 1-15), (8)atmosphere, oceans, and fresh water (Chapters 9, 17, and 18); and (9)toxic chemicals and hazardous wastes (Chapters 19, 20, and 22).

Because the first five clusters are cross-sectoral, they will be reviewed annually by CSD. Other clusters are reviewed on a periodic basis so that each cluster will be reviewed at least every three years. Matters relating to health, human settlements, and fresh water, as well as toxic chemicals and hazardous wastes, were considered in 1994. The 1995 schedule called for review of land, desertification, forests, and biodiversity chapters. Chapters on atmosphere, oceans, and all kinds of seas are scheduled for 1996. In 1997, the CSD will undertake an overall review of the progress achieved in the implementation of Agenda 21 in order to prepare recommendations to a special session of the General Assembly on whether to amend Agenda 21. Agenda 21 is expected to be a document that will continue to guide international action into the next century.

3. Other Sustainable Development Activities

3.1. NATIONAL SUSTAINABLE DEVELOPMENT PROGRAMS

Several nations have initiated significant national sustainable development programs. Included among the more active nations in Agenda 21 implementation efforts are the Netherlands, Canada, and the United States.

The Netherlands National Environmental Policy Plan (NEPP) is a comprehensive strategy for sustainable development that examines the economic and social concerns of maintaining a healthy environment. The NEPP looks at not only specific sources of pollution but also their relationship to relevant ecological, social, and economic systems. The NEPP contains specific programs that relate to climate change, acidification, eutrophication, diffusion, waste disposal, human living conditions, water resources, energy conservation, and natural resources management. Perhaps the most important feature of the NEPP has been the attempts that it makes to integrate environmental and economic considerations.

In 1987, the Canadian National Task Force (NTF) on the Environment and the Economy adopted the Conservation Strategy. The NTF report charged the federal government as well as territorial and provincial governments to adopt conservation strategies. The NTF report also set up multiple stakeholder or roundtable planning processes at the national and provincial levels to develop the conservation strategies. There are now 13 roundtables in Canada, including the national roundtable. Roundtable membership includes representatives from government and labor, farmers, academics, native people, and citizens at large. Roundtables vary from province to province in structure and focus.

In the United States on June 14, 1993, President Clinton created the President's Council on Sustainable Development (PCSD). The function of the PCSD is to make recommendations to the president about how to move the United States toward sustainable development. The PCSD is a broad-based advisory body with 25 members comprised of representatives of government, industry, environmental organizations, and native Americans. The primary goal of the PCSD is to make specific policy recommendations for a national strategy for sustainable development that can be implemented by public and private sectors. The PCSD has developed committees to work on clusters of issues. These clusters include sustainable communities, energy, natural resources management, and ecoefficiency.

3.2. SUBNATIONAL SUSTAINABLE DEVELOPMENT PROGRAMS

Chapter 28 of Agenda 21 expressly calls for states and local levels of government to develop Agenda 21 implementation plans by 1996. Section 28.2(a) states expressly, "By 1996, most local authorities in each country should have undertaken a consultative process with their local populations and achieved a consensus on a local Agenda 21 for the community."

If national governments are to take Agenda 21 seriously, they must encourage states, provinces, and local governments to become involved in Agenda 21 implementation. This is so because states, provinces, or local governments legally often have the dominant role in matters that need to be considered in implementing Agenda 21. For instance, in the United States, states, rather than the federal government, are the most important actors on many aspects of water and air regulation, utility regulation, transportation planning, and land-use control. In addition, most of Agenda 21 is as relevant to state, provincial, or local governments as it is to national governments.

In the United States, several states have begun sustainable planning or held initial meetings on sustainability. These states include Minnesota, Virginia, Maine, New Mexico, New York, North Carolina, Washington, Kentucky, Iowa, Florida, and Pennsylvania. Several cities throughout the world also have begun sustainable development programs; examples in the United States include Seattle, Washington, and Jacksonville, Florida.

3.3. THE NEED TO EXAMINE THE LIMITS OF SCIENCE, ECONOMICS, AND LAW IN SUSTAINABLE DEVELOPMENT DECISIONMAKING

Because new sustainable development programs will surely create considerable conflict between those persons who are deriving a living from unsustainable practices and laws that limit or prohibit the unsustainable behavior, there will be even greater pressure on national political leaders to look to scientists, economists, and other experts to define which human behaviors or activities are unsustainable. Therefore, in the years ahead, government experts will use the languages of science, economics, and law to frame the public policy questions that must be faced in implementing Agenda 21 and sustainable development programs.

In fact, evidence of this increasing need to rely on complex scientific, legal, and economic procedures and analyses is apparent in many sections of Agenda 21. For example, Agenda 21 expressly calls for the use of such complex scientific procedures as toxicological risk assessment as the appropriate tool for determining which hazardous substances are harmful and calls for the use of biotechnology to solve food scarcity problems. If the international community urgently needs to transform unsustainable development practices and put the human community on a path toward sustainable development, concerned citizens and policymakers must critically examine the policy tools of science, economics, and law. This examination is necessary for two reasons.

First, the international community must examine current capabilities of science, economics, and law to determine the ability, limitations, strengths, and weaknesses of each discipline in assisting decisionmakers in moving toward sustainable development programming. For instance, such an examination should seek to determine whether scientific understanding about causes of global warming is sufficient to guide decisionmakers about what changes in human behavior are necessary to avert human-induced climate change. If science is uncertain about the causal relationship between human action and environmental degradation, decisionmakers will have to decide how they should act in the face of scientific uncertainty to achieve the goals of sustainable development. If the world waits until the scientific proof is in, the world is making an ethical judgment that favors the status quo.

Similarly, current economic practices, such as gross national product calculations or cost-benefit analyses, should be examined to determine whether the variables that these practices measure are appropriate indicators of sustainable national health. In a similar vein, existing environmental laws should be examined to determine whether prescriptive rules contained in these laws are consistent with sustainable development goals and are effective in changing nonsustainable behavior. That is, examination of the policy tools of science, economics, and law is necessary to determine whether existing policy tools are capable of guiding future efforts to transform nonsustainable behavior or whether existing policy tools need to be modified or supplemented to include other considerations about, new measures of, or approaches to sustainable development.

Second, the problems that must be faced in implementing Agenda 21 call into question much of the world view that has been dominant during the period of world industrialization. Because humans are on a track that can lead to the destruction of much of life on earth, the world community is urgently challenged to develop an ethic that will recognize the duties that people have to care both for other humans and for future generations and other forms of life with which we share this planet. If a new worldwide sustainable development ethic is needed to support sustainable development programs, the policy languages of science, economics, and law need to be examined to see whether existing practices are consistent with a sustainable development ethic. If the policy languages of science, economics, and law are not ethically neutral or if important ethical positions are often hidden in scientific, economic, and legal languages, it is critically important that the ethical dimensions of science, economics, and law be understood in implementing sustainable development programs.

4. The Purpose of This Book

This book examines the role, capabilities, and certain strengths and weaknesses of existing scientific, economic, and legal tools and ethical considerations in the context of certain sustainable development problems. This analysis is necessary to understand the problems that follow from the application of current scientific, economic, and legal methods and practices to sustainable development problems. This analysis is also necessary to determine whether sustainable development problems create important new challenges and problems for government so that, where appropriate, new tools or approaches may be designed either to overcome limitations or take advantage of the strengths of current scientific, economic, and legal capabilities.

First, the book examines the role of science, ethics, economics, and law as generally applied to sustainable development decisionmaking (Chapters 2, 3, 4, and 5). Sustainable development problems often raise complex scientific, economic, and legal questions, but these questions differ greatly depending on the problem. For instance, questions about the predictive capability of the ecological sciences are more central to the problem of biodiversity than they are to the problem of desertification of dry lands. Problems of toxicology and the movement of hazardous substances through air and water are central to the safe management of hazardous wastes but are not central to problems of deforestation even though hazardous air pollution is a significant threat to some forests. Therefore, the initial chapters on science, economics, law, and ethics treat some of the more general questions that arise about the use of existing scientific, economic, and legal tools in sustainable development decisionmaking. This analysis seeks to describe in a thematic way the capability, limitations, strengths, and weaknesses of each discipline to assist decisionmakers in moving toward sustainable development implementation; it also seeks to understand the need to integrate these disciplines in sustainable development problem analysis and decisionmaking so that sustainable development ethical goals can be achieved.

Next, the book explores the use of science, economics, law, and ethics in the context of some specific global sustainable development problems. These problems are: (1)biodiversity, (2)climate change, (3)water resources, (4)hazardous waste, and (5)nuclear waste (Chapters 6, 7, 8, 9, and 10). This analysis attempts to describe the strengths and weakness of various disciplines when they are applied to concrete sustainable development problems. Although these topics represent only a portion of the sustainable development problems, lessons learned from the analysis of these five problems are often relevant to other sustainable development questions.

5. References

Brown, D.A., ed. 1994. *Proceedings on Ethical Dimensions of the United Nations Program on Environment and Development, Agenda 21*. Earth Ethics Research, Harrisburg, PA.

Gowdy, J.M. 1994. Progress and Environmental Sustainability. *Environmental Ethics* 16: 41-56.

Johnson, S.P., ed. 1993. *The Earth Summit. The United Nations Conference on Environment and Development (UNCED)*. Graham & Trotman/Martinus Nijhoff, London.

Shearman, R. 1990. The Meaning and Ethics of Sustainability. *Environmental Management* 14: 1-8.

(WCED) World Commission on Environment and Development. 1987. *Our Common Future*. Oxford University Press, New York and London.

Chapter 2
THE ROLE OF SCIENCE IN SUSTAINABLE DEVELOPMENT AND ENVIRONMENTAL PROTECTION DECISIONMAKING

John Lemons[1]
Donald A. Brown[2]

1. Agenda 21 and Science

Those designing sustainable development implementation schemes will inevitably look to scientists to help them understand sustainable development problems. Scientists have already made important contributions to the understanding of many serious environmental problems, such as the causal relationship between certain synthetic chemicals and destruction of the ozone layer. If scientists had not identified the relationship between upper atmospheric ozone concentrations and releases of chloroflorocarbons, government decisionmakers would not have agreed to action limiting their production. However, although causes and effects of some environmental problems are understood well, others are not, such as the timing and magnitude of climate change caused by greenhouse gases. If global environmental problems are serious, there is an obvious and urgent need to increase scientific understanding of which human actions cause environmental degradation and how nations can proceed with needed development programs without causing further environmental damage.

Assuming an urgent need to increase scientific understanding of the causes of environmental degradation, the authors of Agenda 21 included two chapters that deal expressly with scientific issues. Chapters 31 and 35 identify programs and methods of ensuring more effective involvement of the scientific community in sustainable development decisionmaking. In addition, many other chapters in Agenda 21 call for an expanded role for science in resolving problems of sustainability and environmental protection.

Despite the fact that Agenda 21 clearly calls for an expanded role of science to promote sustainable development and environmental protection, because of certain perceived limitations with the predictive capability of science and other limitations of the scientific method, there is continuing debate about the role science should play in sustainable development decisionmaking. The debate is comprised of several controversies. First, there is acknowledged need to increase scientific research and involvement of the scientific community in sustainable development decisionmaking. Second, there is controversy about how decisions should be made in the face of scientific uncertainty and the role of science in dealing with issues of uncertainty. Third, issues about the role of science in sustainable development arise when certain value-laden dimensions of scientific methods and tools are considered. Finally,

---

[1]Department of Life Sciences, University of New England, Biddeford, ME 04005, U.S.A.; [2]Bureau of Hazardous Sites and Superfund Enforcement, Pennsylvania Department of Environmental Resources, Commonwealth of Pennsylvania, 400 Market St., Harrisburg, PA 17101-2301, U.S.A.

J. Lemons and D. A. Brown (eds.), Sustainable Development: Science, Ethics, and Public Policy, 11–38.

because of current practices and controversies associated with decisionmaking in the face of scientific uncertainty, the adequacy of science used in environmental impact assessment creates controversy. This chapter will deal with these issues in turn.

Our purpose in this chapter is not to discuss in detail the role and capabilities of science in specific sustainable development problems, for these are matters dealt with in later chapters. Rather, we describe in a general way frequent issues and controversies about the role of science in sustainable development decisionmaking. Because Agenda 21 does not seem to reflect a recognition of some of these problems, it is important the decisionmakers and those designing sustainable development implementation schemes understand these controversies. Consequently, we identify some broad philosophical and practical issues associated with the role of science in sustainable development problems.

2. The Need to Increase Scientific Understanding of Sustainable Development Problems

As stated above, Agenda 21 recognizes that there is an urgent need to increase both the role of scientists in sustainable development decisionmaking and scientific understanding of sustainable development problems. Agenda 21 presumes that the international community needs better scientific understanding of which human actions create adverse environmental effects and what actions can be taken to eliminate or ameliorate environmental degradation. There are two chapters in Agenda 21 that are premised on the notion that increased scientific involvement in sustainable development problem identification and program implementation is critical to the success of changing unsustainable behavior.

Chapter 31 is entitled the "Scientific and Technological Community." The purpose of this chapter is "to enable the scientific and technological community. . . to make a more open and effective contribution to the decision-making processes concerning environment and development." Chapter 31 makes a series of recommendations about how to strengthen the role of science and scientists in sustainable development decisionmaking. These recommendations include: (1)identifying how scientific activities could be responsive to sustainable development needs; (2)creating regional scientific cooperatives; (3)increasing scientific inputs to government processes; (4)strengthening scientific advice to decisionmakers; (5) improving the dissemination of scientific research results; (6)improving links between private and government scientific research; (7)increasing the role of woman in science; and (8)developing and implementing information technologies to enhance dissemination of scientific information.

Chapter 35 is entitled "Science for Sustainable Development." This chapter focuses on the role and use of science in supporting the prudent management of the environment and development needs of humanity. Generally, Chapter 35 calls for better use of existing scientific information through interdisciplinary studies and greater research to increase understanding of sustainable development decisions. Specifically, it recommends the development of four program areas: (1)applying the best scientific information to sustainable development problems, (2)enhancing scientific understanding of linkages between human activities and environmental change, (3)improving long-term scientific assessment, and (4)building scientific capacity and capability. For each of these four program areas Chapter 35 makes many specific recommendations about how to increase the use of science and strengthen the scientific base of sustainable development decisionmaking.

Problems identified in Chapter 35 are of two types, scientific and political problems. For example, the problem of understanding ". . . global atmospheric chemistry and the sources and sinks of greenhouse gases. . . ." is a more traditional scientific problem for which scientific methods and tools can be employed to investigate. In contrast, the problem of intensifying ". . . research to integrate the physical, economic and social sciences to better understand the impacts of economic and social behavior on the environment and of environmental degradation on local and global economies...." is a political problem about funding increased interdisciplinary research pertaining to sustainable development and environmental protection. Moreover, the recommendation to increase interdisciplinary research among physical, economic, and social science is a recognition that better understanding of unsustainable practices is not limited to traditional "hard" scientific information such as physics, chemistry, and biology. The call for interdisciplinary research that includes the economic and social sciences is acknowledgement that the human dimensions of the causes of environmental change are, in addition to information supplied by the "hard" sciences, knowledge that is necessary to change unsustainable practices. Therefore, one of the controversies about the role of science in sustainable development decisionmaking is which scientific research projects should be funded given limited research dollars. Should a nation's limited research money be invested in the biological or social sciences? Should sustainable development problem-solving look to technological solutions to problems of sustainable development such as those that reduce greenhouse gases through treatment of emissions or to social solutions such as programs that encourage energy conservation?

In sustainable development problem-solving, the methods and tools of science have to be applied both to scientific problems in the traditional sense, and to interdisciplinary problems that attempt to integrate the "hard" sciences with social sciences, ethics, and law. Because there is more experience with the application of science to traditional scientific problems among scientists, they often focus on the more traditional scientific and technological solutions in dealing with environmental problems at the expense of integrating the hard sciences with the "softer" social sciences.

Shrader-Frechette (1982) calls this tendency for scientists to focus on technical solutions the fallacy of unfinished business. This fallacy arises out of the assumption that technical and environmental problems have only technical, but not social, ethical, or political solutions. The fallacy of unfinished business is the practice of formulating questions about environmental issues in such a way that the answers necessarily are limited to technical solutions for the problems identified. One example is whether to store radioactive waste in salt mines, or in deep drilled wells, in solidified ceramic form or as a liquid inside double-walled steel tanks. Although the question posed in this way requires a technical solution, the more intractable problem is not the technical one of what storage technique to adopt, but the ethical and social one, such as what risk we can impose on future generations and how we ought to determine the acceptability of a given risk. We have not answered questions such as these in part because we have been asking, not wrong questions but incomplete ones, questions that are epistemologically loaded, questions that presuppose a definition of a given problem for which only an answer in terms of the technological status quo counts as a solution.

As another example, Shrader-Frechette (1982) asks whether we ought to develop coal or nuclear fission in the Ohio River Basin in the United States in order to meet the electric power demand between now and the year 2000? This question was formulated by the U.S. Environmental Protection Agency in its assessment of energy needs for this area. Precisely because the formulation of this technological question allows only for a technical answer in terms either of coal or nuclear fission, it ignores alternative ethical, social, and political solutions to problems such as the conservation of energy.

The propensity to ask the incomplete question is built into the way that we develop new technology and solve technological problems. When a technology goes wrong, someone schooled in that technology is brought in to fix or analyze that technology. Likewise, when a technology is under consideration for future use, someone knowledgeable about that technology is often asked to assess that technology. The result is public policy that does not consider other potentially appropriate options.

Finally, there is the problem of time pressure. The technical expert within a government agency is usually under pressure to make a technical assessment of alternatives in a timely fashion and to ignore any arbitrariness in so doing (Brown 1987). This pressure tends to force the technical expert to analyze only what is genuinely predictable and calculable. Since a known technology is usually taken to be more predictable than alternatives that depend upon social or moral considerations, the technical expert usually focuses his or her attention on those alternatives that are most easily calculable and predictable, namely alternative technologies.

For these reasons, the degree to which scientists working on sustainable development problems consider nontechnological solutions is a controversy about the role of science in sustainable development decisionmaking. Although it is admittedly extremely important to obtain the best scientific information possible about the consequences of human action on the environment and human development objectives, the types of solutions to sustainable development problems considered is an important issue for the success of sustainable development programs.

3. Two Methodological Approaches to the Use of Science in Sustainable Development Problems

Generally speaking, there can be said to be two overall roles for science applicable to broad forms of environmental problems: (1)predictive and (2)holistic. The identification of these two roles should be understood to refer to orientations of thinking and planning that influence the way in which problems are perceived and methods to resolve them are selected. The ensuing discussion about the roles of science necessarily is simplified for purposes of discussion; pure versions of each as applied to broad environmental problems probably are rare.

The predictive scientific approach is grounded in the belief that scientific knowledge consists of reasonably certain "facts" obtained by rational and objective methods and tools (Bella et al. 1994). With respect to sustainable development and environmental protection problems, this belief assumes that: (1)there should be a strong emphasis on data acquisition per se, (2)knowledge of scientific facts will lead to the solution of most problems, and (3)the scientific method as commonly understood has great credibility to discern facts about the natural world. The predictive science approach establishes the necessary goal of science to include: (1)formulating hypotheses and conducting observations to test them, (2)developing an understanding of processes and linkages among variables, and (3)developing reasonably certain predictions. Ideally, this approach tends to be reductionistic, analytical, and empirical where possible in order to increase the likelihood of accurate scientific predictions. Proponents of predictive science believe that scientific information represents objective information suitable for environmental decisionmaking about risks and that it therefore should play a central role in informing debates about broad environmental problems because it reduces the likelihood that decisions will be based on speculative thinking.

Predictive science generally is regarded as an outgrowth of the philosophical movement called positivism. Positivists denied the intelligibility of concepts that were not derived from empirically derived facts and relegated speculation and other forms of untestable thinking to the realm of metaphysics (Abbaganano 1967). Scientists, influenced by positivistic ways of thinking, were taught to eliminate as much as possible all speculation from their scientific descriptions. Only observable, testable facts and mathematical relationships between facts were acceptable as scientific tools; all matters of value were to be purged from consideration (Brown 1987). This positivistic view of science assumes that: (1)all knowledge is founded on experience; (2)concepts and generalizations only represent the particulars from which they have been abstracted; (3)meaning is grounded in observation; (4)the sciences are unified according to the methodology of the natural sciences and the ideal pursued in knowledge is the form of mathematically formulated universal science deducible from the smallest number of possible axioms; and (5)values are not facts grounded in observation and therefore cannot be included as a part of scientific knowledge (Held 1980).

Obviously, the predictive view of science demands a rigid distinction between science and values because values interfere with the search for the truth, for truth, it is assumed, can only be tested through the use of testable hypotheses and observations. To arrive at truthful description, the values positions of the scientist must be excluded from the analysis to prevent a distortion of the scientific description of the truth. Only those things that can be empirically verified count as facts and values questions cannot be empirically verified. From this axiomatic base it follows that: (1)"good science" must not be biased by the values of the scientist, (2)good science is value-neutral, and (3)the good scientist is one who rigorously and consistently purges his or her projects from the biases or subjectivity that values positions will create. "Good science" is therefore antithetical to values discourse and the scientist must be trained to exclude all values discourse from their scientific endeavors. As Maxwell (1987) notes:

> Ideas, in order to be capable of objective rational appraisal, must be entirely factual in character, capable of being true or false, and thus potential contributions to knowledge. Thus religious views, ideologies, social and political policies, personal philosophies, which intermingle judgments concerning facts and values in an essential way, are incapable of objective, rational assessment and have no place within the intellectual domain of scientific, academic inquiry.

Having been influenced by this value-neutral view of appropriate scientific method, many scientists following the predictive scientific tradition assume that the best way to solve our many environmental problems is through the application of scientific procedures from which all values considerations have been rigorously excluded.

By way of contrast, a more holistic science is predicated upon the belief that most broad environmental problems may not be amenable to the application of the predictive science approach and that such an approach in fact is embedded with numerous types of value-laden assumptions, evaluations, judgments, and inferences. Proponents of holistic science reject the beliefs that: (1)problem-solving should reflect an emphasis on data acquisition per se, (2)data will solve problems, and (3)the scientific method is objective and value-neutral. Alternatively, the holistic science approach emphasizes: (1)adequate formulation of problems so that data will contribute to public policy goals, (2)that most results from scientific studies will not yield reasonably certain predictions about future consequences of human activities and that broad environmental problems therefore should be considered to be "trans science" problems requiring research directed toward useful indicators of change rather than precise predictions, and (3)the need to evaluate and interpret the logical assumptions underlying the empirical beliefs of scientists with a view toward ascertaining more fully the validity of scientific claims and their implications.

Funtowitz and Ravetz (1991) argue for a holistic approach to science which they have called "post-normal" science. They state that in environmental matters, where facts are uncertain, values in dispute, stakes high, and decisions urgent, scientists often need to follow methods that might not be appropriate in other scientific endeavors. In deciding such matters, decisionmakers will have to apply values to factual findings. That is, the norms that scientists should follow in serious public policy matters are different than those that should be followed in scientific research matters where important public policy matters are not at stake.

While holistic science is not easy to characterize, it seeks a broad and integrated view of problems and places more emphasis on professional judgment, intuition, and is less bound by analytically derived empirical facts. Proponents of the application of holistic science to problems of sustainable development and environmental protection maintain that its claims are more amenable for practical public policy purposes compared to the claims of predictive science approaches. In part, evaluating the relevance of predictive and holistic science approaches to problems of sustainable development and environmental protection can be based on the consideration of several factors: (1)how the approaches lead to different definitions and conceptual analysis of the problems, (2)the capabilities of the different approaches to yield reasonable certain predictions about the future consequences of human activities on ecosystems and humans, (3)subjective values embedded in the approaches and their implications, (4)the ability of information obtained from the approaches to fulfill burden of proof requirements, (5)the role of the approaches in different decisionmaking perspectives, and (6)how the approaches are used in environmental assessments.

To the extent that the predictive science approach relies on reductionistic analytical science, it leads to different formulations of problems compared to a more holistic approach. Typically, the predictive scientific approach defines problems in a narrow manner by isolating and studying selected variables under controlled conditions in order to obtain accurate and credible results. The approach also leads to formulating environmental problems in particular ways (Miller 1993). For example, a predictive science approach to chemical wastes might be to identify their gaseous, liquid, and solid components and find a separate solution for each. This might entail the identification of harmful chemical pollutants, assessment and evaluation of their risks, the determination of acceptable levels of risks, the development of appropriate technology to manage and dispose of the pollutants in their various forms, and possibly the development or substitution of different chemicals for use in manufacturing or industrial products and processes. Since each of the parts of the problem likely will have a preferred solution, the overall problem of pollution can be said to be the aggregate of the partial solutions for each of the components of the problem. Normally, this results in an emphasis on effluent treatment.

A more holistic approach would extend the problem of chemical wastes beyond that defined by predictive science approaches. For example, the holistic approach might also examine: (1)the relationships between population and/or economic growth as a causative factor for growth of the variables which lead to chemical pollution; (2)the changes in chemical pollution resulting from new technologies of production; (3)the capacity of ecosystems to assimilate pollutants and the efficacy of human ecosystem management in protecting such capacity; (4)the efficacy and appropriateness of neoclassical and alternative economic methods of valuing resources as a basis for pollution policy and law; (5)the goals and efficacy of laws designed to protect ecosystems and human health from the risks of pollutants; (6)cultural, social, and religious factors which might be root causes of pollution and environmental and human health degradation; and (7)assumptions that are necessary to interpret and evaluate inference gaps caused by scientific uncertainty. The holistic approach results in a conceptualization of the problem of chemical waste as being one which is defined not only as the relationship between exposure to chemical pollutants and health, but rather one which also has to do with whether the problem of chemical pollution fundamentally is one of inappropriate economic and public policy and activities, as well as of the consumption levels of individuals.

4. Scientific Uncertainty and Values

Predictive and holistic science approaches also lead to different roles for the use of scientific information in making predictions about the future impacts of human activities on environmental or human health attributes. Ideally, decisionmakers would like reasonably sound scientific information on which to base decisions. Yet, as will be set out more fully below as well as in subsequent chapters, decisions about environmental and sustainable development controversies often will have to be made in the face of pervasive scientific uncertainty. The sources of scientific uncertainty include: (1)limitations of available analytical tools; (2)the complexity and indeterminancy of ecosystems; and (3)the need to make value judgements at all stages of problem identification, analysis, and solution implementation (Lemons, in press).

4.1. SCIENTIFIC UNCERTAINTY CREATED BY ANALYTICAL TOOLS

Decisionmakers must choose from several different types of studies used to investigate the scientific dimensions of sustainable development and environmental problems. Site-specific studies often are conducted in order to understand the responses of particular environmental attributes to different perturbations. These studies are based on the recognition that responses may be specific to the unique conditions existing at a specific site at a given time. Statistical models are used to test or generate hypotheses and descriptions of the responses of environmental attributes to environmental change without making assumptions about the underlying factors responsible for particular environmental attributes and their status. Mechanistic models are used by scientists to predict the consequences of environmental perturbations to environmental attributes. These models are based on the assumption that scientists know the underlying factors responsible for particular environmental attributes and their status and how they are affected by specific perturbations. Comparative studies also are used to understand environmental attributes by attempting to describe and answer questions about patterns and responses of environmental attributes by acquiring data across specified gradients, regions, or larger geographical areas and making statistical inferences from the data acquired. These studies provide information useful for baseline studies to evaluate future change, as well as information which can document large-scale environmental impacts. Various other forms of modeling are used to ascertain responses of environmental attributes to perturbations. Most models focus on either the structural or functional aspects of environmental attributes. Theoretical approaches also are used by scientists to develop generalizable or universal laws which can be used to explain or predict events in a deductive fashion. Theoretical researchers may subject their predictions to experimental confirmation or falsification and, based on the results, accept, modify, or discard the general principles or laws. Historical studies commonly are applied in geological, atmospheric, and paleoecological problems and they often provide useful information in understanding natural phenomena as a function of past conditions. A case study approach is another method used to study environmental attributes in a scientific manner. A goal of such an approach is to test, clarify, amend, and evaluate specific examples or cases. Typically, informal causal, inductive, retroductive, and consequentialist inferences are used to provide meaning to a particular example or case.

All of the aforementioned scientific approaches have limited capabilities to yield reasonably certain predictive knowledge for environmental problem-solving (see, e.g., Shrader-Frechette and McCoy 1993). Because site-specific studies are viewed best as being relevant to the particular study area and its conditions at the time of study, most of the conclusions from such studies are based on inductive reasoning and normally are considered to have heuristic as opposed to predictive value. While many researchers believe that statistical models allow scientists to make general statements about the responses of environmental attributes to perturbations or environmental changes, these models often cannot be used to make accurate specific predictions or to establish the causal connections between perturbations and the responses. Most mechanistic models fail to serve as a basis for reasonable predictions about future responses because of the existence of numerous complexities involved in environmental systems and the lack of understanding concerning them.

Comparative studies generally do not yield reasonably certain predictions because they do not provide mechanistic understanding of causal events and because they may fail when environmental conditions extend beyond the range of a study's conditions. While scientific models focusing on lower levels of hierarchical organization might have some predictive capabilities, those focusing on higher levels (e.g. ecosystems or earth-atmosphere systems) primarily have heuristic value. In part, the limited predictive capabilities of many models stems from the difficulties of understanding the complexity of ecosystem attributes and the linkages between them, as well as from the difficulties of verifying and validating the models. Although theoretical approaches designed to provide generalizable knowledge (e.g., the stability-diversity hypothesis in ecology) have been utilized by many scientists, they increasingly are being criticized as yielding little practical information for environmental decisionmaking. Historical studies often provide information about natural phenomena, but their results are applied better to interpretation of past rather than future events. While the case study approach can allow the formulation of rough or imprecise generalizations, it does not allow for the discovery of general theories or empirical laws that provide precise prediction for environmental problem-solving.

4.2. SCIENTIFIC UNCERTAINTY AND COMPLEX SYSTEMS

Large uncertainties also are inevitably inherent in assessments of biological or ecological systems regardless of the scientific study approach utilized. For example, the stochastic state of ecosystems over time decreases the ability of scientists to derive data for pertinent ecosystem criteria with certainty. Unexpected human intrusions or mismanagement often are responsible for unpredictable changes in species or ecosystems. Recent advances in chaos theory have called into question whether it is even possible to make long-term ecological predictions with any certainty. Several sources of subjectivity also increase uncertainty in predictions about species and ecosystems. For example, people have to decide which ecosystem parameters are more important to base judgments on, often with little or no empirical information available. Assumptions have to be made, often without direct empirical evidence, whether ecosystem parameters should be considered independently or synergistically, and whether threshold values for environmental or health impacts exist, and if so, what such values should be. In addition, a lack of empirical data cannot be separated entirely from practical limitations imposed on environmental scientists. Decisionmakers require information in a relatively short time period, and at reasonable cost. These factors constrain the focus of most ecological studies to lower levels of hierarchical organization, the short-term, small spatial areas, and measurement of relatively small numbers of parameters. Accordingly, adequate knowledge is difficult to obtain for practical reasons as well as the scientific. As a result, environmental science is much softer and less predicative than is realized by many people. Insofar as the predictive science approach is based on reductionistic, analytical, and empirical methods, it attempts to understand complex ecosystem and human health systems by isolating one or a few variables for study under controlled and reproducible conditions and measuring their response to experimental conditions or perturbations. The predictive science approach seeks to establish research programs to document the understanding of the biogeochemical and other natural processes that influence the earth.

Scientists are trained to deal with uncertainty arising from complex technologies by assigning probabilities to various scenarios in an attempt to determine the probability of remote but potential consequences. For example, they might be taught to quantify uncertainty by such methods as fault tree analysis. Where probability of any component of an analysis is known, for instance reliable historical evidence that shows the frequency of valve failure in a nuclear power plant, scientists may use this objectively derived failure rate to determine the probability of this consequence. However, the problem of assigning probabilities is much more difficult where no reliable empirical evidence exists for these consequences. In these cases, scientists develop probabilities for scenarios by using more subjective means such as making analogies from known evidence. Where probabilities of environmental consequences are derived from subjective methodologies, troublesome environmental ethical questions may arise because of the lack of grounding of such predictions in actual experience and the consequent need to determine how conservative these predictions should be.

Certain environmental problems raise questions of such fundamental scientific uncertainty that any attempt to deal with the uncertainty through probability analysis may be not much better than untutored speculation because it is sometimes difficult to describe even subjective probabilities of various scenarios or to predict the consequences of various scenarios. In fact, in an article on the recent development of the science of ecology, one commentator reports that the attempt to develop general predictive mathematically based laws in ecology, a great hope of ecologists in the 1970s, came to nothing in the 1980s (Sagoff 1988). Ecology has failed to develop predictive laws because ecological systems are so inherently complicated that all the small and assumed insignificant variables can easily overwhelm the ecological system and confound the mathematical models, as well as because of the fact that we simply do not understand much about the structure and function of ecosystems. And so the science of ecology has yet to develop a mathematical foundation that has been accepted in the scientific community. Similarly, risk assessors also are faced with many difficulties in predicting the consequences of, say, toxic chemicals on human health (Cranor 1993). These types of considerations lead to the conclusion that projections of the environmental or human health consequences of human activities are often extraordinarily speculative or incomplete.

While the application of the predictive science approach has been successful in the physical sciences as well as in molecular biology, it presents what is almost an intractable problem due to the fact that it attempts to understand complex systems by isolating a few variables so that they can be studied under narrowly controlled and simplified conditions. This approach becomes problematic because when dealing with more complex systems the understanding of the interactions between variables which determine the way in which individual variables express themselves are not able to be discerned. In addition, the use of a reductionistic analytical approach often is not able to discover consequences resulting from indirect or synergistic actions. By dividing research tasks into more easily manageable components, the predictive science approach increases the likelihood that serious future consequences are overlooked. Consequently, in the overwhelming number of cases, the predictive science approach leads to conclusions of negligible impacts of human activities to ecosystem or human health because the pervasive scientific uncertainty inherent in complex environmental problems precludes establishing cause and effect relationships between the activities and impacts at the 95 percent confidence level. Further, by simplifying and reducing parts of complex environmental problems to a more manageable scale, the predictive approach ends up studying a scientific problem which may be very different from the more complex environmental problem from which it stems. For example, in attempting to provide answers to problems of aerial spraying of pesticides in New Brunswick forests in Canada, researchers utilized strict scientific research guidelines and norms in order to attempt to discern whether there was a cause and effect relationship between pesticide exposure from spraying and human health effects. While this problem theoretically is amenable to scientific investigation, it served to focus attention away from more fundamental questions concerning the misdirection and redesign of resource policy (Miller 1993).

4.3. SCIENTIFIC UNCERTAINTY AND ETHICS

As a result of the above problems, decisionmakers making environmental and sustainable development judgments often will be faced with pervasive scientific uncertainty at all stages of problem identification, analysis, and solution. In the face of scientific uncertainty they must decide whether they will err on the side of minimizing false positives, generally referred to as a type I statistical error, or on the side of minimizing false negatives, a type II statistical error. A false positive leads to a decision that errs on the side of environmental protection, while a false negative errs on the side of a desire to not impose unnecessary costs for environmental protection. Therefore, in the face of scientific uncertainty decisionmakers must decide what is the right thing to do, which is a prescriptive rather than a descriptive question. Insisting on high levels of scientific proof before government action may be taken is a prescriptive rule that puts the burden of proof on government decisionmakers and protects the status quo. Such a rule may prevent protective government action where there is a reasonable basis for concern but where science is uncertain about the consequences of certain human activities. Therefore, the standard of proof that should be required of regulatory action is fundamentally an ethical question, not a scientific one.

When scientists are concerned about the search for valid conclusions in normal scientific endeavors statements about cause and effect relative to the consequences of human activities ideally are based on standard norms governing acceptability of scientific evidence. The standard norm followed in such normal scientific endeavors is to accept scientific findings at a 95 percent confidence level. In other words, reductionistic analytical approaches that follow standard scientific norms seek to minimize type I errors which lead to the acceptance of false positive results.

Because of the pervasive nature of scientific uncertainty in environmental matters, there is a tension between the disciplinary norms of good science and good regulation. Government officials cannot wait until all desired scientific information is available prior to deciding on regulatory approaches. Unlike the approach in scientific areas where judgment may be suspended until the scientific proof is in, government officials are expected to act in a timely manner. Very often government officials are expected to make decisions on environmental matters on extremely limited data applied to weak or nonexistent theory. As a result effective regulation may sometimes require government agencies to adopt crude but administrable decision strategies that do not incorporate a high degree of scientific sophistication (Latin 1988).

However, those opposed to regulation can always criticize the regulation on the basis of lack of scientific sophistication. Because many government decisions can be challenged on scientific grounds, those who want to avoid regulation will be successful if legitimizing environmental regulation is limited to scientifically proven information or theories. Politically speaking, decisions initiated to protect the environment in the face of scientific uncertainty often are perceived by some to be irrational because they are said to be without a scientific basis that compels or supports the decisions. Because decisionmakers often must be sensitive to the economic and developmental consequences of decisions to protect the environment, it is likely that many will be reluctant to propose or approve protection measures which might slow or conflict with economic development in situations where scientific uncertainty exists (Brown 1987).

One of the international agreements reached at the Earth Summit in Rio de Janeiro in June of 1992 was the Rio Declaration on Environment and Development. Principle 15 of the Rio Declaration states:

> In order to protect the environment, the precautionary approach shall be widely applied by States according to their capabilities. Where there are threats of serious or irreversible damage, lack of full

> scientific certainty shall not be used as a reason for postponing cost-effective measures to prevent environmental damage.

The precautionary principle establishes the international norm that nations should take steps to protect the environment before potentially harmful effects of a given behavior are proven fully. The precautionary principle suggests a departure from the traditional scientific norms that err on the side of minimizing false positives, a type I statistical error. The precautionary principle assumes that the burden of proof should be shifted so that proposed actions with potential serious consequences are assumed to be harmful until proven benign.

Holistic scientific approaches rarely produce results which are acceptable according to traditional scientific norms. Such approaches recognize the pervasive uncertainty surrounding problems of sustainable development and environmental protection and also recognize that the problems include social, political, economic, and ethical dimensions not amenable to scientific analysis. For example, holistic approaches might focus on improving the understanding of responses of ecosystems and human health to human activities, and on the monitoring and analysis of responses so that sensitive indicators of ecological and human health can be identified. In this sense, holistic approaches focus less on obtaining reasonably certain predictions in the traditional scientific sense, and more on the identification of reliable indicators of ecosystem and human health change.

Typically, proponents of holistic approaches advocate a minimization of type II errors which lead to acceptance of false negative results. In other words, they would favor a higher chance of accepting false positive results and a lesser chance of accepting false negative results which would imply no environmental harm. This approach is more consistent with precautionary approaches called for in the Rio Declaration and Agenda 21. However, the holistic approach has been criticized by some scientists and public policy experts as being too speculative and riddled with bias (see, e.g., Peters 1991). Despite the shift in research focus of the holistic science approach, it does not ensure that complex problems will be understood before serious environmental consequences arise.

For all of the above reasons, the choice between the holistic or predictive approach, as well as any choice made in the face of scientific uncertainty, must be understood to raise ethical questions as well as scientific ones which ultimately must be addressed in law and public policy.

5. Additional Value-Laden Dimensions of Science

Mayo and Hollander (1991), Cranor (1993), Miller (1993), and Shrader-Frechette and McCoy (1993) have presented excellent analyses of how and why numerous value-laden judgments, evaluations, assumptions, and inferences are embedded in scientific methods of ecosystem and human health risk identification, assessment, evaluation, and management, as well as in more basic research methods of environmental assessment, ecology and geohydrology as applied to natural resources problems. Their conclusions are that many of the value-laden dimensions of scientific methodology and information not only are unrecognized by many scientists and environmental decisionmakers, but that the failure to recognize the existence of the value-laden dimensions casts serious doubts about even the best and most thorough of so-called scientific and technical studies used to inform decisions about natural resources problems. In other words, unless the value-laden dimensions of scientific and technical studies and information are disclosed, the positions of decisionmakers will appear to be justified on objective or value-neutral scientific reasoning, when in fact they will be based, in part, on often controversial and conflicting values of scientists and decisionmakers themselves. This situation implies that new definitions of scientific and technical rationality be adopted for sustainable development problems which would be reflective of populist, ethical, and holistic considerations.

5.1. SEPARATION OF FACTS AND VALUES

One of the most common ways in which value issues are hidden in environmental public policy making develops out of the expectation that the technical analysts can isolate the facts under dispute and employ them in accord with the values goals of environmental enabling legislation. The separation of fact and value issues, however, is often difficult, if not impossible. Philosophers who have been concerned about how people understand facts, a topic studied within a branch of philosophy called hermeneutics, have come to realize that what one sees is usually a product of cultural tradition; there are no acts of pure perception that are not dependent on prior value choices. In this context, the decision about which facts to focus on in the analytical stage of research cannot avoid value questions (Brown 1987). For example, should the environmental impact analysis of a dam consider protection of the habitats of deer and elk or should it consider potential destruction of the habitats of skunk or coyote? The decision of what to study is frequently a question of value, not of fact.

The decision about what resources will be used in performing the analyses of the facts is ultimately a value decision for which there is usually no objective standard. Taylor (1984) reports that the determination of what methods and resources the U.S. Army Corps of Engineers will employ to predict the environmental impacts of a project often results from a negotiated settlement between the environmental analyst and the project manager, and frequently depends upon such nonscientific criteria as the amount of budget money that is available to perform the analysis.

Moreover, many of the facts that the analyst attempts to collect for use in policy calculations are not susceptible to purely objective analysis. For instance, it is impossible to determine the visual impact of a water diversion project upstream from a dam, for there simply are no objective criteria for beauty or ugliness. Separating facts from values is also often impossible because the gathering of the facts must rest on hard-to-test or even presently untestable assumptions about the way the world works. The facts at issue in environmental disputes are very often nothing more than guesses based on high-level speculation. Likewise, how facts are arrayed by technical experts are often not policy neutral. For example, simply knowing that some technological activity will result in the deaths of some people who would not have otherwise died prematurely does not tell us whether the activity is murder, killing, allowing some people to die, or even saving lives (in the event some other lives might be saved by the technological activity) (Mills 1985).

No matter how neutral the scientific work is, there may be no neutral description of it that can be incorporated into policy discussions. The psychological literature on decisionmaking shows that people's preferences often are determined by the way a choice is described to them and can change under different descriptions that appear to be equivalent (Davos 1988). Analysis of environmental facts usually requires prior value decisions about level of detail, burden-of-proof, and quality of data. Thus, a value choice is implicit in almost every choice the technical analyst makes.

For example, consider that the use of safety factors in water quality regulations as a means of extra protection for human or environmental health usually are imposed because of poor scientific understanding on the quality of data on exposure. Implicit in the choice of safety factors is an asymmetric cost function with health costs rising more steeply than do overtreatment costs. Implicit in the magnitude of a safety factor are significant uncertainties in health impacts and a steeper cost function for health effects from undertreatment than for overtreatment. When these issues remain implicit in the use of safety factors (as they normally are) the real issues and areas of knowledge and uncertainty are obscured for decisionmakers and for the public. Often, these issues remain implicit or hidden because safety factors and cost factors are described in quantitative terms pertaining to risks or cost-benefit calculations. This increases the likelihood of the misuse of conclusions by decisionmakers who do not understand the basis for deriving safety factors. Other approaches can make the distinctions between science and values or costs explicitly known through the procedures for deriving safety factors. These approaches allow decisionmakers and public observers of the process to understand the basis for decisions about safety factors, such as whether scientific uncertainty was large or whether the costs of overtreatment were much greater than the costs of undertreatment.

5.2. THE NEED TO SYNTHESIZE RESEARCH METHODS AND INFORMATION

The value-laden dimensions of environmental science also stem from other sources. For example, many problems require synthesis of environmental research methods and information from a variety of disciplines. However, synthesis introduces several sources of subjectivity into environmental research. As Cairns (1991) notes, maintenance of ecosystem health can only result in approximations of natural rather than natural assemblages of organisms and their predisturbance abiotic environment. However, the actual biotic and abiotic characteristics of managed or protected ecosystems can vary significantly in terms of their comparisons with predisturbed conditions. Some of the differences between predisturbed and restored ecosystems are determined by practical constraints of management and use as well as by ecosystems' capacities to achieve predisturbed conditions. Other differences are due to selection by research scientists and decisionmakers of biotic and abiotic ecosystem components targeted in management or protection goals. Carpenter (1990) and Carter et al. (1994) also note how ecological, economic, and cultural parameters must be selected and synthesized with one another to promote the goals of sustainability. The selection of indicators of sustainability and the manner and extent in which they are synthesized are based on human judgment as opposed to objective criteria.

Subjectivity also is introduced in the synthesis of environmental research by the choice of integration models and by the unavoidable weighting by the models of their factor components. Modeling the interactions of biotic and abiotic variables with each other is difficult because few models exist, and there often is not a consensus as to which models are most appropriate for a given problem. Further, weighting is implicitly or explicitly performed by the coefficients of mathematical integration models. It is subjective because little basis can be established for nonjudgmental differentiation of the importance of some variables and of their interactions. For example, general circulation models used to project global climate warming vary widely in the variables used, their interactions, weighting of variable coefficients, and whether feedback systems influencing the fate and rate of carbon dioxide transport exist (see Chapter 7).

5.3. METAPHYSICAL ASSUMPTIONS EMBEDDED IN SCIENTIFIC METHODS

The very use of scientific methodology for finding the facts about environmental problems may not be value-neutral in some cases. For example, many people assume that scientific descriptions always picture the way nature really is. Most people in our culture assume that science has historically provided better and better descriptions of the way the world actually exists. That is, science has developed over time increasingly more accurate pictures of ultimate reality. Each scientific advance has made our understanding clearer with succeeding scientific developments building on and therefore clarifying preceding positions.

Many, although not all, philosophers of science have concluded that, contrary to this common understanding in our culture, scientific procedures cannot be expected to produce better and cumulatively more truthful descriptions of the way the world works. Rorty (1979) notes that there is no evidence that science develops better and more accurate mirrors with which to view nature. Kuhn (1962) maintains that scientific progress has actually proceeded with paradigm shifts by demonstrating that the history of science does not support the conclusion that scientific theories start small and grow on each other. Briggs and Peat (1984) argue that for most scientists major theories or paradigms are like spectacles which scientists put on to solve puzzles. Every now and then a paradigm shift occurs in which these spectacles get smashed and scientists put on new ones that turn everything upside down, sideways, and a different color. Once the paradigm shift takes place in any scientific field, a new generation of scientists is brought up wearing the new glasses and accepting the new vision as natural or true.

Kuhn (1962) does not deny that the history of science is marked by improvements in its scope, precision, and consistency of its laws. Science clearly has proven very powerful in developing increasingly more robust laws that have proven over time useful to humans in manipulating nature. The question is, however, whether science has progressed by accumulation of theories that are increasingly more accurate or better mirrors of nature. In other words, the predictive success of any scientific law is no guarantee of its metaphysical accuracy. Kuhn denies that science has developed increasingly clearer pictures of nature because each paradigm shift radically changes the view of what nature is all about. For instance, the Newtonian world of bodies reacting to certain forces is a different world than the Einsteinien world of space-time and energy-matter. Einstein's world is different in turn from Hiesenberg's quantum world. The models created by these theories actually create different world views. If the models created by different current theories in physics actually create different views of what nature is, how can we believe that science actually describes what nature really is? Philosophers of science are divided into at least three major groups on these issues: (1)realists, philosophers who believe that there is an independent reality and that this reality is discovered by the application of scientific methods; (2)pragmatists, philosophers who are not concerned with whether scientific statements are true or false about ultimate reality but are concerned primarily with whether scientific theory helps humans solve problems; and (3)relativists, philosophers who believe that scientific truths are nothing more than the social perspective of the scientist embedded within a scientific age (Briggs and Peat 1984).

If different scientific approaches create different views of what nature is, scientific descriptions are never value neutral. In addition, some scientific methods and procedures tend to create a view of nature that may lead to a disrespect for environmental values. For instance, the value-neutral positivistic view of science tends to see nature as a material substance. Scientists are trained to report impersonal data from which all subjective data have been removed, to reduce all issues to a scale that can be quantitatively manipulated, to think of nature as a substance with measurable analytic parameters. Many scientists tend to see nature as complexes of material because science itself is structured on a material hierarchy. According to Maxwell (1987):

> Scientific theory tends to be hierarchically organized with what is intellectually and explanatorily fundamental at the bottom, each science becoming progressively less and less intellectually fundamental as we ascend to the top. What this means is that a science at one level presupposes and, where relevant, uses the results of sciences at lower, intellectually more fundamental levels, whereas the reverse is not the case. Theoretical physics does not presuppose or use theories from sociology, whereas sociology constantly uses, even if

> only in an obvious and crude way, theories and results of physics (such as the existence and persistence of gravitation). Or to take less extreme examples, chemistry presupposes physics (especially the theory of atomic and molecular structure and quantum theory) whereas fundamental theoretical physics presupposes and borrows nothing from chemistry (apart occasionally from a piece of chemical technology for instruments, which is another matter altogether).

Because physics is the most fundamental of the sciences in the intellectual hierarchy and because physics understands nature in accordance with mathematically defined rules, much of scientific theory relating to environmental matters tends to reduce nature to the laws that presume that nature can be fully explained according to mathematical relationships. Because science starts with the mechanistic assumptions of physics, it can only discover machines. The law-like descriptions of nature tend to reduce nature to material standing reserves ready for human manipulation. Because science is built on law-like rules, scientific laws assure that nature can only be encountered through rules that allow human manipulation. In fact, the drive for a better scientific understanding of nature may be most motivated by a desire to better control nature. Once science is capable of defining nature in terms of empirical laws and regularities, other views about the meaning and importance of nature may be seen as irrational. In this way, the uncritical use of scientific discourse may tend to trivialize important environmental values such as the beauty of nature or respect for animals. Thus, the very use of scientific terminology may lead to unconscious devaluation of the natural environment (Everdon 1987).

One manifestation of the materialistic metaphysical undergirding of much scientific discourse found in the writings on environmental decisionmaking is the tendency to cast all environmental resources into categories of utility. Trees, rivers, animals, and plants are understood in terms of their value to humans. If science sees nature as homogeneous matter that performs certain functions, nature is likely to be understood only as valuable as its current use to humans. If a tree is seen as an oxygen producing machine and a wetland as one of nature's septic tanks, then once alternative ways are identified to meet the human need fulfilled by these natural machines, no intrinsic value is recognized. One manifestation of this problem is the fact that the debate about endangered species of plants and animals is often exclusively focused on the uses of these plants and animals to humans. For all of these reasons, scientific factual descriptions must be understood to introduce questions of value into environmental and sustainable development decisionmaking.

5.4. SCIENCE AND THE BURDEN OF PROOF

The assumptions that a scientist makes about the burden of proof is an important ethical question in decisionmaking about environmental and sustainable development problems. Many environmental laws authorizing governmental regulation put the burden of proof for showing environmental impact or risk on the governmental agencies responsible for rule-making, others allow decisionmakers to take action in the face of mere threats (see Chapter 5). Even when the burden of proof is not assigned by law, scientists often assume that they have the burden of proof in determining whether a proposed action will be environmentally harmful, because scientists have often been taught to be silent in the absence of proof. Because of pervasive scientific uncertainty in environmental and sustainable development matters, it is difficult for the party with the burden of proof to sustain its position. That is, scientists are very skilled in exposing technical weaknesses in an adversary's position when the adversary has the burden of proof. As a result, environmental decisionmakers often are troubled about what decisions to make when their knowledge likely is to be viewed to be inadequate, but when they are under pressure to take action.

One consequence of the placement of the burden of proof is that those advocating environmental protection measures must demonstrate with reasonably certainty that there is a scientific need for protection and that recommended solutions have a reasonable chance of success. Sometimes placement of this burden of proof imposes a requirement for reasonably certain scientific knowledge that may not be possible to meet. If decisionmakers postpone decisions to protect ecosystems or human health until reasonably certain scientific data are available, then such decisions should be understood to be judgments that favor the status quo. Because scientists are taught to refrain from forming conclusions in the absence of sound scientific information, and because often there may be a critical need to take actions to prevent environmental destruction where scientific information is not conclusive, scientific norms may be inconsistent with public policy and ethical principles. In other words, the scientific norm that a scientist refrain from speculation in the absence of reasonable certainty may conflict with public policy and ethical rationales (such as the precautionary principle) for protecting human health or the environment. Thus, the placement of the burden of proof is an important ethical choice that will determine when environmental and sustainable development programs may be implemented.

6. Scientists and Decisionmakers

Decisions about sustainable development problems usually entail the use of expert opinion and advice on scientific matters as well as information obtained through citizen participation. A fundamental dilemma surrounding problems of sustainable development is how to balance the need for expert scientific knowledge with the need to involve the public in the decisionmaking process. In other words, to what extent should scientific controversies about sustainable development problems be openly discussed in the public participation phases of decisionmaking?

As noted previously, proponents of scientific and technical rationality sometimes assert that decisionmaking about problems of sustainable development properly ought to be left to those people with expert knowledge. Persons holding this view apparently believe the proper role of science is to provide factual information to decisionmakers and to leave controversies about the factual information to those competent in evaluating the scientific issues. Any doubts about the validity or authority of scientific analyses and their resolutions are said properly to be left to members of the scientific community. Consequently, the conclusions of scientific analyses do not become part of broader public policy debates such as those which might pertain to such issues as what level of risk is acceptable. Practically speaking, proponents of scientific and technical rationality hold the view that the scientific and technical problems of managing large-scale and complex problems of sustainable development are enormous and that the public cannot be expected to grasp the many scientific and technical issues inherent in understanding and resolving the problems. Further, they perceive that the fundamental differences people have about how problems should be handled generate endless debate and controversy. This implies that while people and local governmental representatives with different interests may review and comment on scientific and technical documents, they would not be brought into the actual decisionmaking process regarding the complex scientific dimensions of problems of sustainable development and environmental protection.

Proponents of the so-called democratic perspective maintain that a fundamental issue for sustainable development decisionmaking is the relationship of citizenry to institutions of power. Holders of the democratic perspective believe that resolving scientific problems of sustainable development ought to be opened to citizen participation and be informed by concerns such as questions of distributive justice, concepts of freedom, and centralized versus decentralized decisionmaking. Proponents of the democratic perspective strive to utilize knowledge from holistic scientific approaches because they attempt to integrate science with other disciplinary knowledge and values in the formulation of public policy. Scientific information is viewed as one element under consideration in the decisionmaking process, but its review and evaluation should not be left to scientific experts entirely. The democratic perspective requires that the general public and decisionmakers become literate in the basic epistemological issues of evidence, uncertainty, and hypothesis testing. It also requires that scientists learn to make concise and articulate defenses of why the best available evidence supports one conclusion instead of another, and that scientists and decisionmakers understand the value-laden dimensions and implications of scientific methods and scientific uncertainty. On this view, the validity of any decision about sustainable development is determined less by particular values or goals that the decision may hold and more on a broad based and open decisionmaking process. The democratic perspective requires a high level of citizen participation in decisionmaking even on issues that are scientifically complex. It also requires a relationship of trust between agencies and the public insofar as environmental protection is concerned.

7. Science and Environmental Assessment

Chapter 35 of Agenda 21 assumes that environmental assessments should be the major tool in providing factual information in sustainable development decisionmaking. A number of countries have adopted laws requiring the assessment of environmental impacts of government actions as part of a decisionmaking process (See Chapter 5). In the United States, for example, the National Environmental Policy Act (NEPA) is an important environmental law that requires careful thinking about consequences of human action because it forces federal agencies to identify and assess the environmental and social consequences of proposed activities and their alternatives in an integrated, systematic, and publicly open manner.

Because NEPA has now been in existence for about 25 years and has served as a model for environmental legislation in other nations, a review of the strengths and weaknesses in the use of science under NEPA can provide information on the role such legislation can play in fulfilling Agenda 21 goals to improve the environmental assessment process in sustainable development decisionmaking.

7.1. SOME GOALS OF NEPA

Over the years, NEPA has been beset by debates over whether it should be implemented and judicially reviewed as a statute that requires merely alterations in an agency's procedures for considering the consequences of development projects, or as a law that defines and mandates substantive changes to protect the environment better. Generally speaking, the U.S. Supreme Court characterizes NEPA as procedural (Rodgers 1990). One consequence of this characterization is that federal agencies sometimes focus more on the procedural aspects of NEPA requirements and less on the substantive requirements, which would require a greater use of scientific information.

Despite the U.S. Supreme Court characterization that NEPA is fundamentally a procedural act, it can be argued that NEPA mandates rational and intelligent decisions based on the use of sound science that fosters protection of environmental values (Lemons et al. 1990). Boggs (1993) maintains that the debate about whether NEPA should be viewed as procedural or substantive establishes a false dichotomy that is both unnecessary and harmful to goals of environmental protection. There are several aspects of NEPA which support this point of view.

The NEPA mandates creating links between knowledge and action by authorizing and directing federal agencies to utilize a systematic, interdisciplinary approach which will ensure the integrated use of the natural and social sciences and the environmental design arts in planning and in decisionmaking. Agencies must develop methods and procedures to ensure appropriate consideration of presently nonquantified environmental amenities and values, and initiate and utilize ecological information in the planning and development of resource-oriented projects. These directives are to be met by agency preparation of detailed environmental impact statements (EIS) which will accompany all pending decisions that might significantly affect the human environment. The requirement for knowledge utilization presents difficult problems of implementation since information must be considered in good faith and must be factually sound.

The NEPA's policy statement can be considered to be clear and straightforward. For example, it declares that the government, through its agencies, will use all practicable means to fulfill the responsibilities of each generation as trustee of the environment for succeeding generations. This is implementable language which creates obligations for federal agencies analogous to those of a charitable trust. Accordingly, courts might then apply more stringent standards of "reasonableness" in reviewing NEPA cases. However, too often agencies and courts have ignored NEPA's policy statement (Caldwell 1989).

Section 102 of NEPA creates direct links between knowledge and action by requiring not just pro forma knowledge creation, but also effective use of that knowledge (Blumm 1990). In other words, the EIS is not to be considered as an end in itself, but rather a means to making better decisions and as an aid to establishing the link between what is learned through the NEPA process and how the information can contribute to decisions which further national policies and goals to protect the environment. Section 102 mandates the preparation of EISs and directly follows NEPA's statement of policy. This implies that policy is to govern agencies' actions (Yost 1990, Sagoff 1992). This goal is made explicit in Section 102(1) which directs that "to the fullest extent possible" not only the requirements for the creation and use of EISs, but also all "the policies, regulations, and public laws of the United States shall be interpreted and administered in accordance with the policies set forth in this Act..." By this line of reasoning, the linkages between knowledge, action, and policy provisions of NEPA establishes a substantively moral framework that should be followed by agencies. In other words, agencies must use sound science to make not only informed decisions, but also must make decisions that are well-reasoned and consistent with the policy provisions of NEPA to promote environmental values. Scientific practitioners can make important practical and theoretical contributions to this task.

7.2. ASSESSING THE STATUS OF SCIENCE IN ENVIRONMENTAL IMPACT ASSESSMENT

Although EISs are one of the most important documents produced during the consideration of agency projections and actions, the quality of scientific information contained in them is questionable (Lemons 1993, 1994). One way to assess the quality of EISs is to conduct postaudits which determine the actual impacts and outcomes of projects or decisions for which an EIS has been prepared. The environmental analysis document written during a project's decision process plays an important role in a postaudit, because the information contained in it must be used as the benchmark against which actual impacts of a project are measured. Consequently, scientific precepts play a role in the postauditing process: (1)insofar as the testing of hypothesis or predictions with valid empirical data are involved, and (2)because the postaudit allows environmental analysts to validate their EIS forecasts and refine their methods and models of impact prediction (Culhane 1993).

A number of post-EIS audits have been conducted for completed projects for which EISs had been prepared (Bassin 1986, Culhane et al. 1987, Culhane 1993). The conclusions of these studies were: (1)that most EISs do not contain minimally satisfactory data or information on physiographic and biological impacts to enable postaudits to be conducted; (2)that predictive accuracy of EISs was low; (3)that the variance in forecast accuracy was significant and unsystematic; (4)that no noteworthy patterns, functions, or variables emerged that explained either accurate or inaccurate forecasts; and (5)that most EISs do not provide adequate scientific information to guide monitoring of critical impacts during project implementation or to use in ameliorating adverse impacts when necessary. Consequently, impact predictions must be viewed as having considerable more uncertainty than those based on more thorough and rigorous scientific experiments.

In the United States, the EIS process and documents are monitored by the Council on Environmental Quality (CEQ) (Clark 1993). The CEQ regulations governing the process emphasize scientific quality by recognizing that it is not just high-quality, state-of-the-art science that NEPA mandates but also a systematic, interdisciplinary, integrated use of the natural and social sciences with an emphasis in ecology. The CEQ regulations are intended to force greater environmental awareness and more careful planning by federal agencies.

Malik and Bartlett (1993) assessed the extent to which federal agencies emphasize and adopt precepts of scientific quality as required by CEQ procedures by examining: (1)the quality of the scientific content and methodology of an agency's impact analysis efforts, and (2)the emphasis given the use of science in the formal rules of that agency and the specificity of those rules. They applied 18 criteria for evaluating these factors in 670 out of the 684 EISs filed by 27 federal agencies and departments in 1991. The criteria applied were fundamental because of their legal, administrative, and scientific legitimacy. The results of the assessment are not promising; no agency or department meets all criteria and most do not even meet a small number of them. According to this study, most agencies do not require even minimal attention to scientific precepts and methodology in the implementation of NEPA.

Utilizing a different approach, Ensminger and McLean (1993) reached a similar conclusion by surveying NEPA practitioners and asking them to rank their responses to 11 NEPA issues in order of importance. According to the survey responses, principal deficiencies of EISs are: (1)the tendency to use them as decision-implementation rather than decisionmaking documents; (2)the lack of effective planning and follow-up concerning mitigation measures identified by the NEPA process; and (3)the inclusion of too much detail that makes it difficult to determine what impacts are to be considered as significant.

While most discussions of NEPA center on the EIS, environmental assessments (EA) are the tool most frequently employed by federal agencies. The CEQ regulations for implementing NEPA define an EA as a concise public document designed to provide sufficient evidence and analysis for determining whether an agency proposal or activity requires the preparation of an EIS or a finding of no significant impact (FONSI). An EA must briefly discuss the need for the proposal, possible alternatives as required by Section 102(2)(e) of NEPA, and the environmental impacts of those alternatives. In order to determine whether EAs are facilitating effective NEPA compliance, the CEQ surveyed 52 federal agencies which have generated thousands of EAs each year (Blaug 1993). The survey results indicate that most EAs are not used as envisioned by the CEQ implementing regulations in three significant respects: (1)agencies rarely use an EA to determine whether an EIS is necessary; (2)agencies prepare EAs which are frequently lengthy and which fail to provide criteria to assess significant impacts; and (3)agencies appear to rely heavily on mitigation measures to justify EAs and FONSI decisions.

Section 1505.5 of CEQ's regulations authorizes agencies to provide for monitoring to assure that their decisions are carried out and specifies that mitigation or other conditions committed as part of the decision shall be implemented by the lead agency or other appropriate consenting agency. In addition, Section 1505.2(c) mandates that a monitoring and enforcement program shall be adopted and summarized where applicable for any mitigation and that it will be a part of the official record of decision for a project. It also states that the record shall indicate whether all practicable means to avoid or minimize environmental harm from the alternative selected have been adopted, and if not, why they were not.

Sections 1505.5 and 1505.2(c) primarily focus on monitoring in conjunction with the implementation of mitigation measures. Mitigation includes: (1)avoiding the impact altogether by not taking a certain action or parts of an action; (2)minimizing impacts by limiting the degree or magnitude of the action and its implementation; (3)rectifying the impact by repairing, rehabilitating, or restoring the affected environment; (4)reducing or eliminating the impact over time by preservation and maintenance operations during the life of the action; and (5)compensating for the impact by replacing or providing substitute resources or environments. Environmental monitoring needs to be used to determine the effectiveness of each of the types of mitigation measures (Smith 1989). Despite the regulations for monitoring promulgated by CEQ, post-EIS monitoring has been given minimal attention in the United States. However, many other nations appear to be more interested in post-EIS monitoring (Canter 1993a), perhaps because: (1)extant environmental monitoring programs may be minimal in scope, particularly in southern hemispheric nations; (2)a greater emphasis is placed on the life cycle of environmental management and not just on obtaining initial project approval via preparation of an EIS as is the case in the United States; and (3)of the recognition of the opportunity to gather environmental data and to use it to increase understanding of environmental stresses.

A number of investigators have pointed out the importance of establishing methods for determining appropriate criteria to assess ecological and human health risks to facilitate decisionmaking under NEPA. McCold (1991) points out that many of the most worrisome environmental problems (e.g. global climate change or loss of biodiversity) are due to the cumulative effects of many minor actions, and this suggests that agencies may be giving too little attention to actions that they do not consider significant. Current NEPA-implementing regulations 40 CFR 1500-1508 may give agencies too little guidance for responding to environmental problems of global or regional concern, and too little direction for ascertaining what magnitude of change is unacceptable. Montgomery et al. (1991) and Perrine and Montague (1991) also review the importance of addressing cumulative impacts of climate change in NEPA reviews, and offer specific methodologies to do so. Southerland (1992) has reviewed available data of major activities adversely affecting terrestrial environments resulting from land conversion, timbering, grazing, mining practices, water management, military, recreational, and other activities. Areas receiving inadequate consideration under NEPA include the loss of old-growth and mature forests, the impacts of habitat fragmentation on biological diversity and wildlife migration, and the degradation of riparian habitats supporting wildlife and endangered species. Hirsch (1993) notes that while EISs sometimes address components of biological diversity, normally they do so by focusing on endangered species rather than by addressing biodiversity problems on a regional, landscape, or ecosystem scale. Although some opportunities exist to improve the consideration of biodiversity in EISs, more specific legislative mandates probably will be required to assure adequate action to minimize losses of biological resources.

Cairns and Niederlehner (1993) discuss the structural and functional attributes necessary to consider for making informed decisions required to support sustained use of ecosystems and long-term productivity. Although measurements of ecosystem structural attributes have a long history of use in environmental assessments, researchers should be cautious in using them to detect adverse impacts. For example, the use of indicator species to detect ecosystem perturbations is problematic for several reasons: (1)the lack of complete knowledge of the responses of most organisms to stress; (2)the occurrence of species for which there is no indicator status can be a major problem, especially where flora and fauna are not characterized well; (3)the responses of organisms to stress is not uniform across types of stress; (4)emphasis on structural attributes confirms the presence of organisms but does not necessarily indicate how well they are functioning; and (5)reliance on measurement of structural attributes is more reactionary than proactive insofar as it tends to record rather than prevent damage.

On the other hand, some ecosystem function attributes are directly relevant to some of NEPA's goals, such as those pertaining to the maintenance of long-term productivity and protection of irretrievable losses of resources. These are aspects of ecosystem function and resilience, wherein the latter term refers to the ability of an ecosystem to return to some approximation of its predisturbance condition. Functional attributes also have the advantage of integrating the effects of stress on all community members and focus on properties that are essential to the sustainability of the ecosystem; they also are generalizable from one ecosystem type to another (Schaeffer et al. 1988). However, there are some problems with using functional attributes: (1)the normal or background variation in functional end points within and between nonimpacted natural systems must be known because significance of a predicted change in function can be judged only in reference to the normal operating range in similar, nonimpacted systems; (2)functional end points may be insensitive to various kinds of subtle or chronic stresses (Schindler 1987); and (3)loss of functional capacity can be compensated by increased activity of another organism. With respect to EISs, not only is it important to predict the magnitude of structural and functional effects of human activities but also their probable duration. Despite this fact, there have been few measurements of resiliency in EISs.

In December 1989, the ministers of environment and health from 29 European countries signed a charter on the environment and health which included a recommendation that environmental assessments should provide greater emphasis on the health aspects of projects (WHO 1989). In the United States, human health impacts generally have been ignored or given superficial attention in most EISs, despite the fact that the CEQ regulations explicitly mandate that the degree to which proposed actions affects public health or safety should be a criterion taken into account in determining impact significance. If assessment of human health risks are to be incorporated effectively into the environmental assessment process, it is necessary to understand the process and uncertainties. Canter (1993b) provides an overview of risk assessment procedures relevant to this process. He divides risk assessment into four major steps: (1)hazard identification, (2)dose-response assessment, (3)exposure assessment, and (4)risk characterization.

Hazard identification is the most easily recognized step in the actions of regulatory agencies (EPA 1984). It is defined as the process of determining whether exposure to an agent can cause an increase in the incidence of a health condition, and it involves characterizing the nature and strength of the evidence of causation. The dose-response assessment is the process of characterizing the relation between the dose of an agent administered or received and the incidence of an adverse health effect in the exposed populations and estimating the incidence of the effect as a function of human exposure to the agent. Many scientific arguments question the validity of the dose-response approach, insofar as there are problems with extrapolating high-dose effects to low-dose effects and in extrapolation of laboratory toxicity data collected on rats and mice to potential responses in humans. Exposure assessment is the process of measuring or estimating the intensity, frequency, and duration of human exposures to a chemical agent currently in the environment or of estimating hypothetical exposures that might arise from the release of new chemicals into the environment (NRC 1983). Exposure assessment often is used to identify feasible prospective control options and to predict the effects of available control technologies on exposure. Risk characterization is the process of estimating the incidence of a health-effect under the various conditions of human exposure described in exposure assessment. It is performed by combining the exposure and dose-response assessments. While the four major components of the risk assessment process are relatively well-defined, they have not been extensively and routinely implemented due to many types of scientific uncertainties pertaining to the use of untested and unverifiable assumptions, modeling errors, natural stochasticity, parameter errors, and the lack of basic information about the health effects of numerous chemicals (Cranor 1993).

7.3. IMPROVING ENVIRONMENTAL IMPACT STATEMENTS

Dickerson and Montgomery (1993) identified several factors influencing the quality of science in EISs: (1)scientific information generally is constrained by limitations on time and resources; (2)they are generally conducted in the context of political controversy; (3)skepticism towards scientific data or conclusions in EISs cannot be resolved by further testing of hypotheses which can be disproved by empirical results; (4)little or no peer review of EISs exists; (5)most federal agencies conduct a relatively small number of EISs on an annual basis, which makes it difficult for them to maintain a high-level of expertise in EIS preparation; and (6)development of more effective methodologies for assessment of global problems such as climate change and biodiversity is needed. They do not find the prevalence of scientific uncertainty in EISs to be particularly problematic. They maintain that the experience of most federal agencies in conducting EISs results in the most egregious projects being dismissed well in advance of the EIS process, and that EIS analysis of projects must now take into account state and federal permitting requirements which provide some assurance that environmental standards will be complied with. They also maintain that the primary purpose of an EIS is to bring together agency thinking and analysis, and a critical audience to focus on project-specific issues as opposed to having a rigorous scientific and technical component serve as the major component of EISs.

On the other hand, NEPA experts have recommended improvements in the EISs. Reilly (1992) suggests that agencies improve their ability to include the full range of scientific activities in EISs, including research, data analysis, assessment, monitoring, and quality assurance. To accomplish this, he recommends that agencies: (1)apply peer review and quality assurance to the planning and results of all scientific and technical efforts that support decisionmaking, (2)improve agencies' use of the nation's best scientists to provide a strong scientific and technical basis for decisionmaking, and (3)consider science early and often in the decisionmaking process. Based on the survey results of Ensminger and McLean (1993), recommendations to improve the NEPA process include: (1)implementing NEPA early in the program planning process; (2)ensuring effective follow-up on EISs and their associated mitigation measures; and (3)clarifying and strengthening the use and role of science in assessing cumulative impacts, in assessing significant impacts, and in the appropriate implementation of EISs. Bausch (1991) notes that many deficiencies of EISs are due to the fact that normally they are conducted on a project-by-project basis without addressing the larger national policy dilemma of how to balance environmental and economic policy objectives.

There has been considerable controversy about how EISs should deal with matters of scientific uncertainty about environmental impacts. The CEQ's regulations include requirements on incomplete and unavailable information (51 CFR 15618,15621). When information on reasonably foreseeable adverse impacts evaluated in an EIS is essential to making a reasoned choice, and costs of obtaining it are not exorbitant, the agency must secure it. However, if this information is incomplete or unavailable, that is the costs of obtaining it are exorbitant or the means of obtaining it are beyond the state-of-the-art, the agency must make clear that such information is lacking. When dealing with scientific uncertainty, an agency must follow four prescribed steps. First, it must state that the information is incomplete or unavailable. Second, it must state the relevance of the missing information. Third, it must summarize the existing credible scientific evidence relevant to its evaluation of reasonably foreseeable impacts. Fourth, it must analyze those impacts based upon theoretical approaches or scientific methods generally accepted in the scientific community. The regulation states clearly that agencies must consider impacts with low probability but catastrophic consequences as long as the analysis is supported by credible evidence that is not based on conjecture, and is within the rule of reason. This rule has been criticized for limiting consideration of catastrophic consequences to situations where there is credible scientific evidence. That is, the regulation prevents discussion of serious environmental impacts if conjecture and speculation must be used in the analysis to fill knowledge gaps. Because of the pervasive nature of scientific uncertainty, the rule is understood to be inconsistent with the precautionary principle which shifts the burden of proof to the proponent of potentially dangerous activities.

Because of the significant amount of scientific uncertainty in predicting the environmental impacts of human activities, opponents of agency decisions have often been successful in challenging agency decisions if they can demonstrate that the agency did not rigorously consider certain impacts or if they can demonstrate that an agency did not follow prescribed steps in dealing with scientific uncertainty. Alternatively, if an agency has followed these prescribed steps, then opponents of an agency's decision will have a difficult time fulfilling burden of proof requirements to overturn that decision.

Other studies have comprehensively reexamined the 25 year commitment to the environmental assessment process (Lemons 1991, 1993). Several themes have emerged from this reexamination. One, effective assessment documents must encourage an integrated approach to the broad range of environmental considerations and be dedicated to achieving and maintaining local, national, and global sustainability. In essence, this principle recognizes conclusions of the World Commission on Environment and Development (1987) and those contained in Agenda 21: (1)that wealthy nations must pursue development while reducing their demands on the environment; (2)that a commitment to sustainability entails recognition that environmental considerations extend beyond biophysical effects by including ways to alleviate poverty; and (3)that guidelines and criteria to scientifically define and measure sustainability must be promulgated.

Two, assessment documents also should identify best options, rather than merely acceptable proposals. This requires critical examination of purposes of projects and comparative evaluation of alternatives. This principle also differs from regulations generally meant only to ensure that proposed undertakings meet established standards for environmental acceptability by requiring development of undertakings that are, relative to other options, most consistent with specific goals of environmental protection as well as with general goals of sustainability. Adopting this principle would require the imposition of standard requirements to examine purposes, needs, and alternatives of proposed projects. However, the implementation of the principle is problematic because the concept of sustainability is ill-defined and methods to assess it are not developed fully (Shearman 1990).

Three, assessment documents should be established in law and must be specific, mandatory, and enforceable. In this regard, the legal developments noted by Herson and Bogdan (1991) are noteworthy. For example, the courts have interpreted cumulative impact analysis requirements liberally, and many are willing to reject NEPA documents on the grounds of inadequate cumulative impact analysis. Consequently, agencies can improve the content and likelihood of developing judicially enforceable EISs by ensuring that EISs do not segment larger projects or ignore closely connected projects under their control or jurisdiction. These types of improvements also would be obtained if decisionmaking was linked to the more specialized controls and incentives used by management and regulatory authorities for different agencies and jurisdictions. The principle would require agreement about matters of significance and adequacy of mitigation measures between different regulatory jurisdictions and authorities.

Four, approvals for projects also should be contingent on enforcement of the terms and conditions of the approvals followed by scientific monitoring of effects. To date, there has been little monitoring of the actual effects of approved undertakings and therefore little basis for judging the accuracy of impact predictions or for improving predictive science.

Five, the EIS process must include provisions for linking EIS documents with the management and regulation of existing as well as proposed new activities. In other words, this principle recognizes that environmental protection would be enhanced if EIS decisionmaking was linked to the more specialized controls and incentives used by management and regulatory authorities for different agencies and jurisdictions. The principle also would require agreement about matters of significance and adequacy of mitigation measures between different regulatory jurisdictions and authorities.

Six, improvements in the use of expert judgment and the role it plays in impact prediction are recommended (Lein 1993). Despite the fact that expert judgment increasingly is utilized as a basis for impact prediction under conditions of scientific uncertainty, there has been little recognition by decisionmakers of recent improvements and formalization of expert judgment approaches for impact prediction. Increasing the understanding of decisionmakers of expert judgment systems would contribute to improving EISs.

Seventh, it is important to improve the methods of deciding when the cost of acquiring additional scientific information necessary to address key issues associated with an EIS is exorbitant under NEPA. Under NEPA regulations, when evaluating potential adverse impacts in the absence of scientific information, an agency must assess whether the costs of acquiring the necessary information is exorbitant and therefore need not be acquired. However, the term "exorbitant" is not defined in NEPA, by the CEQ, or by related case law. Consequently, if a purposeful or high standard for exorbitant is not applied by an agency in seeking additional scientific information, a decision to forego additional research will result. Conversely, if a high standard for exorbitant is applied, then a decision to forego additional research in the absence of scientific information will be made only in cases where doing so is reasonable. Cox et al. (1993) discuss issues of exorbitant costs and make recommendations regarding how to define the term "exorbitant."

As the foregoing discussion about the use of science in environmental assessment suggests, its role can be viewed as being either one of analysis or one of planning, or both. Ideally, an analysis approach relies more on the methods of predictive science where the role of science is to generate reliable information useful in the analysis and assessment of impacts associated with present or proposed human activities and their alternatives. This information is communicated to decisionmakers with the view that it will lead to more rational decisions. This approach is considered distinct from planning and decisionmaking but linked to it through procedures for communication. In other words, the approach is procedural but does not require any particular substantive outcome.

A less common role of science in environmental assessment utilizes planning principles and procedures to determine the order of preference among a set of resource allocation choices. Preferences are based on explicit social norms that act as decision-rules to compare and rank alternative choices and to trade off environmental, economic, and social objectives that might result from alternative future scenarios. The aim in this approach is to use holistic science to facilitate the decisionmaking process by systematically selecting a preferred choice, in this instance one which foster the goals of sustainable development. This approach is more consistent with the interpretation that NEPA or laws like it should require decisions to be consistent with public policy provisions to enhance environmental quality and values. The analytic and planning roles for science in environmental assessment differ in their emphasis. Both approaches benefit from more and better scientific information. However, the former approach emphasizes the scientific analysis of environmental impacts and their alternatives based more on methods of predictive science, while the latter emphasizes a normative policy perspective in which holistic science is used to bring about preferred policy options.

The two approaches can be viewed as being complementary. However, those who favor the former approach likely will make recommendations to improve the use of science in environmental assessment by demanding more and better scientific information. For example, they might recommend the generation of data and identification of impacts on individual environmental components. Those who favor a policy perspective likely will make recommendations to improve environmental assessment by altering the priority of social norms and restructuring planning procedures and institutions. For example, recommendations might require that the identification of interactions between individual impacts on the physical and social environment and that the evaluation and implementation of mitigation techniques be used to promote specific public policy options identified through the planning process.

8. The Role of Scientists

In day-to-day decisionmaking about environmental impacts or risk, environmental protection controversies often are thought of as technical-instrumental problems. To solve such problems scientists or other technically trained personnel use scientific procedures to develop the facts about a particular environmental or human health threat and describe measures that can be taken to prevent or remediate it. For example, if risk associated with toxic substances is viewed primarily as a technical-instrumental problem, science needs to determine whether certain substances create toxic risks to humans or the environment (a question of scientific fact) and if so, what steps can be taken to mitigate against any adverse environmental effects (an instrumental question of means). Since these questions are about facts and means according to conventional wisdom, they are best answered by experts who use what they perceive to be value-neutral scientific procedures as analytical tools to find answers.

Conversely, environmental controversies can be understood as problems that most fundamentally raise ethical questions, questions about what is the right thing to do morally speaking. For example, which environmental amenities should we protect or what should we do with respect to the environment when the technical facts about consequences are uncertain? Ethically speaking, the most important kinds of questions about potential sustainable development controversies might be: (1)What should we do about potential toxic substances before science can specify consequences with certainty?; or (2)Who should have the burden of proof in demonstrating that a particular chemical poses a risk?

Of course, environmental problems usually raise both complex technical-instrumental questions and difficult ethical questions and often the latter are inextricably embedded in scientific reasoning because of the technical and practical need to make untested value-laden assumptions and inferences. Because scientists are trained not to make conclusions in the absence of sound scientific proof, if there is urgent need to take action to prevent environmental destruction where scientific proof is not conclusive, scientific norms may be inconsistent with certain ethical principles. Thus, the scientific norm that a scientist refrain from speculation in the absence of proof may conflict with the goals of precautionary public policies to protect humans or the environment for future generations.

Because there is sometimes a conflict between a scientist's role, qua scientist, which requires him or her to refrain from speculation in the absence of scientific proof and the scientist's role as citizen which requires that he or she speak out in the face of perceived environmental or human health threats, each scientist must decide what role he or she will play in sustainable development controversies. Under conditions of uncertainty, not speaking out might constitute conformance with scientific norms because it minimizes speculation. On the other hand, not speaking out also is tantamount to taking the side of the status quo behaviors or policies responsible for the perceived threats. Speaking out carries the risks and problems of speculation, but it also might carry the advantage of promoting the precautionary approaches recommended in Agenda 21. However, because a scientist's statements, qua scientist, may be confused with his or her role as a citizen, care should be exercised in identifying the basis for statements and conclusions made and, in particular, problems due to scientific uncertainty, inferences made in the face of uncertainty, simplifying schemes, and other values-laden aspects of the scientist's assumptions, inferences, or position.

9. References

Abbaganano. N. 1967. Positivism. In *Encyclopedia of Philosophy*, P. Edwards, ed. MacMillan, NY, pp. 414-419.

Bassin, N.J. 1986. *Analysis of Selected Agencies' Post-Final Environmental Impact Statement Commitments.* EMS-86-002. Environmental Management Support, Silver Springs, MD.

Bausch, C. 1991. Achieving NEPA's Purpose in the 1990s. *The Environmental Professional* 13: 95-99.

Bella, D.A., R. Jacobs, and L. Hiram. 1994. Ecological Indicators of Global Climate Change: A Research Framework. *Environmental Management* 18: 489-500.

Blaug, E.A. 1993. Use of Environmental Assessment by Federal Agencies in NEPA Implementation. *The Environmental Professional* 15: 57-65.

Blumm, M.C. 1990. The National Environmental Policy Act at Twenty: A Preface. *Environmental Law* 20: 477-483.

Boggs, J.P. 1993. Procedural vs. Substantive in NEPA Law: Cutting the Gordian Knot. *The Environmental Professional* 15: 25-33.

Briggs, J. and F. Peat. 1984. *Looking Glass Universe*. Simon and Schuster, NY.

Brown, D.A. 1987. Ethics, Science, and Environmental Regulation. *Environmental Ethics* 9: 331-350.

Cairns, J., Jr. and B.R. Niederlehner. 1993. Ecological Function and Resilience: Neglected Criteria for Environmental Impact Assessment and Ecological Risk Analysis. *The Environmental Professional* 15: 116-124.

Caldwell, L.K. 1989. A Constitutional Law for the Environment: 20 Years With NEPA Indicates the Need. *Environment* 31: 6-11, 25-28.

Canter, L.W. 1993a. The Role of Environmental Monitoring in Responsible Project Management. *The Environmental Professional* 15: 76-87.

Canter, L.W. 1993b. Pragmatic Suggestions for Incorporating Risk Assessment Principles in EIA Studies. *The Environmental Professional* 15: 125-138.

Carpenter, R.A. 1990. Biophysical Measurements of Sustainable Development. *The Environmental Professional* 12: 356-359.

Carter, J., J. Gibson, A. Carr III, and J. Azueta. 1994. Creation of the Hol Chan Marine Reserve in Belize: A Grass-Roots Approach to Barrier Reef Conservation. *The Environmental Professional* 16: 220-231.

Clark, R. 1993. The National Environmental Policy Act and the Role of the President's Council on Environmental Quality. *The Environmental Professional* 15: 4-6.

Cox, D.K., T.M. Keevin, F.T. Norris, S. Gensler, and D.J. Schaeffer. 1993. When Is a Cost Exorbitant Under NEPA? *The Environmental Professional* 15: 145-149.

Cranor, C.F. 1993. *Regulating Toxic Substances*. Oxford University Press, NY.

Culhane, P. 1993. Post-EIS Environmental Auditing: A First Step to Making Rational Environmental Assessment a Reality. *The Environmental Professional* 15: 66-75.

Culhane, P., H.P. Frieseman, and J.A. Beecher. 1987. *Forecasts and Environmental Decisionmaking*. Westview Press Social Impact assessment Services, No. 14, Boulder, CO.

Davos, C.A. 1988. Harmonizing Environmental Facts and Values: A Call for Co-Determination. *The Environmental Professional* 10: 46-53.

Dickerson, W. and J. Montgomery. 1993. Substantive Scientific and Technical Guidance for NEPA Analysis: Pitfalls in the Real World. *The Environmental Professional* 15: 7-11.

Ensminger, J.T. and R.B. McLean. 1993. Reasons and Strategies for More Effective NEPA Implementation. *The Environmental Professional* 15: 46-56.

(EPA) U.S. Environmental Protection Agency. 1984. *Risk Assessment and Management: Framework for Decisionmaking*. EPA 600/9-85-002. Washington, D.C.

Everdon, N. 1985. *The Natural Alien, Humankind and the Environment*. University of Toronto Press, Toronto, Ontario.

Funtowicz, S. O. and J. R. Ravetz. 1991. A New Scientific Methodology for Global Environmental Issues. In *Ecological Economics*, R. Costanza, ed. Columbia University Press, NY, pp. 137-152.

Held, D. 1980. *Introduction to Critical Theory: Horkheimer to Habermas*. University of California Press, Berkeley.

Herson, A.I. and K.M. Bogdan. 1991. Cumulative Impact Analysis Under NEPA: Recent Legal Developments. *The Environmental Professional* 13: 100-106.

Hirsch, A. 1993. Improving Consideration of Biodiversity in NEPA Assessments. *The Environmental Professional* 15: 103-115.

Kuhn, T. 1962. *The Theory of Scientific Revolutions*. University Chicago Press, IL.

Latin, H. 1988. Good Science, Bad Regulation, and Toxic Risk Assessment. *Yale Journal on Regulation* 5: 89-148.

Lein, J.K. 1993. Formalizing Expert Judgment in the Environmental Impact Assessment Process. *The Environmental Professional* 15: 95-102.

Lemons, J., ed. In press. *Scientific Uncertainty and Environmental Problem-Solving*. Blackwell Science, Inc., Cambridge, MA.

Lemons, J., ed. 1991. *The Environmental Professional* 13: 93-184.

Lemons, J., ed. 1993. Science and the National Environmental Policy Act. *The Environmental Professional* 15: 1-160.

Lemons, J. 1994. The Use of Science in Environmental Impact Assessment. *International Journal of Ecology and Environmental Sciences* 20: 303-315.

Lemons, J., D.A. Brown, and G.E. Varner. 1990. Congress, Consistency, and Environmental Law: Nuclear Waste at Yucca Mountain, Nevada. *Environmental Ethics* 12: 311-327.

Malik, M. and R.V. Bartlett. 1993. Formal Guidance for the Use of Science in EIA: Analysis of Agency Procedures for Implementing NEPA. *The Environmental Professional* 15: 34-45.

Maxwell, M. 1987. *From Knowledge to Wisdom*. Basil Blackwell, NY.

Mayo, D.G., and R.D. Hollander (eds.). 1991. *Acceptable Evidence*. Oxford University Press, Oxford, UK.

McCold, L.N. 1991. Reducing Global, Regional, and Cumulative Impacts with the National Environmental Policy Act. *The Environmental Professional* 13: 107-113.

Mills, C. 1985. Faith in Science. *Report From the Center For Philosophy and Public Policy* 5: 1.

Miller, A. 1993. The Role of Analytical Science in Natural Resource Decision Making. *Environmental Management* 17: 563-574.

Montgomery, J.C., B.D. Solomon, J.B. Smith, and C.W. Berish. 1991. Addressing Global Climate Change in NEPA Reviews. *The Environmental Professional* 13: 114-124.

(NRC) National Research Council. 1983. *Risk Assessment in the Federal Government: Managing the Process*. National Academy Press, Washington, D.C.

Perrine, R.L. and D.C. Montague. 1991. Global Climate Change: What Revisions to Environmental Assessment Processes Will Mitigating Strategies Require? *The Environmental Professional* 13: 125-138.

Peters, R.H. 1991. *A Critique for Ecology*. Cambridge University Press, Cambridge, UK.

Reilly, W.K. 1992. Statement before the Committee on Science, Space, and Technology, U.S. House of Representatives, March 19, Washington, D.C.

Rodgers, W.H., Jr. 1990. NEPA at Twenty: Mimicry and Recruitment in Environmental Law. *Environmental Law* 20: 485-504.

Rorty, R. 1979. *Philosophy and the Mirror of Nature*. Princeton University Press, NJ.

Sagoff, M. 1988. Ethics, Ecology, and the Environment: Integrating Science and Law. *Tennessee Law Review* 56: 78-229.

Sagoff, M. 1992. NEPA: Ethics, Economics, and Science in Environmental Law. *Law of Environmental Protection*, Vol. 2, S.M. Novick, ed. Boardman, NY, pp.9-48.4-9-102.

Schaeffer, J.J., E.E. Herricks, and H.W. Kerster. 1988. Ecosystem Health: I. Measuring Ecosystem Health. *Environmental Management* 12: 445-455.

Schindler, D.W. 1987. Detecting Ecosystem Responses to Anthropogenic Stress. *Canadian Journal of Fisheries and Aquatic Sciences* 44: 6-25.

Shearman, R. 1990. The Meaning And Ethics of Sustainability. *Environmental Management* 14:1-8.

Shrader-Frechette, K.S. 1982. Environmental Impact Assessment and the Fallacy of Unfinished Business. *Environmental Ethics* 4: 37-48.

Shrader-Frechette, K.S., and E.D. McCoy. 1993. *Method in Ecology*. Cambridge University Press, Cambridge, UK.

Smith, E.D. 1989. *Future Challenges of NEPA: A Panel Discussion*. CONF-891098-10. Oak Ridge National Laboratory, Oak Ridge, TN.

Southerland, M.T. 1992. Consideration of Terrestrial Environments in the Review of Environmental Impact Statements. *The Environmental Professional* 14: 1-9.

Taylor, S. 1984. *Making Bureaucracies Think*. Stanford University Press, Stanford, CA.

(WHO) World Health Organization Regional Office for Europe. 1989. *European Charter on Environment and Health . First European Conference on Environment and Health.* ICP/RUD 113/Conf. Doc./1 Rev. 2/2803r. Copenhagen.

World Commission on Environment and Development. 1987. *Our Common Future*. Oxford University Press, NY.

Yost, N.C. 1990. NEPA's Promise-Partially Fulfilled. *Environmental Law* 20: 533-549.

Chapter 3
THE ROLE OF ETHICS IN SUSTAINABLE DEVELOPMENT AND ENVIRONMENTAL PROTECTION DECISIONMAKING

Donald A. Brown[1]

1. Ethical Statements Defined and Distinguished from Scientific Statements

Sustainable development controversies can be understood as problems that raise scientific questions about cause or effect or ethical questions, questions about what is the "right" thing to do. This chapter examines the role of ethical reasoning in sustainable decision making.

The use of the term "ethics" in this book is meant to connote the domain of inquiry that attempts to answer the question "What is good?" Ethical statements are propositions of the form that such and such is good or bad, right or wrong, obligatory or nonobligatory. Ethics should be distinguished from the social sciences, such as sociology and psychology, which attempt to determine why individuals or groups make statements about what is good, right, or obligatory. Furthermore, ethics is concerned with prescriptive statements, which attempt to transcend relative cultural and individual positions. Science, as used in this book, is the discipline that attempts to make descriptive statements about the nature of reality through analysis of facts and experience. Science and its derivative technologies attempt to describe through an empirical methodology, facts and relationships among facts, and the laws of nature that govern the universe.

Science aims at value-free descriptions of the laws of nature. However, scientific statements often contain hidden ethical positions throughout analysis of sustainable development problems because of, among other reasons, the unavoidable need to: (1)deal with scientific uncertainty, (2)assign the burden of proof in scientific reasoning, (3)decide what resources will be spent on problem analysis, (4)choose which disciplines will be used in analysis of problems and how to synthesize various disparate disciplines in analysis, and (5)make metaphysical assumptions about the nature of reality.

It is generally accepted that science cannot deduce prescriptive statements from facts. The relationship between facts and ethical positions is of considerable controversy within the philosophical community (Callicott 1982). Although certain linguistic philosophers have held that moral reasoning by individuals does not rely on deductive modes, in which ethical conclusions follow from ethical principles, I believe that it is particularly important in developing public policy that those who make ethical assertions be required to expose ethical premises that support ethical conclusions (Marrietta 1982). That is, one cannot deduce

---

[1]Bureau of Hazardous Sites and Superfund Enforcement, Pennsylvania Department of Environmental Resources, Commonwealth of Pennsylvania, 400 Market St., Harrisburg, PA 17101-2301, U.S.A.

J. Lemons and D. A. Brown (eds.), Sustainable Development: Science, Ethics, and Public Policy, 39–51.

"ought" from "is" without supplying a new minor premise. One cannot introduce an evaluative term, such as "optimal solution," into the conclusion of an argument if the prior premises of that argument are entirely nonevaluative (e.g., dose-response statistics). Although a description of certain facts may suggest an ethical position, one cannot through a description of the facts alone deduce an ethical conclusion. An ethical system such as utilitarianism may provide the minor premise needed for ethical reasoning. For instance, if one concludes that option A will create the greatest happiness, by applying the utilitarian maxim that one should choose the option that creates the greatest happiness, one can conclude that option A is the optimal solution. From a proposition that such and such a problem creates a particular risk, one cannot, however, deduce whether that risk is acceptable without first deciding on certain criteria for acceptability. Therefore, on this largely traditional view of the logic of ethics, science cannot answer ethical questions all by itself.

This is not to say, however, that science is irrelevant to ethics. Ethics is concerned with the ends that should be chosen by people. Science is extremely important in most environmental ethical discussions, because once a particular goal is chosen, science can help evaluate various means that are available to achieve the goal. Science can also analyze which ends are feasible. If a society determines that it is good to build a nuclear power plant, for instance, science can analyze what structures or what types of institutions most effectively and safely achieve the type of power plant desired by the community. Science can also help determine what environmental impacts the community should expect from the power plant. On this view, however, science cannot fully determine whether the power plant should be built, precisely because no amount of descriptive analysis can logically certify a prescriptive course of action. Science is thus obviously fundamental to the description of the sustainable development problems discussed in this book.

In many if not most cases, sound scientific analysis is essential in any attempt to define fully most of the ethical questions considered here. Yet, as was discussed more fully in the last chapter, scientific analysis is rarely value-free.

If we agree that the question of whether society should use nuclear power is essentially an ethical question, while admitting that science can be extremely important in analyzing the facts, and thereby giving content to the ethical question, it must be admitted that there is no generally accepted consensus in the philosophical community about which ethical system to apply to any given problem. Several major philosophical systems attempt to define good, including utilitarianism, Kantian ethics, natural rights, and Rawlsian contract theory, just to name a few. Some philosophers maintain that ethical assertions should be treated as nothing more than the emotive preferences of the person making the assertion on the grounds that they are entirely subjective and relative to the person making the value judgment. Additionally, it is sometimes difficult to determine which facts should be considered and what weight should be given to these facts in any ethical calculus. Because most of the dominant Western philosophical systems make human interests the measure of value, human interests, some critics argue, are the only interests considered in Western ethical systems, with the result that such concerns as the rights of animals are not appropriately included in traditional ethical debate. In the last 20 years, as concern about environmental problems has increased, environmental philosophers have attempted to create new ethical approaches to these complex environmental problems. The next section reviews some of the strengths and weaknesses of some of the more common ethical approaches to sustainable development problems.

2. Types of Ethical Theories

The global environmental crisis has forced a revolutionary reconsideration of ethical theory that dominated public policy debates in the 19th and 20th centuries. This section reviews some of the strengths and weakness of some of the more frequently encountered ethical justifications for policies encountered in sustainable development controversies. The purpose of this section is to sketch out some of the more important issues relating to common ethical justifications for sustainable development policies rather than to treat exhaustively the ethical theories under consideration or to deal comprehensively with the range of ethical positions that might be encountered in discourse about sustainable development.

The discussion focuses on some of the more conventional Western normative theories, because most debate about sustainable development policy is already embedded in traditional discourses of science, economics, and law, which are usually justified by these Western ethical normative theories. In addition, Agenda 21 and most of the international documents on sustainable development use very traditional languages of Western science, economics, and law to describe sustainable development implementation strategies. Therefore, despite growing criticism of traditional ethical theory, most justifications for or against various sustainable policies take the form of traditional Western ethical arguments.

2.1. UTILITARIANISM

Often government officials defend sustainable development or environmental decisions on utilitarian ethical grounds. Utilitarian theory has been particularly influential in economic analysis of environmental policy and regulatory decisionmaking. Because utilitarian theory is the ethical underpinning of free-market theory and welfare economics, understanding the strengths and weaknesses of utilitarianism is particularly important to provide a basis for judging sustainable development policy discourse.

Classical utilitarian theory was developed in the 19th century by Jeremy Bentham and John Stuart Mill. Utilitarian ethical positions assert that those actions are right or good that bring about the best end results. Because utilitarianism makes an action good depending on the results or ends it achieves, utilitarianism is usually classified among "consequentialist" ethical theories. According to utilitarian theory, no act is good or bad in itself; its wrongness depends on the consequences of the action. There are two major forms of utilitarian ethics—act utilitarianism and rule utilitarianism.

Advocates of act utilitarianism assert that an act is good if it brings about the greatest good over bad results; that is, actions are good if they produce the greatest good for the greatest number. Accordingly, each individual must assess the good and bad consequences of his or her actions and choose that action that maximizes good. Advocates of act utilitarianism often eschew rules for human action such as absolute rules on killing and lying because they believe that each situation is different and that it is the consequences of any action that make an action ethical, not conformance with a rule. To determine whether an act is right or wrong, each actor must determine the consequences of each particular act. The right act is that which brings the greatest utility compared with any other alternative, where utility is often defined as happiness, pleasure, or preference satisfaction.

Because environmental policy is most often justified on the basis of cost versus benefit, environmental policy legitimations implicitly rely on act utilitarian ethical theories, although many actual justifications are often only crudely consistent with sophisticated utilitarian theories. A common criticism of act utilitarianism is that it is difficult for any person to determine what is good for another person and particularly difficult to determine with certainty the consequences of certain actions. This criticism is particularly important for environmental policy, where decisions must often be made in the face of pervasive scientific uncertainty about the consequences of action. Another criticism of act utilitarianism is that it is a waste of time or impractical to reassess each action's consequences from the beginning before taking any action.

In response to these criticisms, rule utilitarianism was developed. Rule utilitarianism holds that actors should always follow those rules that bring about the greatest good for all concerned. An act's rightness or wrongness, according to rule utilitarianism theory, therefore depends on whether the actor conforms his or her actions to a rule that if followed by all members of society would bring about the greatest happiness of the group. Rule utilitarianism overcomes an important problem with act utilitarianism in that an actor does not have to calculate the consequences of every action anew before deciding what to do. In environmental policy, rule utilitarianism is often the ethical basis for regulations that establish general rules for action as compared with a process that would make the environmental consequences of each proposed action determine whether any act is permissible. For instance, certain environmental rules require that classes of hazardous substances be treated to the same degree before they may be released into the environment even though there are significant differences in the toxicological properties of individual substances and therefore differences in environmental impacts. Such a rule can be justified on rule utilitarian grounds, because following such a rule creates the greatest utility for all members of society. This is so because it would be too time-consuming and administratively complex to create different treatment objectives and enforcement schemes for each substance that has different toxicological properties. Faced with such rules, regulated parties often argue that treatment obligations should be based on the toxicological properties of each substance, implicitly taking a contrary act utilitarian position.

Both forms of utilitarianism are subject to the following criticisms:

1. The environmental crisis that first received international attention in the late 1960s has forced ethicists to reassess long-accepted ethical theory. Along with other traditional Western ethical positions, utilitarian justifications for environmental policy are challenged for making human interests the measure of value. As will be discussed in more detail in the following section on biocentric ethics, because utilitarianism is usually understood as an attempt to maximize human happiness, it often ignores and undermines the value of nonhuman entities such as plants and animals. Although some philosophers argue that utilitarian calculations could be adjusted to consider the happiness or suffering of any sentient beings, most utilitarian justifications for environmental policy fail to do so. As a result, most utilitarian arguments employed in support of environmental policy are crude utilitarian calculations that many supporters of utilitarianism would reject as incomplete. (For a discussion of the ability to extend utilitarianism to nonhuman entities, see Sharpe 1994.)
2. Utilitarian calculations raise ethical issues that cannot be easily answered from within a utilitarian system (MacIntyre 1977). A utilitarian must decide, for instance, which alternatives will be considered in the utilitarian calculus, which consequences of a given action will be considered, whose assessments of harms and benefits will be allowed, and what time scale will be used in assessing the consequences. The utilitarian analysis therefore often rests upon imprecise judgments of, and prior to, the utility calculus itself.
3. Utilitarian methodology cannot easily accommodate the rights individuals may have to be protected from certain pollutants or to be spared from death-threatening situations. Most contemporary philosophers hold that utilitarian approaches must be supplemented by other ethical approaches, such as the Kantian approach discussed in the next section, which stress such concepts as rights, justice, and due process as fundamental.
4. Utilitarian justifications of environmental policy often assume that questions of value can be reduced to a quantifiable amount. That amount is often money measured in market transactions. Quantification of environmental health and

benefits, however, is often difficult and sometimes impossible. What, for instance, is the value of a human life or of an endangered eagle? These difficult issues are often unsatisfactorily dealt with in utilitarian calculations by determining value by measuring individuals' willingness to pay.

5. Utilitarian theory cannot determine how benefits or costs of subgroups should be distributed among potential winners and losers. That is, utilitarian theory is indifferent in respect to distribution of utility as long as total utility is maximized. Along this line, regulatory decisions that are based on cost-benefit analysis often fail to identify which subgroups in the population will suffer the burden of any decision even though those who suffer from environmental problems are a different group from those who might be asked to pay for the cost of environmental regulation. As a result, most commentators agree that utilitarianism should be supplemented by concepts of distributive justice.
6. Utilitarianism has difficulty in dealing with valuing the impact of environmental problems on future generations. This difficulty is particularly problematic when considering potential environmental impacts that may persist as problems for long periods of time, such as nuclear waste disposal or greenhouse gas buildup in the atmosphere. How should future generations' interests be considered in the calculations, and what present value of these interests should be identified in the utilitarian calculus?

Although utilitarian calculations could be adjusted to take into consideration some of these criticisms, utilitarian justifications supporting environmental and public policy positions rarely deal with these criticisms or, in dealing with them, raise additional ethical questions that can't be decided on utilitarian grounds. For instance, cost-benefit analysis sometimes deals with future generations' interests but discounts future value in a way that raises questions about the rights of future generations.

2.2. RIGHTS AND DUTIES THEORIES

The second most commonly encountered ethical justifications for environmental or sustainable development policy are justifications that ground action or inaction on the notion that certain actions are intrinsically right or wrong. Because such justifications assume that rightness or wrongness turns on some higher standards than the consequences of the action, these justifications often are classified among ethical theories known as nonconsequentialist theories. Nonconsequentialist theories usually speak of rights of individuals to take certain actions or of duties to refrain from action. Theories that ground ethical behavior on duties are usually classified as deontological theories because the word deontological is derived from the Greek word for "duty."

Deontological ethical theories are often encountered in environmental policy discourse in reaction to the limits of utilitarian theory. For instance, because of the difficulty in knowing with certainty the consequences of certain human actions on the environment and therefore determining the rightness of the consequences, persons that support environmental policies often talk of duties to other humans or future generations to refrain from action in the face of uncertainty.

The best-known Western deontological theory is that of the 18th-century German philosopher Immanuel Kant. Kant believed that humans could derive absolute rules of morality based on reasoning alone. Kant held that to determine whether an action was right or wrong, one should look to the rule authorizing the action and ask if logically it could be universalized. If it could not, it would be unreasonable for individuals to give permission to themselves to do things that they could not advocate should be a generally applicable rule for others in society. Thus Kant held that humans are ethical beings because they are rational beings who can freely choose to follow rules that others should be bound by. If one chooses to follow a rule that everyone should follow, one is acting morally. The fundamental ethical duty, called by Kant the "categorical imperative," is to act only in those ways that could be acceptable to all rational beings.

An important corollary of the categorical imperative is the notion that because other humans are rational beings, they should always be treated as ends and never as means. Many constitutional protections, especially those that deal with concepts of due process, are grounded in this Kantian notion that humans are to be respected as autonomous individuals and not be treated as a means for other humans' interests. Such ideas often become manifest in environmental policy when persons assert that individuals or future generations should not have to suffer the pollution caused by another without consent. Many future sustainable development policies will undoubtably have to deal with issues about the scope of individual procedural rights to consent to sustainable development decisions.

Another corollary of the categorical imperative has sometimes become the basis for resisting environmental regulation. That is, because rational beings have the freedom to follow rules that don't transgress the rights of others, actions not transgressing the rights of others are assumed to be prima facie legitimate. For instance, in the face of government regulation that limits use of property for environmental purposes, some assert that their rights have been violated in the absence of proof that their individual use of property has harmed others. As will be discussed in the following chapter on law, property rights theories create important ethical challenges to some sustainable development policies.

The obvious strength of the Kantian approach to environmental policy is that it is an accepted ethical basis for asserting that some actions are wrong without fully knowing the consequences of actions. Most Kantians would assert, for instance, that government lying about pollution levels is always wrong even if the lie resulted in no harm to the person lied to. Similarly, Kantians will argue that humans have a right to a healthy environment undiminished by the actions of another without consent.

Some of the limitations of the Kantian approach to sustainable development problems are as follows:

1. Kantian ethics is difficult if not impossible to apply to most environmental controversies because Kantian ethics is always difficult to apply where a decision involves conflicts between two competing goods. That is, because environmental controversies often involve conflicts between goals that are not objectionable in themselves, such as the use of property for food and shelter versus habitat protection, the categorical imperative is not useful in giving advice about such conflicts. That is, the categorical imperative instructs individuals to act according to rules that can be universalized but gives no advice on how these rules are to be formed.
2. According to Kantian ethics, only humans or other rational beings are intrinsically valuable. This is particularly problematic for environmental controversies because the Kantian ethical system does not contain any basis for giving value to any being that is not rational and therefore provides no basis for asserting intrinsic value for plants and animals. Although some Kantians have attempted to extend rights to nonhumans, most philosophers see rational human beings as the only compelling locus of Kantian morality.

2.3. THEORIES OF JUSTICE

In addition to grounding sustainable development and environmental policy on utility, rights, or duty, proponents of various policy decisions often base positions on grounds of justice or fairness. Theories of justice are particularly important to environmental and sustainable development policy because the other two common theories used as justification for policy, both utilitarian and deontological theories, give no obvious guidance on how goods or bads should be distributed throughout society. Because much environmental and sustainable development policy is grounded on utilitarian justifications, a common criticism of many environmental and sustainable decisions is the failure to satisfy concepts of distributive justice.

Four types of justice claims are encountered in public policy debates. Theories of distributive justice prescribe ways of distributing the benefits and burdens of society. Exchange justice deals with fair exchange of remuneration for products or services. Theories of social justice deal with the duty to be fair to all members of society. The duty of restitutive justice requires that when one harms a moral subject, the person causing the harm must make restitution.

Although questions about these four types of justice may be encountered in public policy debates on sustainable development, understanding issues of distributive justice is particularly important. This is so because environmental policy decisionmaking usually fails to consider distributional effects of proposed actions. For instance, a common justification for environmental policy is some form of cost-benefit analysis, which rarely identifies which subgroups in society will obtain the benefits or who will suffer the burdens of the decision under consideration. That is, cost-benefit-based decisions consider aggregate costs versus benefits, not the fairness of how benefits and burdens will be distributed.

Principles of distributive justice assert that benefits and burdens should be distributed according to concepts of equality or merit or some combination of these two. Principles of distributive justice attempt to resolve tensions between treating people equally and making distributions on the basis of merit or deservedness.

Because of this common failure of environmental policymaking to consider the distributional effects of decisions, a new force has emerged in the last few years, variously called the environmental racism or environmental justice movement. The objective of this increasingly important new force in environmental policy is to see that the distributional effects of environmental policy are disclosed and considered.

Moreover, the very nature of sustainable development problems suggests that questions of when, which, and how benefits of government actions should be distributed will continue to grow in importance. This is so because the urgent need to move toward a sustainable society is based on the conclusion that the world cannot solve environmental problems without solving problems of poverty. Moreover, in solving problems of poverty, the world community can no longer assume unlimited ability to consume natural resources and expand economically. Because the world must solve problems of poverty in the face of limits, issues of fairness of distribution become much more important. Therefore, a fundamentally important issue for sustainable development policy is what is the level of income, education, and other basic goods of society that is entailed by concepts of distributive justice in a world of limits.

Such questions deal with issues of fairness for existing human populations. Questions of distributive justice become even thornier if policy attempts to consider the distributional effects on future generations or nonhumans.

2.4. ANTHROPOCENTRIC VERSUS BIOCENTRIC ETHICS

As a serious academic discipline, environmental ethics began in the early 1970s in reaction to the ever more frequent environmental problems that were getting worldwide attention in the 1960s. Although writers such as Henry David Thoreau, John Muir, Aldo Leopold, and Albert Schweitzer were writing about ethical problems caused by industrial society's disregard of nature in the 19th and early 20th centuries, not until the international media began to give attention to the growing environmental crisis did environmental ethics emerge as a matter for serious study. In the 1960s the world witnessed environmental disasters such as Japan's Minamata mercury pollution, California's Santa Barbara oil spill, Ohio's Cuyahoga River catching fire because of pollution, the destruction of Germany's forests caused by air pollution, and the contamination of food and groundwater by pesticides documented in Rachel Carson's *Silent Spring*. As a result, various challenges to the ethical systems that had dominated Western thought started to arise prominently in debates about the causes of the increasing environmental threats.

The emerging environmental crisis created a powerful challenge to Western ethical systems because ethicists were forced for the first time to consider and articulate the value of nonhuman species of plants and animals. Because utilitarian and deontological ethics, and more prominent Western theories of justice, did not make environmental entities the focus of ethical concern, the emerging environmental crisis became a strong challenge to Western ethical systems. In fact, some concerned with environmental problems charged that Western ethical systems were at least in part responsible for the environmental crisis for their failure to value anything other than human happiness or interests and the consequential devaluing of animals, plants, and ecosystems.

Initially, much of the environmental ethics literature dealt with reforming consequentialist and deontological ethical systems so that nonhuman species would be considered along with humans. These approaches in need of reform were categorized as "anthropocentric," for they relied on human values to proscribe value to nature. From the beginning of the emergence of environmental ethics as an academic discipline, environmental ethicists began talking about "biocentric" ethics, i.e., ethical systems that make all of life, including nonhumans, the center of value. Since the 1970s several approaches to valuing nature that may be loosely classified as biocentric in orientation have become common themes in the environmental ethics literature. These include: (1)biocentric ethics, which attempt to extend utilitarian and deontological theories to all sentient beings; (2)ecocentric theories that make entire ecosystems or environmental communities the center of value; and (3)deep ecology, which holds that humans, nonhumans, and biotic communities are so intrinsically related to each other that it is a mistake to consider them separately. Other challenges to Western ethical approaches have come from non-Western ethical perspectives such as Buddhism and ecofeminism.

Much, if not most, of the literature in environmental ethics continues to be concerned with how humans should value nature. Should humans regard nature as inherently or only as instrumentally valuable? Should we regard nature as spiritually empowered or as a wild force to be subdued? These questions are bound to continue to be central to public policy controversies in the years ahead. Although, with the emergence of the problem of sustainable development with its assumption that environmental problems cannot be cured by increasing the economic pie, issues of distributive justice are likely to vie with questions of methods of valuation as the common focus of debates in environmental ethics and public policy.

2.5. THE ROLE OF RELIGION

Of course, in addition to making value judgments in conformance with various ethical systems, many persons throughout the world look to religious or cultural traditions to find normative rules that define appropriate relationships between humans and nature. Some have argued that because of the urgency of the need for political and personal transformation to avert widespread environmental destruction in the next century, only a radical change in values can bring about the behavioral change needed to protect life on earth. As a result, calls for a new "environmental or sustainable development ethic" have been growing. These calls have come recently from mainstream religious, political, and scientific organizations. Some religious writers have argued that the world's religions must become a major force in implementing a new sustainable development ethic (Hamed 1994, Tucker 1994).

Some religious leaders have argued that only a truly religious transformation can bring about the needed shift in behavior, a cosmological paradigm shift that enables humans to see themselves as part of, rather than apart from, the web of life (Tucker 1994). They argue that such a change in vision is necessary to allow the world and its plants and animals to become reenchanted, to restore a sense of the sacred in nature that was lost during the industrialization period of human history. As a result, much of the literature in environmental ethics deals with the strengths and weaknesses of the ethical underpinnings of an environmental ethic in various religious traditions, including Christianity, Judaism, Buddhism, Taoism, Hinduism, Islam, Native American, Jainism, and others (Tucker and Grim 1993).

3. Distributive Justice and The Good Life

As stated above, the environmental crisis has been viewed to create a serious challenge to dominant Western ethical systems because of their failure to consider the ethical relationship between humans and nature. If the assumptions made in *Our Common Future* and Agenda 21 about the worsening global environmental crisis and the concurrent need to eliminate poverty are correct, the problems that must be faced in implementing Agenda 21 call into question additional aspects of the world view that had been dominant during the period of world industrialization. The global environmental crisis challenges previous assumptions about international and intranational distributive justice and also forces government to think in a new way about the ethical ends of government.

In the world of limits envisioned by Agenda 21, humans must develop programs that are based on a fair international and domestic economic order. That is, because many of the global environmental problems such as the greenhouse effect have been caused largely by the developed nations, the global environmental crisis raises new and sometimes troubling questions of distributive justice concerning the duties of the developed world to assume burdens of preventing future environmental damage, assist the developing world in moving toward sustainable development, or compensate the developing world for past damage.

If the global environmental crisis creates development limits, the international community cannot be considered moral if it protects the powerful nations while neglecting the developing nations (Barahona 1994, Heredia 1994). Moreover, if the global environmental problems are to solved by new technologies, the poor nations cannot rely on expensive technology to solve environmental problems because they are already struggling to survive. Therefore, increases in technology costs mean the poor lose unless the richer nations accept responsibility for the pollution that they have created. Many economists from Western countries argue that the solution to global environmental problems lies in making sure that all human activities are forced to internalize environmental costs fully. However, full costs are high when people are poor and low when the ability to pay is high (Ott 1994). Therefore, development can only be sustainable when equity is made the leading goal (Heredia 1994). When we see earth as a commons with a limited carrying capacity, the question of environmental protection remains inseparable from a need to create an equitable economic international order (Paden 1994, Rolston 1994). Thus, an important question that must be paid attention to in implementation of Agenda 21 is "Who pays for environmental protection?"

Moreover, consumer lifestyles in the developed world demand new, convenient, disposable goods with more and more consumers to respond to ever-increasing purchasing compulsion (Quinn and Petrick 1994). There is a vast gulf between the wants of most of the developed world and the needs of those in the developing world (Westra 1994). For these reasons, arguably the developed world must both modify its consumptive behavior and assist the developing world in moving toward sustainable development.

Therefore, the global environmental threats that influenced the development of Agenda 21 raise unprecedented issues of international distributive justice. Consequently, the international order has been dramatically transformed by Earth Summit documents that have added two new rights to the list of recognized universal rights: (1)the right to an equitable international order, and (2)the right to an environment with health and dignity (Rolston 1994).

Also, in a world of limits, if nations can no longer assume they are able to solve problems of poverty by simply expanding economically, the global environmental crisis raises urgent new questions of distributive justice between rich and poor within nations. That is, sustainable development problems will force nations to consider anew questions of social justice within their borders.

In addition to these questions of international and intranational distributive justice, sustainable development decisionmaking in a world of limits will force governments to consider other ethical questions about the role of government. For example, when water is scarce, governments have a duty to find "the most valuable use" (Priscoli 1994). The most valuable use of water of course raises questions about relative distribution, reallocation, sustainability of existing supplies, social unrest, and governments' search for the good life. In a world of limits, therefore, not only do sustainable development decisions raise ethical questions, but also these decisions must be understood to be positions taken about ethics and values that challenge existing assumptions about the role of government.

4. The Ethical Assumptions of Agenda 21

Most commentators agree that Agenda 21 is a historically important but flawed attempt to move the international community toward solving the twin problems of environment and poverty. That is, if the international community desperately needs to adopt an ethic that respects other forms of life and future generations, serious problems with Agenda 21 must be recognized during future years of implementation. For instance, in Agenda 21 there is no concern for the preservation of the natural environment for itself, that is, for reasons that transcend human purposes or goals (Katz 1994, Sagoff 1994). Katz argues, for instance, that rather than exhibiting a respect for nature, Agenda 21 is a compromise among the rich and poor nations, industrial workers, private individuals, and indigenous people.

In a similar vein, Agenda 21 follows a narrow anthropocentric approach to the twin problems of environment and development; that is, in Agenda 21 only human beings are of ultimate concern (Weiss 1994). There is little evidence in Agenda 21 that moral duties may be owed to the natural environment, to animals, plants, and ecosystems, and that these things may possess a value of their own independent of their usefulness to humans (Weiss 1994); nature in Agenda 21 is only valuable for its potential use to humans as a resource (Paden 1994). On a similar theme, the Agenda 21 chapter on biodiversity, Chapter 15, pays little attention to the loss of biodiversity but includes a strong emphasis on the development of biotechnology (Sagoff 1994). Sagoff points out: "There is nothing wrong, of course, with wanting to promote biotechnology. But what has this got to do with—and why is this the central chapter on—ending the mass extinction of species and the destruction of their habitats?" Sagoff concludes: "[Agenda 21's] authors apparently assume that the principle reason to protect biodiversity is to maintain an enormous inventory of raw materials for eventual economic applications, for example, in biotechnology." Sagoff goes on to note that Agenda 21 dismisses all noneconomic reasons for protecting biodiversity, for example, religious, ethical, and cultural values, even though these may provide strong grounds for conservation.

5. Theoretical Versus Applied Ethics

As stated above, calls for a new sustainable development have recently come from many throughout the world, including philosophers and religious, environmental, and political leaders. There are several reasons, however, to be concerned about pinning the world's hopes exclusively to the call for creation of a new sustainable ethic that will guide the day-to-day practices of human life.

First, such calls for a new sustainable ethic sometimes seem to assume that an ethic can be created by simply calling for its creation, without understanding how ethical positions arise out of existing social practices and needs or within existing ethical belief systems. For example, because any person struggling to survive is likely to be influenced in his or her view of "right" or "wrong" by the day-to-day forces against which he or she must struggle, no simple call for a sustainable living ethic is likely to be greatly influential until dire threats to survival are eliminated. In order to survive, the poor must sell whatever is marketable (Weir 1994).

Second, merely calling for a sustainable development ethic may be useless unless ethical discourse is integrated into the languages in which sustainable development problems are discussed. Agenda 21 is expected to be largely implemented by national and local governments that translate the general principles of Agenda 21 into specific programs and laws. The most likely response of these governments is to assign these laws and programs to government agencies staffed largely by engineers, scientists, lawyers, economists, and other experts who are expected to implement laws and manage sustainability problems. As is discussed more fully in the chapters on science, economics, and law, norms of these disciplines often conflict with a sustainable development ethic, and the languages of science, economics and law often hide important ethical positions. Therefore, calls for a new ethic must be supplemented by ethical analysis of day-to-day sustainable decisions and integration of ethics into the scientific, economic, and legal languages in which these decisions are discussed.

Along this line, several commentators have criticized Agenda 21 for its over-emphasis on technological rationality, science, and technical solutions to solve the twin problems of environment and development (Landen 1994). One commentator described technological rationality as that which relies on better techniques to solve human problems while ignoring discussion of human ends or goals, that is, value rationality (Heyd 1994). One of the problems with over-reliance on technological rationality is that it withdraws from analysis of power relations (Heyd 1994). Because technological rationality doesn't consider power relations, differences in power are never subject to public scrutiny. Further, technological rationality also tends to treat nature as a mere storehouse of resources for development, thus ignoring other ways of valuing the nonhuman environment. In this way, technological rationality may increase the tendency to treat nature as a mere commodity available for human use. For this reason, it is critically important that the ethical dimensions of science and technology be understood in implementing Agenda 21.

Third, formulations of new sustainable ethical principles tend to be at such a level of abstraction that they may not be helpful in resolving the kinds of concrete conflicts that will come up in real sustainable development problems. For instance, although many may agree that humans should not treat animals as commodities available for any human use, does this rule prevent farmers from killing wolves who attack cattle? Do humans have a duty to protect every individual of any animal species, or is the duty to protect the species when it is threatened with extinction by human behavior? Should humans protect ecosystems or communities of plants and animals within ecosystems? Are modifications of ecosystems permissible if the integrity of the ecosystem is maintained? Do humans have a duty to restore ecosystems that have already been degraded by human activity, and if so, to what level of its prior state? General principles of sustainable development will probably not be helpful in resolving these and other concrete sustainable development problems. In other fields of applied ethics such as biomedical ethics, philosophers have concluded that understanding specific contexts and interpretation of background facts is often more important than having ethical theories that facilitate making deductive ethical judgments (Winkler 1993). If other fields of applied ethics have found that the understanding of specific facts, consequences, and historical contexts of proposed actions are extraordinarily important when making ethical judgments, applied ethics in sustainable development decisionmaking will require even more attention to contextual details. This is so because sustainable development decisions will often involve making decisions in the face of pervasive uncertainty about consequences and making judgments that will resolve conflicts among a variety of human, animal, and environmental interests and concerns.

For these reasons, those concerned with the ethical dimensions of sustainable development decisionmaking must be particularly concerned with integrating ethical discourse into the details of day-to-day sustainable development decisions.

6. References

Barahona, R.G. 1994. Ethical Questions Embedded in Biodiversity Provisions of Agenda 21. In *Proceedings on Ethical Dimensions of the United Nations Program on Environment and Development Agenda 21*, D.A. Brown, ed. Earth Ethics Research Group, Harrisburg, PA, pp. 11-16.

Callicott, J.B. 1982. Hume's Is-Ought Dichotomy and the Relation of Ecology to Leopold's Land Ethic. *Environmental Ethics* 4: 16-74.

Hamed, S. 1994. Seeing the Environment Through Islamic Eyes. In *Proceedings on Ethical Dimensions of the United Nations Program on Environment and Development Agenda 21*, D.A. Brown, ed. Earth Ethics Research Group, Harrisburg, PA, pp. 95-114.

Heredia, R.C. 1994. The Ethical Implications of a Global Climate Change: A Third World Perspective. In *Proceedings on Ethical Dimensions of the United Nations Program on Environment and Development Agenda 21*, D.A. Brown, ed. Earth Ethics Research Group, Harrisburg, PA, pp. 121-128.

Heyd, T. 1994. Agenda 21 and the Limits of Technological Rationality. In *Proceedings on Ethical Dimensions of the United Nations Program on Environment and Development Agenda 21*, D.A. Brown, ed. Earth Ethics Research Group, Harrisburg, PA, pp. 129-138.

Katz, M. 1994. Sustainable Development and Imperialism: Ethical Reflections on Agenda 21. In *Proceedings on Ethical Dimensions of the United Nations Program on Environment and Development Agenda 21*, D.A. Brown, ed. Earth Ethics Research Group, Harrisburg, PA, pp. 153-156.

Landen, L. 1994. Environmental Decisions as Human Decisions: The Appropriate Role of Science, As Revealed by Looking at the Atmosphere. In *Proceedings on Ethical Dimensions of the United Nations Program on Environment and Development Agenda 21*, D.A. Brown, ed. Earth Ethics Research Group, Harrisburg, PA, pp. 157-166.

MacIntyre, A. 1977. Utilitarianism and Cost/Benefit Analysis: An Essay on the Relevance of Moral Philosophy to Bureaucratic Theory. *Values in the Electric Power Industry*, K.M. Sayre, ed. Notre Dame University Press, South Bend, IN.

Marrietta, D.E., Jr. 1982. Knowledge and Obligation in Environmental Ethics: A Phenomenological Approach. *Environmental Ethics* 4: 15-62.

Ott, K. 1994. Ethical Questions Embedded in Water Resource Provisions of UN Agenda 21. In *Proceedings on Ethical Dimensions of the United Nations Program on Environment and Development Agenda 21*, D.A. Brown, ed. Earth Ethics Research Group, Harrisburg, PA, pp. 217-234.

Paden, R. 1994. Free Trade and Sustainable Development: The Moral Basis of Agenda 21 and Its Problems. In *Proceedings on Ethical Dimensions of the United Nations Program on Environment and Development Agenda 21*, D.A. Brown, ed. Earth Ethics Research Group, Harrisburg, PA, pp. 235-246.

Priscoli, J.D. 1994. A Perspective on Some Emerging Ethical Dilemmas in Water Resources Management. In *Proceedings on Ethical Dimensions of the United Nations Program on Environment and Development Agenda 21*, D.A. Brown, ed. Earth Ethics Research Group, Harrisburg, PA, pp. 247-260.

Quinn, J.F., and J.A. Petrick. 1994. Agenda 21: Biodiversity and Responsible Land Use Planning and Management: Economic, Legal, Scientific and Ethical Implications of Modernist, Post-Modernist and Universalist Environmental Philosophies. In *Proceedings on Ethical Dimensions of the United Nations Program on Environment and Development Agenda 21*, D.A. Brown, ed. Earth Ethics Research Group, Harrisburg, PA, pp. 261-266.

Rolston H., III. 1994. Environmental Protection and an Equitable International Order: Ethics After the Earth Summit. In *Proceedings on Ethical Dimensions of the United Nations Program on Environment and Development Agenda 21*, D.A. Brown, ed. Earth Ethics Research Group, Harrisburg, PA, pp. 267-284.

Sagoff, M. 1994. Biodiversity and Agenda 21: Ethical Considerations. In *Proceedings on Ethical Dimensions of the United Nations Program on Environment and Development Agenda 21*, D.A. Brown, ed. Earth Ethics Research Group, Harrisburg, PA, pp. 289-300.

Sharpe, V. 1994. *Ethical Theory and the Demands of Sustainability.* An unpublished paper presented at the American Chemical Society Meeting on Ethics and Risk Assessment, Washington, DC.

Tucker, M.E. 1994. The Role of Religion in Forming an Environmental Ethics. In *Proceedings on Ethical Dimensions of the United Nations Program on Environment and Development Agenda 21*, D.A. Brown, ed. Earth Ethics Research Group, Harrisburg, PA, pp. 315-320.

Tucker, M.E. and J.A. Grim. 1993. *Worldviews and Ecology.* Bucknell University Press, Lewisburg, PA.

Weir, J. 1994. Who Can Save the Earth? Agenda 21 and Professional Expertise. In *Proceedings on Ethical Dimensions of the United Nations Program on Environment and Development Agenda 21*, D.A. Brown, ed. Earth Ethics Research Group, Harrisburg, PA, pp. 339-354.

Weiss, S.D. 1994. Ethical Issues in Toxic Waste Export. In *Proceedings on Ethical Dimensions of the United Nations Program on Environment and Development Agenda 21*, D.A. Brown, ed. Earth Ethics Research Group, Harrisburg, PA, pp. 355-382.

Westra, L. 1994. Ecosystem Integrity and Agenda 21 Science, Sustainability and Public Policy. In *Proceedings on Ethical Dimensions of the United Nations Program on Environment and Development Agenda 21*, D.A. Brown, ed. Earth Ethics Research Group, Harrisburg, PA, pp. 382-392.

Winkler, E.R. 1993. From Kantianism to Contextualism: The Rise and Fall of the Paradigm Theory of Bioethics. In *Applied Ethics: A Reader*, E.R. Winkler and J.R. Coombs, eds. Blackwell, Cambridge, p. 342.

Chapter 4
THE ROLE OF ECONOMICS IN SUSTAINABLE DEVELOPMENT AND ENVIRONMENTAL PROTECTION

Donald A. Brown[1]

1. Introduction

In the last few years, some economists who are members of a movement in academic economics known as ecological economics have begun to examine and criticize the use of many traditional economic methods of analysis often used in environmental decisionmaking (see, e.g., Costanza 1991). These criticisms have often focused on traditional economic approaches to valuing environmental entities, the use of cost-benefit analysis in environmental decisionmaking, and systems of national accounting such as gross national product (GNP). Yet for the most part, economic analyses in environmental and sustainable decisionmaking continues to be based on more traditional economic methods and approaches that have been criticized in the growing ecological economics literature. This chapter examines some of the more controversial issues that arise in the application of traditional economic analysis methods to sustainable development and environmental controversies.

As the 20th century ends, most developed and developing nations assume that a major function of government, if not the most fundamental and important one, is to provide citizens with a healthy economy and opportunities for meaningful employment. Although great differences exist among nations in the degree of state ownership or control of the means of production, for most nations economic policy is the cornerstone of both domestic and foreign policy. Because of the success of Western democracies in achieving wealth and the demise of communism, most nations have adopted market economies in which the means of production are privately held.

Although humans throughout history have always exchanged goods for value in one form of market or another, which goods were produced and who produced them were determined by social norms embedded in religious and cultural traditions. Especially after the fall of communism, the acceptance of market economies throughout the world has been so rapid that the relationship between markets and society has been altered drastically (Brown 1994). That is, where economic systems were once submerged in general social relations, now markets provide the framework of society. The prevalence of the international markets has so invaded national governments' and individuals' choices that most political discourse in many parts of the developed world is almost exclusively about how to organize citizens' lives and corporations so that they can be players in the international economy. Where jobs

[1]Bureau of Hazardous Sites and Superfund Enforcement, Pennsylvania Department of Environmental Resources, Commonwealth of Pennsylvania, 400 Market St., Harrisburg, PA 17101-2301, U.S.A.

J. Lemons and D. A. Brown (eds.), Sustainable Development: Science, Ethics, and Public Policy, 52–63.

and roles in society were often determined by custom, religion, and other social forces, roles played by individuals and employment opportunities within nations are now often largely determined by international market forces. One consequence of this social transformation is that issues about the means of achieving economic efficiencies, often referred to as instrumental rationality, tend to dominate public discourse about the ends and purposes of government or values rationality. An example of this phenomenon drawn from the United States is that political discourse tends to be more and more exclusively devoted to economic issues such as job creation rather than questions about what is the proper function of government in attaining the good life for citizens or in protecting the environment.

Perhaps because issues of efficiency and instrumental rationality dominate government policymaking, when policymakers recognize that particular value questions have to be considered in environmental and sustainable decisionmaking, the values are often discussed in terms of economic considerations, in costs and benefits, efficient markets, or changes to GNP. Yet some environmentalists and philosophers argue that environmental and sustainable development decisions are political questions that cannot be reduced to economic questions (Sagoff 1988). This chapter reviews these and some other of the more important issues that arise out of the use of these economic tests in development of sustainable development and economic policies.

Those who support market approaches to government economic policies often argue that successes of Western democracies in achieving unparalleled levels of prosperity, lengthened life, and stable social and political systems is strong evidence for economic policies that minimize government interference in market mechanisms. Moreover, according to proponents of unrestricted markets, market economies have achieved a maximization of individual liberty and wealth without the need for inefficient and liberty-reducing government steering mechanisms. Given this record of success, it is argued, a heavy burden of proof must be met by proposals for significant interference with market mechanisms.

Some sustainable development programs proposals would require significant intervention in market mechanisms—for example, those that propose to limit the use of land for human settlements so as to maximize energy efficiency and minimize urban sprawl. Therefore, some sustainable development program proposals are likely to be strongly opposed on economic grounds both by those who have direct economic interests and by economists who believe in minimum interference with market mechanisms.

Conversely, some economists and environmentalists (e.g., Daly and Cobb 1989) have argued that markets, left up to their own devices, most often fail to protect human health and the environment because of the inability of market exchanges to: (1)internalize damages to animals, plants, and the natural environment; (2)provide for certain public goods; (3)anticipate environmental damage; and (4)properly value environmental entities. Economists argue that these failures occur because: (1)interests of those not represented in the market exchanges are affected by the market exchange; (2)no persons own the environment and therefore no market in environmental entities is created; (3)environmental impacts of market exchanges are not understood by participants to the market exchange; (4)the market methods of assigning value are inconsistent with nonmonetary values of environmental entities; and (5)markets assume unlimited natural resources.

As discussed more fully below, government economic policy must ultimately decide how it will reconcile potential conflicts among the need for public goods, desires to correct market failures, the need to use nonrenewable resources, the desire to make national industries competitive in international markets, and wishes to extend the benefits of a healthy economy to all citizens. Thus, in the implementation of sustainable development policy that attempts to attain both environmental protection and development objectives, a nation's economic policy is both relevant to development targets and potentially in conflict with environmental protection goals. As a result, the economic policy pursued by any nation is of central importance to achieving sustainable development objectives. That is, for instance, the percentage of a nation's GNP devoted to environmental protection, the creation of public environmental goods such as parks or wilderness areas, and the elimination of poverty will affect both the speed and the ability of a nation to put itself on a sustainable path. Similarly, a nation's tax policy is also a strong potential tool in implementing sustainable development programs. In other words, any nation may tax activities that are unsustainable and thereby create economic incentives to move toward more sustainable behavior. Similarly, tax policies that subsidize unsustainable behavior, such as mineral depletion allowances, can work to undermine sustainable development goals.

To resolve potential conflicts among environmental, economic, and social goals of any nation, Chapter 8 of Agenda 21 calls for integration of economic, environmental, and social planning. Implicit in this call for integration is the expectation that nations should choose economic policies that do not conflict with social and environmental objectives. This chapter reviews some of the conflicts between neoclassical economic policy designed to expand economic growth and the environmental and social goals of Agenda 21.

2. Ethics, Efficiency, and Sustainable Development

2.1. ARGUMENTS FOR THE USE OF MARKET MECHANISMS IN SUSTAINABLE DEVELOPMENT POLICYMAKING

Several arguments are commonly made in support of the use of market mechanisms in environmental and sustainable development policymaking. These include claims that market mechanisms should be preferred because markets: (1)maximize efficiency, (2)promote liberty, (3)provide benefits to those that most deserve them, and (4)result in the most mutual advantage to citizens. The following discussion considers the most common arguments for the use of market mechanisms, namely, the arguments about efficiency and liberty (Buchanan 1985).

2.1.1. *Efficiency*

Economists are concerned with how people can best satisfy desires and needs given limited amounts of labor, resources, and material that are available to satisfy human wants. Thus, economics has been concerned with which government programs or policies will maximize human satisfaction in the face of the inability to produce unlimited amounts of goods. This branch of economic theory is generally referred to as welfare economics.

In pursuit of welfare maximization, economists often recommend a strategy generally referred to as Pareto optimization. A Pareto optimal solution is one that provides the best mix of results in any decision so as to create maximum human satisfaction (Goodland and Ledec 1994). A premise of Pareto optimally is the notion that an economic system is good and right if its distribution of consumer and other goods is such that it maximizes human satisfaction. A solution is Pareto optimal if there is no alternative that makes one person better off and no one worse off. In this way, Pareto optimal solutions can be said to maximize human satisfaction. Supporters of market economies argue that under ideal conditions, markets will lead to Pareto optimal solutions because persons will continue to trade in markets until no one will be made better off by the trade (Daly and Cobb 1989). Markets are therefore understood to be efficient in maximizing human welfare.

Because neoclassical economics assumes that an ideal economic system is one that is efficient in maximization of human satisfaction, many economists strongly support markets and oppose interference with market mechanisms because interferences reduce efficiency and therefore decrease human welfare. These economists also assert that market mechanisms should be applied to sustainable development problems because they will result in the most efficient and best use of scarce resources, thereby minimizing waste of energy and other precious natural resources.

Because economists support markets due to their ability to maximize human satisfaction, market-based decisions are usually understood, as a matter of ethics, as utilitarian justifications for public policy. However, as is discussed in more detail below, implicit in the economist's prescription of markets is the notion that welfare will be determined solely by individual human preferences registered as prices in markets. Because many utilitarians argue that utility or human happiness should be measured by factors other than human preferences, market economic theory is generally understood to be, at best, a form of utilitarianism, known as preference utilitarianism.

2.1.2. *Liberty*

Another common justification for market-based decisions is that in voluntary exchange markets, individual liberty is enhanced. That is, in markets, individuals are free to choose which products to buy. Because individual choices determine what will be produced, economic power is separated from political power. For this reason, in market economies, power is less concentrated in political organizations than in political systems where governments make economic decisions (Brown 1994). Liberty is also enhanced in market economies because for markets to work, there must be decentralization of employment activities. Based upon the assumption that some degree of individual liberty is desirable in sustainable societies, proponents of market economies therefore argue that markets are the best hope for protection of individual liberties.

2.2. CRITICISMS OF THE USE OF MARKET MECHANISMS IN SUSTAINABLE DEVELOPMENT POLICYMAKING

2.2.1. *Failure to Cover Market Externalities*

In addition to the support of market mechanisms, market economy governments also assume the role of correcting market failures, where market failures are understood to be costs to society that do not show up in economic exchanges and are not borne by the parties to the exchange. Programs directed at fixing market failures are often identified as programs that internalize the externalities, that is, as programs that force certain costs to be implicitly considered within market decisions that would otherwise be imposed on others than the participants in the market exchange. Externalities are sometimes defined as unintentional side-effects of production and consumption that affect a third party positively or negatively (Turner et al. 1993).

The classic example of market failures is pollution because, although the prices of industrial commodities cover costs of labor, materials, and a return on investment, they usually fail to include costs of environmental damage. Although market failures are understood to occur whenever prices fail to cover costs not considered in an exchange, market failure is a particularly serious problem in environmental matters because of the resultant environmental and human health impacts. For example, the price charged for steel does not cover costs of the damage to the atmosphere caused by carbon dioxide emissions from the steel manufacturing process. Therefore, the price of steel does not internalize all the costs of making steel.

Some market failures are caused by governments inappropriately intervening in markets. Particularly troubling to some economists are economic policies that subsidize prices of scarce resources, because these policies encourage waste of those resources. For example, government subsidies of water for farmers in arid areas such as the U.S. Southwest encourage the waste of a valuable resource.

Because of the common failure of markets to cover costs of environmental damage in economic exchanges, most economists agree that these market failures should be corrected. How they should be corrected, however, is a matter of considerable controversy. Some economists support regulatory mechanisms designed to minimize or eliminate the damage. These regulatory mechanisms often take the form of statutes or regulations that prohibit or minimize environmentally damaging activities through the setting of standards. Such regulatory approaches are often referred to as "command-and-control" approaches to environmental regulation when they specify specific standards. Other economists argue for market solutions such as effluent taxes or emissions trading regimes to correct the market failures.

For many economists, effluent taxes and emissions trading policies are preferable to command-and-control regulatory mechanisms because economically based approaches: (1)create economic incentives to reduce pollution; (2)allow more flexibility than regulatory approaches in achieving pollution reduction strategies and thereby increase efficiency of pollution reduction efforts; and (3)reduce transaction costs of environmental compliance, because more limited technical information needs to be given to regulators than is required in command-and-control regulatory schemes.

Those that support command-and-control regulatory approaches often point to: (1)the inappropriateness of putting market-based prices on environmental entities, (2)a variety of ethical problems with preference utilitarianism that is the ethical basis for the market-based approaches, and (3)the need to supplement market incentives with legally enforceable standards because of the need to assure that environmental entities do not suffer long-term damage.

Proponents of effluent taxes argue that the way to control pollution is to tax emissions of pollutants at a level that will correct for market failures. The major problem with effluent taxes is setting the tax at the right level. From an economic perspective, if the tax is set too low, there will be too much environmental damage; but if the tax is too high, production that is needed to meet development goals will be curtailed. Moreover, setting the tax at the right level requires sophisticated understanding of how the activity that is being taxed will affect the environment and what is the value of the environmental entity that may be damaged by the activity under consideration. Because of the pervasive nature of scientific uncertainty in environmental matters, it is therefore difficult to know what is the right level of emissions and the right price to attach to these emissions, information that is necessary to set the tax. If the tax is based upon the willingness of individuals to pay for the environmental entity, from an ethical perspective there is a problem of using market valuations to determine what the environmental entity is worth.

Similarly, emissions trading policies work by allowing parties to buy, sell, and bank rights to discharge pollutants. For instance, if the government determines that 5,000 pounds of sulphur dioxide may be released before ambient air quality standards are met, an allocation among discharges is first made to distribute the 5,000 pounds, and then parties may sell or bank reductions below their allocation. As in the effluent tax situation, there are problems in setting the initial allocation, both in determining levels that will assure protection from environmental damage and in determining how to distribute the allocation among the parties. From an ethical perspective, emissions trading policies are sometimes thought to be problematic in that they grant to those that have financial resources the right to pollute.

Although many economists support some government intervention in markets to correct for market failures, not all economists agree on which problems create market failures. Pollution that kills whales and eagles may be accepted as a market failure by some economists who deny that a market failure exists when development destroys the habitat of a skunk or coyote. Is the use of prime farmland for urban development or the filling in of a wetland for a parking lot a market failure? Economists tend to define pollution as environmental degradation that decreases human welfare or base the determination on some other value judgement. Because willingness to pay defines human welfare according to economic welfare theory, markets only fail when pollution damages environmental entities that humans are willing to pay for. However, because others argue that environmental entities should be recognized to carry nonmarket values (see discussion below), some argue that market failures should be seen as a fact that follows from a value judgment rather than an economic calculation (Sagoff 1988, Brown 1994).

2.2.2. *The Propensity of Market Valuation to Treat Environmental Entities as Commodities*

A serious limitation to neoclassical economic theory applied to environmental problems is how value is assigned to environmental entities. Because individual preferences measured in economic exchanges are what determine value in neoclassical economic theory, economists usually assign value to environmental entities by measuring the willingness of humans to pay for those entities. Thus, the value of scenery or the noise of an industrial plant is determined by comparing real estate prices of affected properties with prices of nonaffected properties. The value of fish killed by an oil spill is the price per pound of fish at a local fish market. In this way, economic tests tend to treat animals, plants, and other environmental entities as commodities whose value is determined solely by their use to humans. Therefore, a strict market approach offers no limits as to what should be sold in the market, because value is determined solely by the prices individuals are willing to pay.

Some philosophers assert that nonhuman sentient beings and other environmental entities have a right to exist that transcends their use value to humans. From this perspective, the fate of environmental entities should not be determined by human subjective preferences. In addition, some economists and philosophers argue that values of the environment should be understood to transcend market preferences (Sagoff 1982). If animals or plants have intrinsic value or other nonmonetary value, market valuations may understate or ignore these values. Because market-based prices only measure the strength of human desires, they do not reflect values that are not dependent on subjective human preferences.

By way of contrast, for many matters, society has determined that certain conduct is so destructive of important nonmonetary values that the behavior should not be tolerated, even if some persons are willing to pay for the conduct. Child labor laws, for instance, represent a value judgment about the undesirability of exploiting children that is antithetical to free market ideology. Similarly, the use of certain drugs and the right to torture animals are activities that societies have decided are not available for sale. Because of the desire to give each person in a democracy an equal amount of political power, it is always wrong to pay for votes. Thus society has demonstrated that there are some values that should not be jeopardized by markets. Certain values held by individuals are derived not from their role as consumers but as citizens who hold values such as honesty and courage, virtues which are diminished by unfettered market theories of value. Following this line of thinking, a common criticism of market-based valuation applied to environmental and sustainable development decisionmaking is that the market treats things that should not be for sale as tradable commodities and thereby undermines other important societal values and interests.

In the last decade, economists have derived a variety of techniques that attempt to overcome some of the limitations of market-based evaluations applied to natural environments. These techniques include, among others: (1)the replacement cost technique (RCT), which determines the value of damaged resources on the basis of the cost to restore the damage, and (2)the contingent valuation method (CVM), which attempts to determine the willingness of persons to pay for the existence of environmental entities (Turner et al. 1993). Despite the benefits of RCT and CVM over other market techniques of assigning value, both RCT and CVM have limitations as methods for determining the value of environmental entities.

The RCT approach cannot be applied to environmental damages for which restoration is not feasible. For instance, it is often infeasible to restore river sediments contaminated with hazardous substances, because attempts to restore the river bed usually make the contamination worse due to inevitable spreading of contamination through necessary stirring of contaminated sediments. In a similar vein, many ecological systems once damaged cannot be completely restored. A site mined for coal by strip mining, for instance, can never be put back into premining conditions despite the ability successfully to grow a vegetative cover that eventually makes the site aesthetically pleasing. Also, an extinct species of plant or animal is extinct forever; it cannot be restored. RCT is also challenged by some economists because it sometimes results in damage estimates that are believed to be grossly disproportionate to the real value. For example, full restoration of a 10-acre site contaminated by hazardous substances in Douglasville, Pennsylvania, has been estimated at $500 million, an amount that some assert is grossly disproportionate to the site's actual value.

The CVM is a technique that attempts to determine the value of environmental entities by asking individuals what they are willing to pay not to protect environmental entities and resources in an unconsumed and undamaged state. Thus, CVM is understood to be a way of getting at existence value. Criticisms of CVM are, however, many. Like other economic tests of value, CVM is based on individual subjective preferences and not on intrinsic or other nonmonetary values. CVM has also been strongly criticized because it is so subjective that it is viewed to be not reliable. When CVM has been used in environmental litigation, it has been often vigorously attacked as being so untrustworthy that it should not be admitted into evidence.

2.2.3. *The Failure to Produce Public Goods*

Another limitation of unfettered markets in the implementation of sustainable development policy is the inability of markets to provide certain public goods that people care about that are not sold in markets (Schrecker 1984). Such public goods as defense, police, health, education, and parks are goods that markets, left up to their own devices, cannot readily create. Even the strongest supporters of free markets usually recognize some role for government in producing these public goods, because they recognize that markets may not produce them. However, the scope of which "goods" should be considered as "public" and be produced through public sector financing or other forms of market intervention is, of course, a matter of considerable controversy. Are wilderness areas, wetlands, uncontaminated water supplies, habitat for endangered species, scenic vistas, and/or a nation's biodiversity public goods?

Programs that attempt to increase a nation's storehouse of public goods sometimes conflict with economic policies focused on wealth maximization and development. For instance, money raised from private-sector taxes to finance public parks makes the private sector less competitive in an international market compared with nations that do not subject their private sector to equivalent levels of taxation. Government economic policy must therefore resolve conflicts between desires to increase public goods and wealth-maximizing market policies.

2.2.4. *Ethical Limitations of Preference Utilitarianism*

As stated above, market-based decisions are usually understood, as a matter of ethics, as utilitarian justifications for public policy. As stated in Chapter 3, utilitarian justifications for sustainable development policy can be criticized from a variety of ethical perspectives. Utilitarian justifications for market-based economic policy have been criticized on the following grounds:

1. Utilitarian justifications for environmental policy are challenged for making human interests the measure of value, thereby ignoring arguments about the intrinsic value of environmental entities. Economic analyses not only make human interests the only measure of value but also assume that individual preferences expressed as willingness to pay are equal to human interests. Economic analysis is therefore sometimes referred to as preference utilitarianism. Preference utilitarianism is further criticized as conflating individual interest with societal interests. In preference utilitarianism, each preference is given equal weight within a quantitative assessment.

   Some philosophers have argued that preference utilitarianism is fundamentally incompatible with utilitarian theory because true utilitarianism requires maximization of happiness, not the quantitative maximization of sheer preferences (Sagoff 1988). That is, utilitarianism requires that desires be ranked according to their ability to promote happiness, while preference utilitarianism assumes that every acted-upon desire is equal in its ability to satisfy happiness. By equating what is desired with what is valuable, preference utilitarianism denies that some desires should be desired more than others.
2. Utilitarian calculations raise ethical issues that cannot be easily answered from within a utilitarian system. A utilitarian must decide, for instance, which alternatives will be considered in the utilitarian calculus, which consequences of a given action will be considered, whose assessments of harms and benefits will be allowed, and what time scale will be used in assessing the consequences. The utilitarian analysis therefore often rests upon imprecise judgments that are prior to the utility calculus itself. Preference utilitarianism, the often-cited justification for economic-based public policy, ignores these problems by assuming that whatever individuals want is good, and therefore there is no need for individuals to consider alternatives to nor consequences of whatever they are willing to pay for.
3. Utilitarian methodology cannot easily accommodate rights, individuals, or other sentient beings that may have to be protected from environmental threats. For instance, there is nothing in utilitarian theory that prevents someone from polluting another's drinking water with health-threatening hazardous substances if total human happiness is increased by the activity that causes the pollution.
4. Utilitarian justifications of environmental policy often assume that questions of value can be reduced to a quantifiable amount. For economic theory, that amount is money measured in market transactions. Quantification of environmental health and benefits, however, is often difficult and sometimes impossible. What, for instance, is the value of a human life or of an endangered eagle? The utilitarian need to quantify the value of life in monetary terms may lead to undervaluation of the life when considered from a nonmonetary or intrinsic value perspective.
5. Utilitarian theory cannot determine how benefits or costs of subgroups should be distributed among potential winners and losers. That is, utilitarian theory is indifferent in respect to distribution of utility as long as total utility is maximized.
6. Utilitarianism has difficulty in dealing with valuing the impact of environmental problems on future generations. This difficulty is particularly problematic when considering potential environmental impacts that may persist for long periods of time, such as nuclear waste disposal or greenhouse gas buildup in the atmosphere. Economic theory deals with these issues by discounting future values, a procedure

that is particularly problematic when applied to environmental or sustainable development policy.

7. Preference utilitarianism makes willingness to pay the measure of value but ignores ability to pay. In environmental affairs, ignoring ability to pay creates several distortions in the valuing process, including questions about whether individuals with limited ability to pay would pay more to avoid environmental threats if they had greater ability to pay.

2.2.5. *The Problem of Discounting for the Future*

Economic theory usually has difficulty assigning values to environmental events that take place in the future, especially the distant future (Spash 1993). Neoclassical economic theory assumes that value will be applied to future environmental events in the same way the market assigns value to future events. Because persons will pay less now for something that will be received in the future than for the same thing in the present, economists typically discount the value of future events. Discounting allows decisionmakers to compare the present value of both costs and benefits so that decisions that need to be made in the present can be made under a common quantitative measure. Economists discount future events because if someone waits a year for another person to pay them $1,000 that is now owed, it means that the person will have to forgo the interest that the money could earn if it were invested now. Therefore, money is worth less in the future that it is now. Accordingly, economists discount the value of future events. Economists defend such discounting on the basis that this is how real people value future events in markets. In other words, the economists' position is in part an empirical claim about how individuals value future events.

The amount of the discount rate is a matter of considerable controversy. The present value of $100 twenty years from now is $37.70 at a 5 percent rate or $14.90 at 10 percent. The higher the discount rate, the lower the present value. Because many environmental decisions are based on cost-benefit analysis where the present value of future benefits is compared against present costs, high discount rates applied to future benefits may greatly distort the importance of those benefits. Because future environmental benefits may extend hundreds of years into the future, present value of long-term future benefits will be extraordinarily small if the benefits are discounted over the entire period of concern. High discount rates discourage projects with long-term benefits, while promoting projects with long-term costs.

Discounting environmental events in this way can be criticized in a number of ways:

1. Discounting assumes that only contemporary individuals count in assigning values to future events, thereby ignoring the rights of future generations. This assumption frames our responsibility for the future in terms of our own returns on investment, not on the rights or interests of future generations.
2. Discounting, like other economic techniques that attempt to maximize welfare, ignores questions of distributive justice because it is indifferent to how costs and benefits will be allocated among subgroups. The particular distributive justice problem created by discounting is, of course, intergenerational distributive justice. That is so because discounting assumes that the environmental entities that may be affected in the future by a current environmental decision are resources of those making the decision in the present. The farther in the future the environmental concern, the less is the overlap between those who make the decision and those who suffer the consequences.
3. Discounting, like other economic quantitative methods of analysis, translates future values into present market prices, thereby ignoring nonmarket-based values such as issues of intrinsic valuation of other sentient species or other environmental entities.

Discounting could, by way of example, lead to the economically rational extinction of species, even in the absence of other market failures.

3. Limits of Cost-Benefit Analysis

Governments often require that proposed new environmental regulatory programs pass a cost-benefit analysis (CBA) before being adopted. A CBA is often used as a decision rule in environmental matters because decisionmakers assume that governments should choose options that maximize or at least improve human welfare or utility. Therefore, according to some economists, government should choose those sustainable development options for which benefits most exceed costs (Leonard and Zecekchauser 1983).

Like other welfare-maximizing techniques considered in this chapter, CBA can be criticized for: (1)valuing environmental entities on the basis of willingness to pay and thereby ignoring other, nonmonetary values; (2)failing to consider questions of intergenerational and intragenerational distributive justice; and (3)using discounting to determine the present value of future benefits.

In response to these criticisms, proponents of CBA argue that: (1)nonmonetary considerations could be considered by decisionmakers in conjunction with CBA under other considerations, (2)questions of distributive justice can be built into CBA so that the distributional effects of a decision are considered or so that those who are adversely affected by CBA-based decisions are compensated, and (3)there is nothing inherent in CBA that mandates that a given discount level be applied to future environmental benefits. These proponents of CBA propose, therefore, the use of CBA in a way that is sensitive to some of the criticism of applied economic policy discussed in this chapter.

Another frequent criticism of CBA is that to make the comparison required by the analysis, it is necessary to reduce costs and benefits to a single scale, that is, money (Kelman 1981, Schultz 1994). This need for quantification creates several problems. First, some benefits are extraordinarily difficult to quantify because the entities under consideration are not traded on markets and therefore have no monetary value. For instance, how does one establish the value of human life or increases in respiratory ailments? As we have seen, economists often attempt to solve the problem of the absence of available monetary values for nontraded environmental entities by devising methods such as contingent valuation. However, these attempts to find ways of putting a value on things that do not have market values are subjected to additional criticisms and problems. Second, this need to quantify benefits in terms of money also has the propensity to bias the monetary values at the expense of nonmonetary values. Third, the quantification of benefits of a proposed regulatory action is only possible when the environmental impacts of proposed actions are understood. For instance, Shultz (1994) points out that until recently, CBA of air pollution programs would probably not have considered the effect on depletion of stratospheric ozone because the effects of chlorofluorocarbons and other pollutants on it has only recently been recognized. Because environmental and sustainable development decisions will have to be made in the face of pervasive scientific uncertainty, the ability to quantify benefits will always be constrained by limitations in understanding impacts of human actions on the environment.

4. Problems With Systems of National Accounting

Nations typically attempt to quantify their economic activity through a variety of national accounting practices such as GNP. The GNP is a measure of a nation's total market value of final goods and services produced by its economy during a year. The GNP or its derivative, gross domestic product (GDP), has been traditionally used by governments as an indicator of national economic health and as a target of national policy objectives. Policies that increase GNP are viewed as desirable because they increase human welfare within a nation. Yet maximization of GNP as a national policy goal may be inconsistent with environmental and sustainable development goals for the following reasons:

1. GNP does not include the loss to natural capital, such as wetlands, that occurs in development (Daly and Cobb 1989). Existing methods of computing national health count things such as bullets and bombs heavily while ignoring destruction of natural resources. Sustainable development losses to the natural resources of a nation through development should be subtracted from the measures of productivity calculated in GNP. That is, as losses to forests, wildlife, and soil are created by development projects, accounts of national health should reflect those losses in the same way capital depreciation is subtracted from corporate profit.
2. GNP is an inadequate indicator of sustainability because it counts unsustainable behaviors on the same scale as sustainable behaviors. For instance, GNP counts environmentally destructive projects on an equal scale with projects that enhance energy efficiency. As a result, Henderson (1994) argues that current GNP/GDP practices are responsible in part for felled forests, air and water pollution, exhausted soils, depleted natural resources, and holes in the ozone layer. Therefore, GNP should be reformed to provide separate indicators of sustainable health. Henderson argues that these indicators should not be integrated into dollars that measure productivity, because such attempts to translate environmental values to dollars will wind up hiding controversial ethical issues behind economic calculations.
3. GNP fails to count many important quality-of-life variables while ignoring other factors that decrease the quality of life. For instance, the value of parks, once they are created, is not counted in GNP, while costs of increased stress and disease caused by economic decisions to move businesses out of the country are ignored in GNP. For these reasons, many argue that GNP must be supplemented by quality-of-life indicators.

Because of these problems with GNP, some economists have advocated several reforms (see, e.g., Costanza 1991, Eakins 1992). These reforms have included: (1)adding or subtracting from GNP various elements to produce an adjusted national product (ANP) that is a better indicator of sustainable income, (2)supplementing this ANP with figures for nonmarket production and social and environmental indicators to give a broader framework within which welfare can be evaluated, and (3)combining ANP with other indicators to give an overall index of welfare, which could replace GNP altogether as a social welfare indicator.

5. References

Brown, P. 1994. *Restoring the Public Trust.* Beacon Press, Boston.

Buchanan, A. 1985. *Ethics, Efficiency, and the Market.* Rowman and Allanheld, Totowa, NJ.

Costanza, R., ed. 1991. *Ecological Economics: The Science and Management of Sustainability.* Columbia University Press, New York.

Daly, H., and J. Cobb. 1989. *For the Common Good.* Beacon Press, Boston.

Eakins, P. 1992. *The Gaia Atlas of Green Economics.* Anchor Books, New York.

Goodland, R., and G. Ledec. 1994. Neoclassical Economics and Principles of Sustainable Development. *Applied Ethics*, E. Winkler and J. Coombs, eds. Blackwell, Oxford, U.K.

Henderson, H. 1994. Redefining Wealth and Progress. In *Proceedings on Ethical Dimensions of the United Nations Program on Environment and Development Agenda 21*, D.A. Brown, ed. Earth Ethics Research Group, Harrisburg, PA, pp. 115-120.

Leonard, H., and R. Zecekchauser. 1983. Cost-Benefit Analysis Defended. In *Social Conflict and Environmental Law*, A. Greenbaum, A. Wellington, and E. Baar, eds. Captus Press, Vancouver, Canada, p. 41.

Kelman, S. 1981. Cost-Benefit Analysis: An Ethical Critique. In *Regulation* (Jan.-Feb.): 74-82.

Sagoff, M. 1982. At the Shrine of Our Lady of Fatima or Why Political Questions Are Not All Economic. *Arizona Law Review* 23: 1281-1298.

Sagoff, M. 1988. *The Economy of the Earth*. Cambridge University Press, New York.

Schrecker. 1984. The Limits of Cost-Benefit Analysis. In *Social Conflict and Environmental Law*, A. Greenbaum, A. Wellington, and E. Baar, eds. Captus Press, Vancouver, Canada, p. 47.

Schultz, P.C. 1994. Cost-Benefit Analysis and Environmental Policy. In *Ecological Economics* 9(3): 197-199.

Spash, C.L. 1993. Economics, Ethics, and Long-Term Environmental Damages. *Environmental Ethics* 15(2): 117-132.

Turner, R.K, D. Pearce, and I. Batemen. 1993. *Environmental Economics*. Johns Hopkins University Press, Baltimore.

Chapter 5
THE ROLE OF LAW IN SUSTAINABLE DEVELOPMENT AND ENVIRONMENTAL PROTECTION DECISIONMAKING

Donald A. Brown[1]

1. Introduction—Law and Sustainable Development

This chapter first examines some of the problems with international and national law that will need to be faced in moving toward sustainable development. This introduction is followed by a review of the role of law in sustainable development decisionmaking. Next, the chapter discusses the problems with the science-law and economics-law interfaces and then concludes with a discussion of some additional problems in implementing sustainable development law.

Chapter 39 of Agenda 21 calls for a review of international environmental law "to evaluate and promote the efficacy of that law and to promote the integration of environment and development policies...." International environmental law consists of: (1)bilateral or multilateral treaties, (2)binding acts of international organizations, (3)rules of customary international law, and (4)judgments of international courts and tribunals (Sands 1994). Over a thousand treaties deal with environmental matters. These treaties have been the most frequent method of creating binding international rules relating to the environment (Birnie and Boyle 1992). For the most part, this body of international law prescribes acceptable behavior between nations rather than the behavior of individuals within nations. In many instances, international law acts as a framework for the development of more specific national law. The United Nations currently has 181 member states and another dozen or so that do not participate. Most nations have adopted a body of national environmental law that governs the action of individuals relating to the environment. Despite this large body of international and national environmental law, sustainable development cannot yet be said to be a norm of international law. There are several reasons for this.

First, although many developed nations support environmental controls, the developing nations give priority to development matters and resist international imposition of environmental controls (Birnie and Boyle 1992). As a result, fewer developing states have become parties to binding treaties on environmental matters. Unless the developed nations assist the developing world economically or through the transfer of technology, the developing nations are unlikely to move on the path toward sustainable development.

Second, nations have traditionally resisted international agreements, which are generally perceived to diminish national sovereignty. Sovereignty means that each nation has

[1]Bureau of Hazardous Sites and Superfund Enforcement, Pennsylvania Department of Environmental Resources, Commonwealth of Pennsylvania, 400 Market St., Harrisburg, PA 17101-2301, U.S.A.

J. Lemons and D. A. Brown (eds.), Sustainable Development: Science, Ethics, and Public Policy, 64–76.

exclusive jurisdiction over its territory and the natural resources found there, as well as a corresponding duty not to intervene in the areas of exclusive jurisdiction of other nations. This concept of territorial sovereignty does not coexist comfortably with an environmental order that consists of a biosphere of interdependent ecosystems, which do not conform to the artificial territorial boundaries between nations. Because of threats to sovereignty, nations have resisted agreeing to limit exploitation of natural resources within their territory.

Third, the body of international and national environmental law is often conflicting and piecemeal in a way that allows major environmental insults to go unchecked by the law. For instance, the United States has laws on endangered species but not on biodiversity, and laws relating to mineral extraction contain subsidies that often clash with the goals of environmental regulation (Futrell 1994).

In response to these gaps in international law, Agenda 21 was adopted. It contains 40 chapters focused on solving the twin problems of environmental protection and sustainable development. Each of the 40 chapters includes a statement of objectives, an outline of required activities, guidelines for developing a framework of action, necessary institutional changes, and identification of implementation needs. It calls for the governments not only to adopt new environmental laws and programs but also to commit to significant economic, social, and international institutional reforms. In addition, Agenda 21 creates international expectations that nations integrate environmental, economic, and social planning.

In addition to Agenda 21, the treaties on climate change and biodiverisity and many other regional and international agreements will have an effect on the speed and degree to which the international community moves toward sustainable development. As stated more fully in Chapter 1, some of the most important international agreements include: (1)the Convention on the Prevention of Marine Pollution by Dumping of Waste and Other Material, generally known as the 1972 London Dumping Convention; (2)the 1982 U.N. Convention on the Law of the Seas; (3)the 1986 Vienna Convention for the Protection of the Ozone Layer as amended by the 1988 Montreal Protocol; (4)the 1989 Basel Convention on the Control of Transboundary Movements of Hazardous Wastes and Their Disposal; (5)the Program of Action on World Population agreed to at the United Nations Conference on Population and Environment at Cairo in September 1994; and (6)agreements reached at the United Nations World Summit for Social Development in Copenhagen in March 1995.

In December 1992, the U.N. created the United Nations Commission on Sustainable Development (CSD) to monitor worldwide progress of implementation of Agenda 21. The CSD is comprised of the representatives of 53 United Nations member nations elected by the Economic and Social Council of the United Nations for three-year terms.

Among other duties, the CSD is charged with the responsibility of reviewing national reports submitted to the United Nations regarding each nation's activities undertaken to implement Agenda 21. Governments are asked to submit voluntary annual national reports to the CSD for its review. The CSD is also expressly charged with reviewing national commitments on financial targets set by Agenda 21 and, in particular, the Agenda 21 goal that each developed nation should provide 0.7 percent of gross national product for development assistance to the developing world. Many hope that the CSD will become an important international force to move the world community forward on a sustainable development path.

2. The Role of Law in Sustainable Development Decisionmaking

Much of the hoped-for transition from unsustainable to sustainable development can be accomplished without the compulsion of law. Throughout the world, many architects have designed environmentally friendly buildings, engineers have constructed nonpolluting transportation systems, farmers have adopted low-impact farming practices, and corporations have adopted aggressive pollution prevention strategies, all without the coercion of law.

Perhaps the best hope for the greatest worldwide gains toward sustainable development lies in motivated individuals voluntarily reducing or eliminating unsustainable behavior. Yet the need of individuals and governments to share common and scarce resources will undoubtably create conflict and controversy. Because of the inevitable conflict over resource use, the adoption of environmental protection and sustainable development laws is an important element in moving the world toward a sustainable future. Moreover, much existing law creates incentives for unsustainable behavior, such as laws that subsidize consumption of nonrenewable resources. Therefore, much existing law needs to be amended to accomplish sustainable development goals.

How does the law fit into day-to-day sustainable development problem-solving schemes? According to the model followed in much of the developed world, government agencies staffed largely by technical experts break down environmental problems into "objective" technical problems and a "subjective" policy component. When making a decision, the decisionmaker looks at the guidance contained in the law, then applies the objective technical facts to the decision rule found in the law. For instance, in the United States under the National Environmental Policy Act (NEPA), 42 U.S.C. 4321 *et seq.*, for actions of the federal government that have potential significant environmental impacts, decisionmakers are required carefully to identify environmental impacts of proposed actions. In some cases, laws adopted by legislatures give general prescriptive guidance but leave to administrative officials the duty to develop more specific rules in policies and regulations or through adjudications that are consistent with the general prescriptive guidance in the authorizing legislation. For instance, in the United States, a law such as the Resource Conservation and Recovery Act (RCRA), 42 U.S.C. 6901 *et. seq.*, provides that persons may not treat, store, or discharge hazardous wastes without a permit, but delegates to the U.S. Environmental Protection Agency (EPA) the duty to define through regulations what is a hazardous waste and other more specific rules on the treatment, storage, and discharge of hazardous wastes.

Once legislation or implementing regulations have developed prescriptive guidance, government technicians are understood to apply scientifically derived "facts" to politically derived rules in day-to-day decisions that apply the law. This analysis leads to the conclusion that governments must turn to the law or regulations interpreting the law to determine the applicable prescriptive rule to be applied to sustainability decisionmaking. A closer analysis of most environmental laws, however, reveals that the prescriptive rules contained in many environmental laws are vague. A relevant example in the United States is NEPA which clearly is law that articulates environmental policy goals with a distinctly ethical character. The goals of NEPA state that the act's purpose is to establish a harmonious relation between humans and the environment. NEPA is thus understood to be a law that incorporates an ethically based environmental approach to federal decisionmaking. However, the exact nature of the environmental ethical approach embedded in NEPA is ambiguous, because the goals also include words or phrases that seem to recognize the need to balance environmental concerns with the need to meet "social and economic requirements of present and future generations" and the requirement to use all "practical" means and measures to create and maintain conditions where people and nature can exist in "productive" harmony (Sagoff 1987a). With a law this vague, a decisionmaker, in the face of political pressure, can hide controversial ethical positions behind statements that a decision fully complies with the law. Only if laws give clear prescriptive direction can the law overcome the short-term political forces that work against its implementation.

Several commentators have argued that Agenda 21 is notoriously vague (Rothenberg 1994). Areas of ambiguity, imprecision, or apparent self-contradictory recommendations weaken the force of the document (Westra 1994). The ambiguity stems from the political need to reach agreement at an abstract level of generality when Agenda 21 was drafted

(Blomquist 1994). Because of this ambiguity, Agenda 21 is capable of many interpretations, including those asserting the following: (1)primacy should be given to ecosystem integrity and environmental considerations should get priority over other economic or social considerations such as individual rights, preferences, or fairness; (2)environmental values should be considered and allocated efficiently with other values, including economic and development considerations; and (3)Agenda 21 forces technology and science to develop in appropriate ways so that no conflict exists between environmental protection and development.

It is therefore apparent that Agenda 21 fails to answer some of the difficult public policy questions posed by the potential conflict between development interests and environmental protection considerations and, therefore, does not create a clear prescriptive rule for governments to follow in applying the facts of any individual controversy and in making a decision. Laws that are developed to implement Agenda 21 should set clear lexical priorities between environment and development based on ethical considerations that recognize the value of nonhuman animals and plants and the rights of future generations.

Agenda 21 is generally referred to as "soft law," meaning that it is not binding on signatory nations but rather operates as a set of normative principles that will guide the development of specific laws and treaties in the years ahead. The hard choices, therefore, have been left to the development of the implementing law. Many countries will undoubtedly point to existing laws such as NEPA as legislation that already implements Agenda 21. These countries will argue that it is not necessary to pass new laws because existing laws satisfy Agenda 21's goals. Yet many of these laws do not contain clear prescriptive rules about resolving conflicts between protecting life on earth and desired development. To implement the sustainable development goals of Agenda 21, many existing laws need to be replaced with laws that clearly prohibit environmentally destructive actions while encouraging development that does not harm ecosystems. As one commentator on Agenda 21 has stated, "Neither prudence, nor yet morality, will be served if the new measures we attempt to implement are neither clear, nor new enough, and if they simply represent a half-hearted effort to keep everyone satisfied" (Westra 1994).

3. The Science-Law Interface

3.1. THE PRECAUTIONARY PRINCIPLE

As stated more fully in Chapter 2, sustainable development decisions will have to be made in the face of pervasive scientific uncertainty. Decisionmakers will have to decide in making such decisions whether they will err on the side of environmental protection or refrain from imposing costs in the absence of conclusive proof. One of the international agreements reached at the Earth Summit in Rio de Janeiro in June 1992 was the Rio Declaration on Environment and Development (Johnson 1993). Principle 15 of the Rio Declaration states:

> In order to protect the environment, the precautionary approach shall be widely applied by States according to their capabilities. Where there are threats of serious or irreversible damage, lack of full scientific certainty shall not be used as a reason for postponing cost-effective measures to prevent environmental damage.

The precautionary principle establishes the international norm that nations should take steps to protect the environment before potentially harmful effects of a given behavior are fully proven. The precautionary principle departs from many traditional approaches of law which presume that no harm has occurred until a party can demonstrate damage and causation. For this reason, the precautionary principle represents a fundamental shift in many legal principles that might otherwise apply to environmental controversies. Therefore, environmental law based on the precautionary principle will require governments to focus. on potential threats to the natural environment. Under a precautionary approach to environmental law, the burden of proof is shifted to the party who seeks to undertake potentially harmful activity (Weintraub 1992). Application of the precautionary principle implies a shift of approach from giving contaminants the benefit of doubt to giving the benefit of doubt to human health and the environment (Roht-Arriaza 1992).

3.2. SCIENTIFIC EVIDENCE IN LEGAL PROCEEDINGS

Because environmental decisions must be made in the face of pervasive scientific uncertainty, legal rules on the use of scientific evidence in court proceedings may determine when environmental laws may be enforced or implemented. If rules of evidence restrict the use of scientific evidence in court proceedings to that which is highly certain, enforcing or implementing an environmental law may be impossible in matters where certain scientific evidence is theoretically or practically unavailable. Therefore, rules on the use of scientific evidence in legal proceedings must be understood to be important ethical choices about when an environmental law may be enforced. A review of law on the use of scientific evidence in court proceedings in the United States reveals that rules differ depending on the type of proceeding.

3.2.1. *Tort Actions*

A tort is a wrongful act that causes a recoverable damage. In the United States, legal actions to recover environmental damages or to obtain an injunction to prevent harm from pollution may be brought under a variety of tort theories, including public and private nuisance, negligence, and trespass. In proceedings to prevent environmental harm or recover damages from pollution, federal courts in the United States follow the federal rules of civil procedure.

The U.S. Supreme Court in 1993 changed the standard of admissibility of scientific evidence in all civil proceedings in *Daubert* v. *Merrell Dow Pharmaceuticals, Inc.*, 113 S.Ct. 2786 (1993). In *Daubert*, the U.S. Supreme Court rejected a test for admissibility of scientific evidence that had existed for 70 years in federal courts established in the case of *Frye* v. *United States*, 293 F. 1013 (D.C. Cir. 1923). The *Frye* test allowed into evidence expert testimony deduced from a well-recognized scientific principle or discovery, if the thing from which the deduction is made must be sufficiently established to have gained general acceptance in the particular field in which it belongs.

The *Frye* test, therefore, prohibited the introduction of scientific evidence in civil proceedings unless the scientific evidence had reached high levels of certainty in the scientific community. Under *Frye*, causal evidence of damage to the environment or human health cannot be admitted in legal proceedings unless high levels of certainty of causation have been reached by the relevant scientific community. As a result, persons who have a reasonable basis for concern that they may have been harmed by unwanted exposure to chemicals cannot recover damages if the evidence does not establish causation with a high level of certainty. For example, if someone with cancer could prove that they had been exposed to chemicals that cause cancer in animals, they cannot recover damages without reliable epidemiological proof of causation of cancer in humans. Since epidemiological evidence of causation of cancer in humans is sometimes practically impossible to obtain, such a rule makes certain types of recovery for damages impossible. Although *Frye* is no longer the rule on the admissibility of evidence in federal courts, it is still the test of admissibility in some states in the United States such as Pennsylvania and California.

In *Daubert*, the U.S. Supreme Court liberalized the test of admissibility of scientific evidence. The court announced the following four pronged analysis to assist courts in determining whether the evidence is relevant and reliable:

1. Is the scientific method used by the expert to derive an opinion capable of being tested? That is, is the method capable of being shown to be false? If the method is not capable of being shown to be false, then the method is not scientific and, hence, not admissible. Although there is little case law on environmental matters since *Daubert*, examples of environmental methods of analysis that cannot be verified include environmental models.
2. Has the scientific method been subjected to peer review publication? Publication only strengthens admissibility; nonpublication does not impart inadmissibility. Publication strengthens admissibility by providing public scrutiny of the method, increasing the likelihood that substantive flaws in the method will be detected. Many analytical tests and assumptions used in environmental matters have never been subjected to peer review. For instance, in risk assessment, analysts often make assumptions about the toxicity of chemicals that have not been tested for toxicological properties by drawing analogies from other chemicals that have been tested.
3. Does the method have a known error rate or the existence of outside standards that monitor the method? Obviously, a low error rate will encourage admissibility. Many analytical tests and assumptions used in environmental matters, such as environmental models, have no known error rate.
4. Does the method have general acceptance in the general scientific community? Acceptance by a specific scientific community is not needed for admissibility as it was under *Frye*. A method, however, with only minimal support in the general scientific community may properly be viewed with skepticism.

From the above analysis, it is apparent that although *Daubert* may have liberalized the rule on the admissibility of scientific evidence in civil proceedings, evidence that establishes a reasonable basis for concern about harm but does not conclusively establish causation is not admissible after *Daubert*. The rules on the admissibility of scientific evidence in civil proceedings in the United States are therefore not consistent with the Rio Declaration's precautionary principle.

3.2.2. *Administrative Action*

Laws dealing with environmental matters avoid many of the admissibility problems encountered in tort actions by giving governments power: (1)to take legal action if the government determines that an activity creates a "threat" to human health or the environment or (2)to create standards that can be enforced. Through a grant of power to government to take action where a "threat" of environmental damage exists, the government avoids the problem of showing causation of damage in fact. For instance, the Comprehensive Environmental Response, Compensation, and Liability Act (CERCLA, usually referred to as Superfund) authorizes injunctive action upon a showing of "a threat of imminent and substantial endangerment." In one case, the court held that "the United States need not prove an actual imminent and substantial endangerment, but may obtain relief on proof that the danger may exist" (*U.S.* v. *Conservation Chemical*, 619 F. Supp. 192, 1985). Similarly, if the government has the power to enforce standards directly, the standards are understood to be a definition of environmental harm, thus avoiding evidentiary problems associated with showing actual harm. For instance, under the Clean Water Act, states in the United States set in-stream water quality standards through rule-making, and environmental harm is presumed if government can show that an action caused a violation of water quality standards.

Another way in which environmental law avoids the admissibility problems of uncertain scientific evidence is to limit judicial review of administrative actions to the administrative record created by the administrative agency. For instance, in legal disputes about cleanup remedies under the Superfund law, it was held, "In considering any objections to the Environmental Protection Agency's cleanup of spilled waste oil contaminated with PCBs, the court may only look to the administrative record" (*U.S.* v. *Mexico Feed & Seed*, 729 F. Supp. 1255, 1990). In cases where judicial review is from the administrative record, the court does not call witnesses nor admit evidence but simply reviews the record of public comment about the proposed action prepared by the administrative agency. Because there are no witnesses or evidence heard in such court proceedings, there are no problems of admissibility of scientific evidence. In record review matters, the agency's actions are afforded deference and must be upheld if they are based on relevant factors and are not a clear error of judgment (*Citizens to Preserve Overton Park, Inc* v. *Volpe*, 401 U.S. 402, 1971). In one record review case, the court concluded:

> Where environmental protection statute is precautionary in nature, evidence is difficult to come by, uncertain, or conflicting because it is on the frontier of scientific knowledge, regulations are designed to protect public health, and decision is that of an expert administrator, court will demand step-by-step proof of cause and effect; but administrator may apply his expertise to draw conclusions from suspected, but not completely substantiated, relationships between facts, from trends among facts, from theoretical projections from imperfect data, from probative preliminary data not yet certifiable as fact, and the like. (*Ethyl Corporation* v. *EPA*, 541 F. 2d, 1, 1976)

Although under laws that take a precautionary approach, administrators may decide questions of uncertainty in favor of environmental protection, most laws do not require that decisionmakers resolve questions of uncertainty in this way. Moreover, not all environmental laws clearly prescribe a precautionary approach. Many environmental laws require a finding of harm as a factual prerequisite before taking protective regulatory action. When science is uncertain about the environmental consequences of human action, insisting on high levels of scientific proof before government action may be taken is a prescriptive rule that puts the burden of proof on government decisionmakers and protects the status quo. Such a rule may prevent protective government action where there is a reasonable basis for concern but where science is uncertain about the consequences of certain human activities. The standard of proof that should be required of regulatory action is an ethical question, not a scientific one. If we let scientific standards dominate legal institutions, we are making ethical choices that may be inconsistent with the precautionary principle. Although insisting on rigorous certainty may make sense in criminal cases where society wants to preserve presumptions of innocence, more flexible standards of admissibility might be appropriate in matters where government is expected to act according to the precautionary principle.

Even if a law allows a precautionary approach, there is no guarantee that the law will be implemented in a precautionary manner. Because of pervasive scientific uncertainty in environmental problems, technical experts within government often refuse to act out of fear that they will enrage a legislator who will have them fired if it is discovered that they have imposed insupportable costs upon a constituent. This reluctance is also consistent with most scientific training. The scientist is trained to be very conservative in asserting cause-and-effect relationships. Many traditionally trained scientists will not act quickly if there is uncertainty about the cause of an environmental problem. If a position, once taken, is later discredited by subsequent scientific research, the technical person who suggested the cause-and-effect relationship may suffer peer sanctions for being associated with a faulty scientific hypothesis. Because scientists are taught to be silent in the absence of proof, scientific norms of behavior may be inconsistent with public policy norms. Therefore, some administrators may be inclined to take no protective action until matters of scientific uncertainty are resolved.

If a law authorizes a precautionary approach but government refuses to implement the law in a way that resolves uncertainty in favor of the environment challengers of the government's action are in a weak position in any challenge. Persons who wish to challenge government decisions that will be reviewed on the administrative record need to show that the government action is arbitrary or capricious, contrary to law, or based upon insubstantial evidence in the administrative record (*U.S.* v. *Akzo Coatings of America*, 719 F. Supp. 579, 1989). Because the burden of proof is on those challenging the government action, challengers may have great difficulty in meeting such a burden in areas of significant scientific uncertainty.

Generally speaking, courts tend to defer to the agency's decisions in scientific disputes litigated from administrative records (*Browning-Ferris Industries of South Jersey, Inc.* v. *EPA*, 31 ERC 1088, 1990). However, courts will often overturn agency actions if they feel that the agency has failed to demonstrate in the administrative record that an adequate scientific basis exists for its decision. For instance, courts have upheld appeals of EPA decisions when: (1)the EPA failed to do adequate testing of a substance before listing it as a hazardous waste (*American Mining Congress* v. *EPA*, 907 F. 2d 1188, 1990) and (2)the EPA made a decision about the toxicity of a substance at a site contaminated with hazardous substances without determining whether the substance was present in highly toxic or low-toxic form (*National Gypsum Co.* v. *EPA*, 986 F. 2d 40, 1992).

3.2.3. *Mathematical Models and Environmental Decisions*

Mathematical models are frequently used in developing environmental regulations and in making day-to-day environmental decisions. In the United States, Clean Air Act implementation relies on the use of models in permitting decisions, enforcement, and long-range planning. The Superfund program and permitting under RCRA use models to determine groundwater flow, predict transport of hazardous pollutants, assess risk, rank hazardous sites, and determine natural resource damages. Clean Water Act implementation relies on models to set effluent limitations for discharges into surface waters. Prediction of environmental impacts under NEPA also depends on the use of a variety of different types of models.

These models can never achieve levels of certainty reached in other scientific endeavors because: (1)ecological systems are open systems rather than the closed systems described by the models, and therefore, the models fail to deal with unmeasurable and underdetermined parameters and certain cause-and-effect relationships; (2)ecological models cannot be verified or validated in ways that other scientific processes can be tested; and (3)models usually must make simplifying assumptions for theoretical and practical reasons (Oreskes et al. 1994).

Courts have traditionally deferred to agency expertise in challenges to mathematical models (Case 1982). However, the judicial review of computer models in environmental decisionmaking has been uneven. For instance, in *South Terminal* v. *EPA* (504 F. 2d 646, 1974), petitioners successfully challenged the EPA's use of a model used in the Metropolitan Boston Air Quality Transportation Control Plan. In this case, the court found that the EPA relied on insufficient evidence in constructing and applying an air pollution model. Similarly, petitioners have challenged successfully the coefficients used in air pollution models to represent assumed weather conditions. How courts review models is often a function of the judge's expectation about the degree to which government decisions should be based on sound science. If judges do not understand the inevitable imprecision entailed by the use of any environmental model, they are likely to decide cases in a way that undermines the precautionary principle.

3.2.4. *The Duty of the Government to Speculate About Uncertain Environmental Impacts in Environmental Impact Statements*

Principle 17 of the Rio Declaration states:

> Environmental impact assessment, as a national instrument, shall be undertaken for proposed activities that are likely to have a significant adverse impact on the environment and are subject to a decision of a competent national authority.

What does the law require of preparers of environmental impact statements (EISs) when they are faced with possible but uncertain environmental impacts? In the United States, EISs are required by NEPA for government actions that have potential significant impacts on the environment. In addition, environmental impact assessment requirements similar to NEPA have been adopted by over thirty countries.

Under NEPA, courts take a hard look at the adequacy of the EIS and require that the EIS be a "full disclosure" of the environmental impacts of the proposed action. To enforce the hard look, the courts distinguish between a "substantive" and a "procedural" challenge to an EIS (Mandelker 1994). The duty to reject a project on environmental grounds is generally referred to as NEPA's "substantive" duty, while the duty to prepare an adequate EIS is a "procedural" duty. The U.S. Supreme Court found in *Strycker's Bay Neighborhood* v. *Karlen* (444 U.S. 223, 1980) that NEPA did not impose a substantive requirement that elevated environmental concerns over other concerns. Courts have, however, consistently held that NEPA requires that EIS analyses be based on a "good faith effort" of the agency to identify environmental impacts fully. Thus, NEPA is understood to create procedural duties to examine potential environmental impacts carefully but no substantive duty to refrain from taking actions that have adverse environmental impacts. Thus, successful challenges to EISs under NEPA have almost always followed from failure to examine potential impacts adequately rather than from unwillingness to mitigate adverse environmental impacts. An EIS must fully explain its inquiry, analysis, and reasoning to survive a procedural challenge.

Congress did not address the problem of scientific uncertainty in identifying potential impacts when it passed NEPA. In 1978, the U.S. Council on Environmental Quality (CEQ) adopted regulations (40 C.F.R. 1500-17) that addressed the problem of scientific uncertainty. These regulations provided that if scientific uncertainty existed that could be cured by further research, the agency had to do or commission the research. If the necessary research was exorbitantly expensive or beyond the state of the art, the agency had to make it clear that uncertainty existed and had to include a "worst-case analysis" in its EIS (Reeve 1984). When the full extent of environmental impacts from an agency action was uncertain or unknown, an agency was under a duty to discuss the worst possible consequences and the probability of their occurrence. The worst-case analysis rule received much criticism for delaying decisionmaking and forcing the agency to speculate about low-probability events; the rule was subsequently repealed by the CEQ in 1986 (Fitzgerald 1992). The new rule (40 C.F.R. 1502.22) states that if information is not available, the federal agency must make reasonable efforts, in light of overall costs and the state of the art, to obtain missing information which, in its judgment, is important to evaluating significant adverse impacts on the human environment that are reasonably foreseeable. If the costs of obtaining the information are exorbitant or the means of obtaining it unknown, agencies must: (1)state that the information is incomplete or unavailable; (2)state the relevance of this information to evaluating reasonably foreseeable significant environmental impacts; (3)summarize credible scientific evidence relevant to evaluating impacts; and (4)evaluate these impacts based upon theoretical approaches to research methods generally accepted in the scientific community. Reasonably foreseeable impacts shall include low probability catastrophic impacts if the analysis of the impact is supported by credible scientific evidence, is not based on pure conjecture, and is within the rule of reason (Mandelker 1994).

One commentator sees the new rule as essentially the same as the old worse-case analysis rule (Tutchon 1989). The new rule attempts to limit speculation by requiring that analysis be based on credible scientific evidence. Tutchon argues that since credible scientific evidence allows theoretical approaches or research methods that are generally accepted in the scientific community, the new rule is not much of a change from the original one. However, the new rule may be interpreted by the courts as a limitation on identifying serious possible impacts where scientific theory is weak. If the precautionary principle requires that nations not take actions that have possible serious adverse environmental impacts, the new rule could be understood to be inconsistent with the precautionary principle, if in identifying possible serious impacts, the analyst must rely partially on speculation. The new rule presumes that an action is not harmful unless credible scientific evidence supports that it is harmful. Under a precautionary approach to environmental law, the burden of proof is shifted to the party who seeks to undertake potentially harmful activity to prove that it is not harmful. Because the new rule presumes that actions are safe unless there is credible evidence supporting that it is not, it is arguably inconsistent with the precautionary principle.

4. Economics-Law Interface

Principle 16 of the Rio Declaration states:

> National authorities should endeavor to promote the internalization of environmental costs and the use of economic instruments, taking into account that the polluter should, in principle, bear the cost of pollution.

The Rio Declaration's call for the polluter to pay is designed to assure that the full costs of environmental protection are borne by those creating potential environmental damage. In a similar vein, many economists assert that the solution to environmental problems is to create mechanisms that assure that the full costs of environmental protection be internalized.

How environmental costs should be internalized is a matter of considerable controversy. Some economists support regulatory mechanisms designed to minimize or eliminate the damage. These regulatory mechanisms often take the form of statutes or regulations that prohibit or minimize environmentally damaging activities. They are usually referred to as command-and-control (CAC) mechanisms. Other economists argue for market solutions, such as effluent taxes or emissions trading regimes, to correct the market failures.

Many economists assert effluent taxes and emissions trading policies are preferable to CAC regulatory mechanisms because they: (1)create economic incentives to reduce pollution and (2)allow more flexibility than regulatory approaches in achieving pollution reduction strategies and thereby increase efficiency of pollution reduction efforts. Those who support CAC approaches often point to: (1)the inappropriateness of putting prices on environmental entities and (2)a variety of ethical problems with preference utilitarianism that is the ethical basis for the market-based approaches. (See Chapter 4 for a fuller discussion of some of the limits of market-based mechanisms.)

Although CAC methods have been the dominant tools used to internalize costs, many market-based methods have been put into place and are operating under environmental laws passed in recent years. Among the several types of economic instruments that have been used to implement environmental policy, frequently used approaches have included pollution fees, tradable allowances, deposit-refund systems, and information disclosure requirements (Dudek et al. 1994). However, in most market-based approaches, the government still sets firm environmental goals but defines them in terms of performance standards, leaving the choice of specific response and compliance strategies to each individual enterprise. As a result, many market-based approaches ultimately rely on CAC-type policy choices, in the form of effluent or emission limitations, to determine what level of environmental degradation is acceptable. Because "acceptability" of environmental damage is ultimately a prescriptive concept, rather than a value-neutral descriptive judgment, even market-based approaches demand ethical choices about their implementation.

Environmental laws have differed greatly in the extent to which cost considerations are relevant to performance standard-setting. Four approaches have been followed in environmental law in the United States. They include:

1. Cost-oblivious statutes that set standards on health considerations only, such as national ambient air quality standards under the Clean Air Act.
2. Cost-effective statutes in which the goals of protection are set legislatively, but the selection of means can be made on the basis of efficiency, such as the provision under the Clean Water Act that states that technology used in municipal treatment plants must reflect the most cost-efficient alternative.
3. Cost-sensitive statutes that require that cost be considered among other considerations by government in setting standards. This approach stops short of requiring that the standard be set on a cost-benefit basis. One example of a cost-sensitive approach is the setting of fuel economy standards under the National Energy Conservation Policy Act, which makes cost pertinent in vague and varying ways in relation to numerous energy, health, and environmental decisions.
4. Strict cost-benefit analysis approaches that require that standards be set at a level determined by the cost-benefit analysis. Under the Reagan and Bush administrations, Executive Orders 12,291 and 12,498 prohibited environmental regulations unless they could pass a cost-benefit analysis (Rodgers 1980).

These four approaches represent different resolutions of potential conflicts between efficiency and ethical considerations in environmental law. In such laws, as in other forms of social legislation, some see legislation as an opportunity to promote public values and some instead promote market efficiency by regulating the market. Even if a person takes the position that pollution should be eliminated eventually on ethical grounds, it may be necessary to take cost into account for practical reasons in the short term (Sagoff 1987b):

> That progress toward stated goals must be deliberate, but it need not succeed all at once, is evident in court decisions which recognize that economic "feasibility" is a legitimate factor to be considered in protecting safety and health, and that EPA need not insist upon every possible reduction if it determines such insistence counterproductive. The fundamental idea is to make progress in view of the circumstances, not to insist uncritically upon perfection.

Cost considerations are clearly valid for many types of environmental decisions, such as the timing of environmental compliance or the priority of attacking environmental problems. If some form of economic efficiency is included as a policy choice in legislation, it is important, however, to see such choice as an ethical decision rather than a value-neutral calculation, because no amount of economic analysis can logically prescribe a given course of action. It is also important to understand conflicts between efficiency and sustainability. Efficiency determines whether a particular benefit or resource outweighs its cost; sustainability requires sustained yield (Campbell-Mohn et al. 1993). The assumption implicit in the efficiency objective is that the market will reflect scarcity before a resource is inextricably exhausted; sustainability does not rely on the market to assume the planning function.

Courts have often reviewed cost analysis undertaken in environmental decisions, even when the analysis was authorized by law. Mandelker (1994) concludes:

> The courts will disapprove a cost-benefit analysis if it is so biased or conclusary that it provides a misleading explanation of the basis on which project costs and benefits were determined. One common error in cost-benefit analysis that leads to judicial disapproval is the failure to quantify all the environmental costs of a project....In other cases, the courts disapproved cost-benefit analysis because environmental benefits were not properly quantified.

Although there are fundamental philosophical problems with using willingness-to-pay as a method of measuring the value of environmental resources, courts have been harsh on economic methods that attempt to use alternative methods for valuing resources, such as contingent valuation (see Chapter 4). Courts prefer willingness-to-pay because it is objective to the extent that you can determine values by looking at market prices. Courts disfavor contingent valuation because it is viewed as highly subjective and therefore unreliable. Therefore, if society wants to assure that environmental entities are not treated just like other commodities in cost-benefit analysis, society will have to support laws that authorize other methods of valuing resources such as contingent valuation or replacement value.

5. The Role of Citizens in Moving Toward Sustainable Development Law

A significant problem with Agenda 21, according to one commentator, is that while it is now part of international law, "international law is no law at all" (Rolston 1994). If Agenda 21 is to be implemented, the international community will have to rely on institutions with no enforcement authority such as the United Nations Commission on Sustainable Development. This institution's power stems largely from its ability to report on and disclose the inadequacy of national efforts. Because public pressure that follows from disclosure will be the sole power for changing the behavior of recalcitrant nations, the role of nongovernment organizations is critical to the success of Agenda 21. That is, if Agenda 21 is to be comprehensively implemented, concerned citizens throughout the world will have to monitor and challenge national progress toward implementation of Agenda 21. As this book points out, however, if citizens want to be effective in monitoring national progress, it will be necessary for citizens to become involved in scientific, economic, and legal issues and other details that face nations in implementing sustainable development programs, because it is the scientific, economic, and legal details of worldwide programs that will determine whether nations follow a sustainable development path. A significant limitation on citizen involvement in international compliance with Agenda 21 is the lack of technical expertise to be a credible player in the scientific, economic, and legal languages that will structure debates about sustainable development. If governments are serious about sustainable development implementation, they should fund the expertise to support citizen involvement, for much of the technical expertise that exists in science, economics, and law is employed either by governments that may be recalcitrant or corporations with narrow economic interests.

6. References

Birnie, P.W., and A.E. Boyle. 1992. *International Law and the Environment.* Clarendon Press, Oxford, England.

Blomquist, R.F. 1994. Judging the United Nations Agenda 21 Industrial Pollution Prevention Provisions: An Ethical and Policy Analysis. In *Proceedings on Ethical Dimensions of the United Nations Program on Environment and Development Agenda 21*, D.A. Brown, ed. Earth Ethics Research Group, Harrisburg, PA, pp. 17-58.

Campbell-Mohn, C., B. Breen, and J.W. Futrel. 1993. *Sustainable Environmental Law, Integrating Natural Resource and Pollution Abatement Law from Resources to Recovery*. Environmental Law Institute, Washington, DC.

Case, C.D. 1982. Problems in Judicial Review Arising from the Use of Computer Models and Other Quantitative Methodologies in Environmental Decisionmaking. *Environmental Affairs Law Review* 10: 251-356.

Dudek, D., R. Stewart, and J. Wiener. 1994. Technology-Based Approaches Versus Market-Based Approaches. In *Greening International Law*, P. Sands, ed. New Press, New York, pp. 182-209.

Fitzgerald, E. 1992. The Rise and Fall of the Worst Case Analysis. *University of Dayton Law Review* 18: 1-96.

Futrell, J.W. 1994. *The Transition to Sustainable Development Law*. Environmental Law Institute, Washington, DC.

Johnson, S.P., ed. 1993. *The Earth Summit. The United Nations Conference on Environment and Development (UNCED)*. Graham & Trotman/Martinus Nijhoff, London.

Mandelker, D.R. 1994. *NEPA, Law and Litigation*. Clark Boardman, New York.

Oreskes, N., K. Shrader-Frechette, and K. Belitz. 1994. Verification, Validation, and Confirmation of Numerical Models in Earth Sciences. *Science* 264: 641-646.

Reeve, M. 1984. Scientific Uncertainty and the National Environmental Policy Act—The Council of Environmental Quality's Regulation, 40 C.F.R. Section 1502.22. *Washington Law Review* 60(87): 89-101.

Rodgers, W.H. 1980. Benefits, Costs and Risks: Oversight of Health and Environmental Decisionmaking. *Harvard Environmental Law Review* 4: 191-226.

Roht-Arriaza, N. 1992. Precaution, Participation, and the "Greening" of International Trade Law. *Oregon Journal of Environmental Law and Litigation* 7: 57.

Rolston, H. 1994. Environmental Protection and an Equitable International Order; Ethics After the Earth Summit. In *Proceedings on Ethical Dimensions of the United Nations Program on Environment and Development Agenda 21*, D.A. Brown, ed. Earth Ethics Research Group, Harrisburg, PA, pp. 267-284.

Rothenberg, D. 1994. Say What You Mean! The Undefined in Agenda 21. In *Proceedings on Ethical Dimensions of the United Nations Program on Environment and Development Agenda 21*, D.A. Brown, ed. Earth Ethics Research Group, Harrisburg, PA, pp. 285-288.

Sagoff, M. 1987a. NEPA: Ethics, Economics, and Science in Environmental Law. In *Law of Environmental Protection*, S. Novick, ed. Clark Boardman, New York, pp. 9-48 to 9-102.

Sagoff, M. 1987b. Ethical and Economic Principles in Environmental Law. In *Law of Environmental Protection*, S. Novick, ed. Clark Boardman, New York, pp. 5-9 to 5-79.

Sands, P. 1994. *Greening International Law*. New Press, New York.

Tutchton, J. 1989. *Robert* v. *Methow Valley Citizens Council* and the New "Worst Case Analysis" Regulation. *Environmental Law* 8: 287-294.

Weintraub, B. 1992. Science, International Regulation, and the Precautionary Principle: Setting Standards and Defining Terms. *N.Y.U. Environmental Law Journal* 1: 172.

Westra, L. 1994. Ecosystem Integrity and Agenda 21: Science, Sustainability and Public Policy. In *Proceedings on Ethical Dimensions of the United Nations Program on Environment and Development Agenda 21*, D.A. Brown, ed. Earth Ethics Research Group, Harrisburg, PA, pp. 383-392.

Chapter 6
CONSERVATION OF BIODIVERSITY AND SUSTAINABLE DEVELOPMENT

John Lemons[1]
Pamela Morgan[1]

1. Introduction

Biological diversity, or biodiversity, is a term that is now part of the vocabulary of policymakers, academics, lawyers, and laypeople, and yet its full meaning is often not understood. Although Agenda 21 does not define biodiversity, the 1992 Convention on Biological Diversity defines it as "the variability among living organisms from all sources...and the ecological complexes of which they are a part; this includes diversity within species, between species and of ecosystems" (Johnson 1993).

There are many hierarchical levels of biodiversity, ranging from the genetic level to the landscape level, with each subsequent level supporting the next. At the smallest scale is genetic diversity, which includes the variety of genes within a population or a species. Genetic variation is necessary within a population in order for it to maintain reproductive vitality, resistance to disease, and the ability to evolve or to adapt to changes in the environment. Populations contribute to species diversity, which involves not only the number of different species (the variety component) but also how the total number of species is divided up (the relative abundance component) (Odum 1994). Add to species diversity the interactions among these species to form the next level in the hierarchy, community diversity. Biodiversity also includes abiotic processes, and at this larger scale is then considered ecosystem diversity. Clusters of interacting ecosystems are often looked at on an even greater scale, the landscape level. These different levels of biodiversity include not only the structural diversity in each but also the variety of functional processes occurring at each scale.

Biodiversity is valued for a variety of reasons. Biological resources are used by humans in products such as food, pharmaceuticals, and fiber. Biodiversity also supplies humans with what have been called "ecosystem services," which includes the maintenance of atmospheric gases, climate control, and nutrient cycling. Other values are not related to consumptive resource uses, such as aesthetic and recreational values. Some also argue that biodiversity has intrinsic value, separate from its current or potential uses by humans.

Because of the variety of values associated with biodiversity, worldwide concern has arisen over its loss, which has primarily been studied in terms of species extinction. Although the process of extinction is natural, the rate of extinction has increased since the arrival of humans on earth. It is difficult for scientists to determine to what extent the loss of species has been caused by human impacts. The natural rate of extinction, or extinction that would occur in the absence of human influence, is estimated from the fossil record. This background

[1]Department of Life Sciences, University of New England, Biddeford, ME 04005, U.S.A.

J. Lemons and D. A. Brown (eds.), Sustainable Development: Science, Ethics, and Public Policy, 77–109.

rate of extinction is best known for birds and mammals, where the current extinction rate has been calculated to be 100 to 1,000 times greater than the natural extinction rate (Primack 1993). The often-cited total number of species lost per year is 100,000 (Groombridge 1992). The cause for this elevated rate of extinction is attributed to a variety of human influences: (1)the introduction of exotic species, (2)destruction of habitat, (3)overexploitation of species due to hunting and deliberate extermination, (4)habitat fragmentation, (5)pollution, and (6)the spread of disease.

The overall objective of Agenda 21's recommendations to conserve biodiversity is "...to improve the conservation of biological diversity and the sustainable use of resources." A summary of more specific objectives recommended for governments, nongovernmental organizations, the private sector, and financial institutions is: (1)press for early entry into force of the Convention on Biological Diversity; (2)increase knowledge of the status of biodiversity and the sustainable use of resources and dissemination of this knowledge; (3)protect biodiversity both *in situ* and *ex situ*; (4)analyze relevant costs and benefits of conserving biodiversity, with particular attention to socioeconomic aspects; (5)develop national strategies to incorporate the conservation of biodiversity into plans for development; (6)recognize and foster traditional methods of conserving biodiversity employed by indigenous people and women; (7)ensure the fair sharing of benefits derived from biological resources between the sources of those resources and those who use them; and (8)explore the potential of biotechnology to conserve biodiversity for agricultural, health and welfare, and environmental purposes. Other recommendations pertain directly to the disciplines of science, public policy, economics, and ethics. In general, science is called on to develop methods and technologies for the conservation of biodiversity and the sustainable use of biological resources. More specifically, it should establish baseline information on the status of biodiversity at the genetic and ecosystem levels and develop efficient sampling and evaluation methods for these surveys. Long-term research on the importance of biodiversity in ecosystem functioning and the ability of ecosystems sustainably to produce resources needed by humans is encouraged. Along with this, science will have to play a role in achieving the objective of promoting sustainable production systems such as agroforestry, traditional methods of agriculture, and range and wildlife management.

The recovery of endangered species and the restoration of damaged ecosystems also will demand the expertise of scientists. Maintenance and recovery of minimum viable populations (MVP) of species in their native surroundings and the *in situ* conservation of ecosystems and natural habitats is recommended. Finally, the *ex situ* conservation of biological diversity (especially genetic) is addressed. Improvement and diversification of methods for *ex situ* conservation plus the promotion of national efforts to maintain gene banks are recommended, and where possible, genetic resources should be stored and maintained in the country of origin.

The conservation of biodiversity is not solely a scientific problem; it also involves public policy, economic, and ethical decisions regarding how to promote it within the context of sustainable development objectives. For example, Agenda 21 identifies the need to alleviate poverty, to reduce inequities between developed and developing countries, and to accommodate the socioeconomic needs of present people while providing for the future. Agenda 21 also refers to the values of biodiversity and encourages a greater understanding of these values, although they are not clearly defined. Following, we discuss: (1)the goals of sustainable development and conservation of biodiversity and potential conflicts between them, (2)guidelines for management of biodiversity, (3)the status of scientific knowledge about biodiversity and the implications of scientific uncertainty, (4)the linkages between different sustainability problems, and (5)value-laden and ethical dimensions of promoting sustainable development and conserving biodiversity.

2. Goals of Sustainable Development and Conservation of Biodiversity

In 1987, the World Commission on Environment and Development (WCED 1987) proposed that development that "...seeks to meet the needs and aspirations of the present without compromising the ability to meet those of the future" be defined as "sustainable development." This definition for sustainable development has become widely accepted in the international community. During the 1992 United Nations Conference on Environment and Development, U.N. Secretary-General Boutros-Ghali proposed a similar definition for the participants of Agenda 21, which was that sustainable development is that which "...meets the needs of the present as long as resources are renewed or, in other words, that does not compromise the development of future generations" (Johnson 1993).

Chapter 15 of Agenda 21 recognizes that biodiversity should be conserved to meet three overall goals: (1)to promote social sustainability in order to reduce poverty and to obtain a more equitable distribution of goods and social welfare, (2)to promote economic sustainability in order to maintain natural resources capital (i.e., biodiversity) for the benefit of future generations, and (3)to promote environmental sustainability in order to improve human welfare by protecting the sources of biological resources for human needs. Each of these forms of sustainability is assumed to be required for and not to be in conflict with the other forms. While Agenda 21 contains laudable goals, it does not seem to recognize fully the potential for basic conflicts between its goals and the principles that govern species and ecosystems. Specifically, the conflicts relate to those between the needs of humans and the requirements of supporting ecosystems, as well as to the implications to biodiversity of living in a sustainable manner, assuming it is possible to do so.

2.1. THE NEEDS OF HUMANS AND ECOSYSTEMS

With regard to the first set of conflicts, living in a sustainable manner requires that social and economic development not exceed the carrying capacity of ecosystems. Carrying capacity is a characteristic defined for the population of a given species, and in the case of Agenda 21, this is the human species. The carrying capacity for humans depends on numerous and complex interactions between the environment, lifestyle aspirations, technological development, and economic and social organization (Daily and Ehrlich 1992). Goodland et al. (1993) note that humans can live sustainably by limiting population, limiting affluence, or improving technology by reducing the intensity of the throughput of resources used. Agenda 21 broadens a more narrow definition of the human carrying capacity to include the capacity of ecosystems to support healthy populations of other species while maintaining their capacities to support sustainable uses. In this sense, sustainable development and conservation of biodiversity should be considered in a common context and as essential parts of an indispensable process, which is achieving a decent future for humanity.

Despite Agenda 21's use of a more broad definition of the human carrying capacity, the fact that sustainable development and conservation of biodiversity are separate processes creates potential conflicts. Generally speaking, humans use ecosystems in relatively earlier successional stages as opposed to more mature ecosystems because the former are more productive. However, the use of earlier successional ecosystems presents conflicts with goals of conserving biodiversity for three reasons. One, the maintainence of earlier successional ecosystems requires intensive management and inputs of energy and materials that contribute to resource use and pollution, which threaten species or ecosystems directly or indirectly. Two, earlier successional ecosystems are less biologically diverse than more mature ecosystems; hence, within-habitat diversity is decreased when humans utilize earlier successional ecosystems. Three, maintenance of early successional ecosystems occurs at the expense of more natural ecosystem types and therefore reduces the total biodiversity. One result is that there are fewer species that can then serve as colonizers for disturbed habitats.

There also may be a basic conflict between sustainability goals and the maintainence of ecosystems. To the extent that Agenda 21 recommends the maintenance and use of relatively stable ecosystems for structural purposes even under sustainable conditions, it seems to imply that they exist in an equilibrium state by virtue of its emphasis on those particular species that have utility or potential utility for humans. This is problematic because the developmental pathways of ecosystems often are nonlinear and discontinuous and require nonequilibrium conditions to maintain their integrity and biodiversity (Kay 1991). According to nonequilibrium ecosystem theory, the general types of structural responses of natural or managed ecosystems to environmental changes include: (1)temporary changes followed by a return to post-disturbance conditions, (2)continued operation but with changes in structural attributes (e.g., reduction or increase in the number of species), (3)continued operation but with the emergence of some new structural attributes which replace or add to existing structures (e.g., new species or food web pathways), (4)development of an ecosystem with new structural attributes, and (5)ecosystem collapse with no regeneration. Nonequilibrium theory also suggests that when humans attempt to maintain ecosystems in particular states (even if according to criteria of sustainability), the specific outcomes of these types of general responses are likely to be both variable and unpredictable because there is no inherent single state for which ecosystems can be managed in the long term. Further, the responses of ecosystems to future changes is a function of both immediate environmental change and historical changes (Bonnicksen and Stone 1985). The difficulties of understanding the nature and consequences of immediate environmental change combined with the fact that it often is impossible to ascertain historical changes and their consequences makes it even more difficult to predict responses of ecosystems to change. Consequently, management of ecosystems for sustainability requires considerable adaptability.

Robinson (1993) has presented an excellent analysis of some of the potential problems or conflicts among the different forms of social, economic, and ecological sustainability and their potential implications to biodiversity. For example, he notes that there are many population levels at which species can be harvested sustainably, and the extent to which a species can be harvested depends in large part on whether it exhibits density-dependent compensation by increasing its rate of growth. Species that exhibit density-dependent compensation tend to be found in relatively earlier successional and less biologically diverse ecosystems. Consequently, it becomes more feasible for humans to utilize species from such ecosystems because their production can be maximized despite the fact that to do so requires the maintenance of earlier successional ecosystems which have less biodiversity. Alternatively, it can be problematic for indigenous or other peoples to utilize species from mature ecosystems for commercial exploitation because they do not appear to contain species with high enough densities and rates of population growth to support more than relatively minimal exploitation. The human use of species also is problematic for the overall conservation of biodiversity because harvest of one or more species can have significant and unpredictable ecological ramifications throughout the community (Larkin 1977, Redford 1992). For example, managing particular species populations for sustained harvest often leads to a shift in relative abundance of coexisting species, the extent of which will depend upon the tightness of coupling of harvested species to others in the food web. Two conclusions emerge from these considerations: (1)although species can be exploited to meet the goals of social and economic sustainability, the fact that they can be exploited does not say anything about the level of exploitation; and (2)the use of any species is likely to result in the loss of some biological diversity.

Conflicts between sustainable use and biodiversity at the community or ecosystem level also exist. Theoretically, communities or ecosystems can be sustainable at different levels of intensity of management and use. However, ecosystems managed at more intensive levels will be less biologically diverse than those managed at less intensive levels, although the former will be able to support more people with a higher quality of life. Conflicts pertaining to ecological sustainability also derive from the fact that, in part, one of the reasons for conserving biodiversity is because ultimately it is the source of ecosystem services that humans value as well as upon which other species may depend. Ecosystem services derived from the functional consequences of biodiversity are used at the local, regional, and global levels (although there is overlap among the different levels). At the local level, examples of ecosystem services include the cycling of nutrients, retention of eroded sediments and protection of water quality, and production of food. At the regional level, examples of ecosystem services are the production of wood and fiber. Examples of ecosystem services at the global level include maintenance of global nutrient cycles and regulation of atmospheric gases. Potential conflicts may arise among different forms of sustainability by virtue of the fact that on the local (or regional) level, ecosystems will be exploited to meet the goals of social and economic sustainability, whereas at the global level, ecosystems may be required to be preserved or used at lower levels of exploitation in order to contribute to the maintenance of global ecosystem services.

Recently, Westra (1994) has proposed that ecosystems be managed according to concepts of ecological integrity. Such concepts include the abilities of ecosystems to: (1)maintain optimum operations under normal environmental conditions; (2)cope with changes in environmental conditions; (3)continue the process of self-organization on an ongoing basis; and (4)continue to evolve, develop, and proceed with the birth, growth, death, and renewal cycle. Ensuring the maintenance of ecological integrity would require the establishment of large core areas of nature relatively undisturbed by humans to provide for the protection of biological diversity. While the establishment of such core areas would likely represent the best guarantee that biological diversity would be conserved, it remains to be seen whether this would conflict with the goals of social and economic sustainability. For example, developing nations are not likely to support policies calling for total exclusion of human resource extraction from large areas in the absence of major increases in tourism, foreign debt relief, or other direct subsidies from developed nations. Even in developed nations, support for preservation of large land areas is controversial (Lemons 1987). In the event a conflict occurs among different goals of sustainability, Agenda 21 offers no prescriptions about whether sustainable ecosystems should be managed to promote higher levels of biodiversity or a higher quality of life for more people.

2.2. SUSTAINING BIODIVERSITY AND SOCIOECONOMIC SUSTAINABILITY

Agenda 21 promotes environmental, social, and economic sustainability and rightly considers each to be required for a sustainable future. Despite the importance of all forms of sustainability, they may be in conflict with each other at times. Lonergan (1993) recognizes three ways in which impoverishment has been acknowledged as a key cause of environmental degradation: (1)where poverty is the underlying cause of environmental degradation; (2)where poverty is a proximate cause of environmental degradation, but where national/international policies/institutions are the major determinants; and (3)where environmental degradation is caused primarily by overconsumption and affluence but is reinforced by poverty. Each of these causes of environmental degradation is a function of the cumulative actions of the various users of resources who have diverse interests and needs, levels of wealth and power, and access to public policy decisionmaking. Although different resource users may have interests in sustainability, they may differ in their concerns for the environmental, social, or economic aspects of sustainability as well as in their assessment of how quickly plans for sustainable development should be implemented. For sustainability to be achieved, the rights and interests of different user groups have to be defined so that decisions can be made concerning when and how the needs of different user groups are met and when losses of biodiversity are judged to be acceptable. This means grappling with the question of how to specify and balance the rights and interests of indigenous peoples, local communities, regions, nations, the international community, and multinational corporations. Grappling with this question is highly problematic insofar as many local indigenous peoples and the 20 percent of humanity which is the most impoverished often are accorded little or no role in decisions about resource use. For all of the different resource users, concepts of ecological sustainability need to be used to define acceptable impacts of resource use on biodiversity, while concepts of social and economic sustainability must be used to define acceptable measures to obtain equitable improvements in social and economic well-being.

As a more specific example, consider the case of local or indigenous people who may support extractive reserves in tropical forests as a means to sustainable development. Their ability actually to use such reserves at a given level of intensity is dependent upon different user groups at the national and international levels finding and maintaining viable commercial markets for reserve products. Consequently, if national or international market demand or financial conditions change, then the use of extractive reserves on a sustainable basis may not be maintained despite the willingness of local people to use them in this manner. If the commercial products from extractive reserves do not meet the socioeconomic needs of either local people or more economically powerful users who will have different needs and interests in sustainability, then sustainable management of reserves will fail despite the fact that the resources are or could be managed in an ecologically sustainable fashion. If such management fails, then the result likely will be continued overexploitation of the local environment in order to meet the needs of local people if they are impoverished. This line of reasoning also suggests that the majority of the world's most impoverished or marginalized people will have to be acknowledged to have rights and interests in matters of sustainability. If their rights and interests are not met, then they will continue to overexploit marginal lands, which leads to concomitant losses of biodiversity.

Conflicts also can exist because the time scales for assessments of ecological and social or economic sustainability can be different. For example, the needs of impoverished local people may require rapid improvements in their living conditions, and business decisions must reflect appropriate rates of return for investments even under conditions of sustainability. The time scales for assessments of improving protection of or for utilizing biological resources may be slower than those for making decisions about social or economic sustainability, because assessing the damage to biological resources often requires relatively long time periods, especially under conditions of scientific uncertainty.

3. Guidelines for Management of Biodiversity

Nations' laws prescribe various goals and levels of protection for biodiversity. Federal laws and policies in the United States that can be used to conserve biodiversity include: (1)wildlife statutes, (2)agencies' legislative mandates, (3)laws that create liability mechanisms, (4)laws that guide federal actions, (5)laws and policies to compel or encourage conservation by private parties, (6)specific laws to conserve natural resources, and (7)federal executive orders and policies.

In the United States, examples of wildlife statutes include: (1)the Lacy Act (16 U.S.C. 701), which promotes the restoration of game and wild birds to areas where they have become extinct; (2)the Federal Aid in Wildlife Restoration Act (16 U.S.C. 669e[a][1]), whose purposes include land acquisition for wildlife rehabilitation and wildlife habitat enhancement for mammals and birds; and (3)the Federal Aid in Fish Restoration Act (16 U.S.C.

777c[a][1]), which promotes restoration and enhancement of fishery resources for sport or recreation. Examples of agencies' legislative mandates include the National Park Service Organic Act (16 U.S.C. et. seq.) and the Forest Service Organic Act (16 U.S.C. et seq.), which established the overall management goals of those agencies. An example of a law that creates liability mechanisms is the Comprehensive Environmental Response, Compensation, and Liability Act (42 U.S.C. et seq.), which contains provisions for the recovery of monetary damages by government organizations for natural resource injuries resulting from the release of hazardous substances. Laws that guide federal planning include: (1)the National Environmental Policy Act (42 U.S.C. et seq.), which is intended to promote harmony between humans and the environment, to prevent or eliminate environmental damage, and to enhance the quality of renewable resources; and (2)the National Forest Management Practices Act (16 U.S.C.A. 1600-1614.), which requires the U.S. Forest Service to assess renewable resources and to develop a National Renewable Resources Program, including plans for integrated and interdisciplinary management of national forests. An example of a law that directs private parties to conserve natural resources is the Surface Mining Control and Reclamation Act (30 U.S.C. 1201 et seq.), which established a nationwide program to mitigate or prevent the adverse environmental effects of surface mining for coal. Examples of laws to protect and conserve biological resources include: (1)the Endangered Species Act (16 U.S.C. 1531 et seq.), which is designed to conserve threatened and endangered fish, wildlife, and plants; (2)the Marine Mammal Protection Act (16 U.S.C. 1361 et seq.), which authorizes actions to conserve and replenish those stocks and species that are endangered or severely depleted; (3)the Fish and Wildlife Conservation Act (16 U.S.C. 2901 et seq.), whose purpose is to provide states with financial and technical assistance to conserve nongame wildlife; (4)the Fish and Wildlife Coordination Act (16 U.S.C. 661-667e), which requires a federal agency or permitee to consult with the Fish and Wildlife Service prior to controlling or modifying a body of water in order to prevent or lessen potential damages to natural resources; (5)the Soil and Water Resources Conservation Act (16 U.S.C. 2001 et seq.), whose purpose is to conserve soil, water, and related resources; (6)the North American Wetland Conservation Act (16 U.S.C. 4401-4413), which is designed to conserve and restore wetland ecosystems and maintain healthy populations of migratory birds in North America; and (7)the Clean Air Act (42 U.S.C. 7401 et seq.) and the Clean Water Act (33 U.S.C. 1251 et seq.), whose purposes are to prevent and control pollution in order to enhance and protect human and ecosystem health. Federal executive orders and policies include: (1)Executive Orders 11514 and 11991, which commit the federal government to provide leadership in protecting and enhancing the quality of the nation's environment to sustain and enrich human life, including resource restoration; (2)Executive Order 11988, which was designed to avoid adverse impacts of flooding associated with floodplain development.

The above laws and policies affect forests, mined lands, wetlands, plants and wildlife, and ecosystems. Berger (1991) posits that in combination with the many other laws affecting natural resources, a sound legal foundation exists for the conservation of biodiversity. On the other hand, while the types of laws mentioned above serve as examples of measures that can be used to promote the conservation of biodiversity, they have limited capabilities to do so. For example, the Endangered Species Act has been criticized because it: (1)primarily protects high-profile individual species rather than overall biodiversity; (2)lacks clearly defined thresholds to delineate endangered, threatened, and recovered species; (3)does not protect metapopulations adequately; (4)does not adequately document many biological determinations and therefore prevents meaningful scrutiny and participation by the public and scientific communities in decisions about protection of species; (5)does not protect habitat reserves sufficiently to sustain recovered populations; and (6)allows or fosters the discounting of uncertain or nonimmediate factors in the decisionmaking process about species' protection (Rohlf 1991). The treatment of biodiversity in environmental impact assessments conducted under the National Environmental Policy Act has been viewed as inadequate on two accounts: (1)biodiversity often is not considered in the assessment process even when there are reasons to do so, and (2)when biodiversity is considered in impact assessments, its treatment is inadequate (Hirsch 1993). Further, most other legislation designed to protect species or ecosystems does not establish criteria for protection of attributes of ecosystem health or integrity.

Legislative mandates guiding the goals and policies of federal agencies usually provide for administrative discretion in the balance chosen between use of resources that potentially impairs biodiversity or other natural resource values and their protection, and administrative decisions therefore have favored short-term consumptive use of resources over long-term protection of biodiversity and other natural resources. This has been shown to be true for agencies like the U.S. National Park Service, which has relatively strong legislative mandates to preserve biotic resources in their natural state (Lemons 1987), as well as for agencies such as the U.S. Forest Service. In other words, it must be recognized that while agency administrators may have the force of law or administrative discretion to conserve biodiversity, in the absence of stronger legislative directives, they will not abandon their own existing statutory authority and responsibility for their individual agencies.

Conserving biological diversity also requires successful management policies and strategies. These are determined, in part, by the legislative mandates governing governmental agencies and conferring protection on biological resources, as well as by the particular visions and capabilities of the agencies and others involved with the development and implementation of management.

A number of studies have critically analyzed the efficacy of management designed to conserve biodiversity (see, e.g., Soulé 1986, Grumbine 1990a, Costanza 1991, Slocombe 1993a). Generally speaking, their conclusions are that most management efforts fail to employ scientific methods that produce usable knowledge that fosters meeting socioeconomic needs while conserving biodiversity. In particular, these efforts also fail to base the management and conservation of biodiversity on ecosystem concepts. Lemons (1987) has documented how management of U.S. national parks promotes visitor use and enjoyment to the detriment of parks' species and ecosystems. Most assessments of environmental impacts to biodiversity do not include the use of sound science or post-project audits or monitoring (Cairns and Niederlehner 1993, Lemons 1994a). Keiter and Boyce (1991) have shown a pervasive lack of management based upon ecological boundaries and interagency cooperation, and Lemons (1986) has documented a lack of scientific information to support federal agencies' decisions. Burkardt et al. (1990), Wellman and Tipple (1990), and Cole (1992) have documented the need for federal resource agencies to bring about organizational change to promote conservation of biological resources, and Hargrove (1989) has pointed out the need to include values in decisions about biodiversity but has found little evidence that rigorous analyses of values are determinants in decisionmaking. Slocombe (1993b) has identified two kinds of obstacles that limit effective management: (1)those related to the ends intended by those using different management approaches, and (2)those related to the theory and implementation of scientifically based management itself. The first type of obstacle impedes integration of ecological and socioeconomic systems and arises from a mix of local, national, and international policies and practices, as well as from complex patterns of land ownership. The second type of obstacle is derived from informational and theoretical uncertainty about using scientifically based management approaches as well as from the different philosophies pertaining to planning and management.

Based upon a review of published papers, Grumbine (1994) identified ten attributes thought to be important to the successful management of biodiversity: (1)hierarchical context, wherein a management focus on any one level of biodiversity is not sufficient; (2)ecological boundaries, wherein management should be based on ecological boundaries and therefore might require working across administrative/political boundaries; (3)ecological integrity, wherein management should focus on protection of populations, species, ecosystems, and landscapes and on the ecological patterns and processes that maintain diversity and evolutionary potential; (4)data collection, wherein management requires research and information; (5)monitoring, wherein managers must know the results of their actions in order to evaluate their success or failure; (6)adaptive management, which assumes that scientific information is provisional and which allows managers to remain flexible so as to be able to respond to new information better; (7)interagency or international cooperation, which allows different groups to work together and to integrate conflicting legal mandates and management goals; (8)organizational change, which requires changes in the way agencies and private sector groups work with each other and share decisionmaking; (9)recognition of the fundamental influences humans have on ecological patterns and processes; and (10)values, which play an important role in attitudes and decisions about biodiversity.

While attributes have been identified as being important for the conservation of biodiversity, they are emphasized to varying degrees by both scholars and resource managers. For example, as might be expected, scientists emphasize the scientific attributes thought to be important for conservation of biodiversity and underestimate the importance of the other attributes. Conversely, public policy analysts tend to ignore information available in conservation biology literature. Both of these groups tend to ignore considerations of values, which largely are dealt with by philosophers. Perhaps more problematically, although these attributes are thought to be important for the conservation of biodiversity, the actions of many governmental agencies do not appear to reflect their utilization.

Based upon our foregoing discussion, achieving sustainable development and the conservation of biodiversity will require a reconciliation of traditional legislative mandates and goals of most governmental agencies that do not have an exclusive focus on conserving biodiversity and that also may have responsibilities for economic and development interests that might compete with the goals of sustainable development. Achieving such a reconciliation will be difficult enough. As we discuss in the following sections, meeting the goals of sustainable development and conserving biodiversity also must take into account: (1)limitations of the predictive capabilities of the science of conservation biology and (2)conflicts over competing values.

4. The Status of Science and Scientific Uncertainty

Scientifically speaking, management goals to conserve biological diversity must include: (1)maintenance of viable populations of native species *in situ*, (2)representation within protected areas of native ecosystem types and their natural ranges of variation, (3)maintenance of evolutionary and ecological processes, (4)long-term policies to maintain the evolutionary potential of species and ecosystems, and (5)accommodation of human use of resources consistent with the above four goals.

A reading of the language contained in Agenda 21 suggests an ambivalence concerning whether scientific understanding of problems of conserving biodiversity is thought to be adequate to serve as a basis for decisionmaking with reasonable certainty or whether uncertainties are so pervasive that few predictive capabilities exist. On the one hand, Agenda 21 calls for the development and use of predictive capabilities of science in decisionmaking and the use of methods to "...identify components of biodiversity important for its conservation and for the sustainable use of biological resources, ascribe values to biological and genetic resources, identify processes and activities with significant impacts upon biodiversity, evaluate the potential economic implications of conserving biodiversity and the sustainable use of biological and genetic resources." On the other hand, Agenda 21 recommends significant improvements in scientific capabilities and the adoption of precautionary management approaches in recognition of scientific uncertainty. Consequently, the expected role of the sciences in conserving biodiversity is not clear from the language of Agenda 21.

Soulé (1985) has proposed functional postulates for the science of conservation biology: (1)many species that constitute natural communities are products of coevolutionary processes; (2)species are interdependent; (3)many species are highly specialized; (4)extinctions of keystone species can have important long-range consequences; (5)introductions of exotic species may reduce diversity; (6)many ecological processes have thresholds below and above which they become discontinuous, chaotic, or suspended; (7)the temporal continuity of habitats and successional stages depends on size; (8)if population densities of ecologically dominant species exceed sustainable levels, they can alter prey populations and other species with whom they share resources; (9)genetic and demographic processes have thresholds below which nonadaptive, random forces prevail over adaptive and deterministic forces within populations, and (10)most nature reserves are inherently disequilibrial for large or rare organisms. With respect to Agenda 21's expected role of science, it is reasonable to ask whether these postulates are sufficiently generalizable and therefore of utility in serving as a basis for predictions, or whether their value is to serve as indicators of ecological change or in the understanding of the complexities of nature.

Sagoff (1988) distinguishes two types of criteria that might be applied to judge the usefulness of the ecological sciences to problems of biodiversity. The first type of criteria should contain descriptions of situations that facilitate prediction and control of species or ecosystems to meet prescribed goals. According to these criteria, a predictive science also will be a prescriptive science when decisions depend upon a rationality designed to meet instrumental goals. An example of such a criterion would be when decisionmakers ask ecologists to manipulate and control ecosystems in order to maximize the sustained yield of a resource. The second type of criteria contains descriptions that help people decide what to do but not exactly how to do it. These criteria assist in the understanding and application of normative concepts and how to apply them rather than assist with the manipulation and control of nature in order to achieve predetermined ends, such as when decisionmakers request the assistance of ecologists in protecting relatively natural ecosystems by mitigating changes that human interventions may cause in them. The distinctions between these two types of criteria may be subtle, but they are important. To assist with the task of maximizing the sustainable yield of a biological resource, scientific information will have to be sufficient to inform management decisions with reasonable certainty. To assist with the task of protecting natural ecosystems by mitigating changes brought about by human intervention, ecological knowledge will have to be sufficient to guide decisionmakers in identifying important ecological structural or functional attributes and in understanding the values that may lead to their conservation or preservation. Consequently, ecological knowledge may be used to gain predictive knowledge for the achievement of instrumental ends, or it may be used to understand the complexities of nature and to help identify its qualities that humans might wish to protect.

Although Agenda 21 may be interpreted as being ambiguous regarding the expected role of science in conserving biological diversity, our interpretation of its recommendations is that a considerable expectation is placed on science to yield reasonably certain predictions to serve as a basis for decisionmaking about biological conservation. To the extent that our interpretation is correct, then the ability to implement Agenda 21's recommendations depends, in part, upon the capabilities of science. Based on a review of the status of scientific knowledge concerning biodiversity and relevant areas of ecology, our view is that its utility as a basis for reasonably certain predictions is constrained by a number of theoretical and practical factors.

4.1. SCIENTIFIC KNOWLEDGE ABOUT BIODIVERSITY

Agenda 21 articulates the need for baseline information concerning the status of biodiversity which would include information about the number of species present on the earth as well as the rate at which these species are becoming extinct. The understanding of the rate of species extinction depends, in part, on the knowledge of the current number of species on the planet.

The known number of named and categorized species is estimated to be between 1.4 and 1.8 million (Groombridge 1992). A more precise number is not available due to differences of opinion regarding how to identify and categorize species and because an official count of the species that have already been named does not exist (May 1992). Estimating the total number of species on the planet has proven to be difficult, and calculated values range from 3 to 30 million (May 1992) to 100 million (Ehrlich and Wilson 1991). One reason for such a discrepancy is that scientists have arrived at these numbers by assuming that a ratio of known to unknown species obtained in a known situation holds in an unknown one. For example, scientists have found that for birds and mammals, there are approximately two to three times as many tropical as nontropical species. This ratio has been used to estimate the total number of insect species from the currently known number of species. If we use 1 million as an approximate known number of insect species and we assume that 60 percent of these are nontropical and that about 40 percent of nontropical species have been described to date (which gives a total nontropical species count of 1.5 million), then using the ratio of tropical to nontropical species given above, an estimate of from 4.6 to 6 million insect species worldwide can be obtained (Groombridge 1992). This type of estimate is dependent, of course, on the validity of the assumption that there is a similarity between tropical to nontropical ratios for bird and mammal species and those for more diverse but less well-known groups. In fact, there now are data from the tropics indicating that the diversity of insects in tropical forest canopies may be at least ten times that of insect diversity in temperate forests. This would increase the estimate of insect diversity to 11.5 million species.

Another approach to estimating the number of insect species was attempted by Erwin (1983), who used insecticidal fogs to enable him to count the number of beetle species in canopies of Amazon tropical trees. His results suggest that insect species are often distributed in very local ways. Erwin collected 1,200 species of beetles from the canopy of a single tree species (*Luehea seemannii*) in Panama, of which 800 were herbivores. Using an estimate that 20 percent (160 species) of these herbivores feed only on this species of tree, he then calculated that because beetles comprise 40 percent of the total of all insect species, there could be as many as 400 species of insects that feed in each tree species' canopy. He further estimated that these canopy species comprise about two thirds of the total number of insects living on each tree species, bringing the total number of insect species per tree species to 600. Finally, assuming 500,000 species of tropical trees, there may be 30 million species of insects in the tropics alone (Primack 1993). As these examples indicate, estimates of the number of species are based on a combination of empirical data, extrapolations, and assumptions which must be understood in order to obtain as accurate a picture as possible of predicted values for species diversity.

An important question that needs to be addressed is: How quickly is biodiversity being lost? Predictions that half the world's species will be lost within the next half century have led some to warn of consequences as severe as the collapse of entire ecosystems (Cairns 1988). The loss of biodiversity usually is discussed in terms of the rate of species extinction. Although most scientists agree that there is a serious extinction problem and that this problem is caused by humans primarily, there is disagreement about how severe the problem really is. Ehrlich and Wilson (1991) have warned that 4,000 species a year may be lost in the tropical forests alone, given that 2 million species live only in the tropics. If the number of species inhabiting only the tropics is raised to 20 million, this raises estimates of the number of tropical species lost per year to 40,000. Mann (1991) disagrees with these estimates and claims that Wilson's and Ehrlich's estimates exaggerate and distort the extinction problem by relying on several questionable assumptions.

Estimates of extinction rates are based upon a number of assumptions, including: (1)the rate of habitat loss, (2)the shape of the species-area curve, and (3)the absolute number of species on earth. With respect to the first assumption, Ehrlich and Wilson (1991) cite sources that estimate the loss of rain forest habitat at a rate of 1.8 percent per year, whereas Mann's sources indicate that the rate of actual forest clearing is 0.5 percent per year. According to Mann, the reason for the discrepancy is that Amazonia includes habitat other than forest, such as savannah and semidesert, and that these are being converted to farmland at a rapid rate. Therefore, the loss of forest habitat is less than commonly believed. The second assumption above is based on the island biogeography model, a theory proposed by MacArthur and Wilson (1967) that describes the relationship between the size of an island habitat and the number of species it supports. This species-area relationship states that the size of the island is proportional to the number of species found there, with larger islands having greater species diversity and smaller islands having less. The theory is based on data from islands of varied sizes and has been empirically validated to some extent (Primack 1993). The species-area relationship has been extended to predict the effect of habitat loss on extinction rates in nonisland situations. Lovejoy et al. (1986) have studied different-sized forest patches surrounded by deforested areas in order to test this assumption in nonisland settings. Some results suggest that a loss of habitat of 50 percent would correspond to a decrease of 10 percent in the number of species. However, the use of island biogeography models in nonisland situations can be deceiving, because whereas islands are surrounded by water, many forest patches are surrounded by secondary forest. Secondary forests, unlike the water surrounding a smaller island, will still support a variety of species. In addition, when considering the number of species that can reside in a particular community type, there is a finite limit to the number of different species that can live there. Consequently, an increase in area corresponds to an increase in the number of species only up to a point. Critics of Ehrlich and Wilson argue that if habitat destroyed is in the upper reaches of the species-area curve, species extinction may be minor or nonexistent (Mann 1991). Even so, Ehrlich and Wilson (1991) claim that "...by the most conservative estimate from island biogeographic data, 0.2 to 0.3 percent of all species in the forest are extinguished or doomed each year." They also argue that many species are found only in very local habitats, such as mountain ridges or woodland patches, and can be eliminated as a result of even a small amount of habitat destruction. These types of arguments not withstanding, a more fundamental problem is that despite some uses of island biogeographical models, many researchers argue that they cannot be generalizable and hence have no predictive capabilities whatsoever (Shrader-Frechette and McCoy 1993). The third assumption, which concerns the absolute number of species on earth, has been discussed above. The current discrepancy in the estimates of how many species inhabit the earth can affect the calculated rate of species extinction by several orders of magnitude.

As described earlier, biodiversity includes many levels from genes to landscapes, and its study may include short-term or long-term research on both or either structural and/or functional attributes. Scientists working to conserve biodiversity recognize this hierarchy and therefore base their solutions to biodiversity problems on a variety of scales. In some cases, problems will require the application of methods at one scale, while in other cases it will be necessary for biologists to work at several scales. The appropriate scale will depend on the research question asked, although conservation strategies will probably be most effective if applied to a range of scales, because there is no a priori single correct choice of scales for a particular biological resource or ecosystem (Noss 1992).

On the structural level, a focus of study may be on individual organisms, populations, species, or ecosystems; alternatively, a functional study relevant to biodiversity problems may emphasize factors such as production, the nature of changes in population densities with predation, or energy flow. A suitable temporal scale may be defined, in part, by factors of climate, physiology, and nutrient cycling and may include the short or long term. Spatial scales may be locally or globally defined and may interlock with temporal scales. Because activities threatening biodiversity take place on a variety of structural, functional, temporal, and spatial scales, ecological studies necessarily must define bounds on the scales of interest; this operationally defines the units of study. However, even when scales of interest are defined, they are done so somewhat arbitrarily. For example, all ecosystems exchange information and material with others up to the global level. Because ecosystems are not closed, even with boundaries established on scales, they cannot be defined unequivocally. From the standpoint of scientific predictions, a major problem is that in many cases the biologically or ecologically relevant scales are not obvious; when scales are known, they often are known better for lower hierarchical levels of organization. Effective management of biodiversity therefore demands that structural, functional, temporal, and spatial scales be chosen carefully; yet it is not often possible to make choices on sufficiently precise data.

Agenda 21 recognizes the issue of scale by including recommendations addressing the conservation of biodiversity at many levels. Traditionally, the conservation of biodiversity has been performed at the species or population level, where scientists have worked primarily to prevent the extinction of threatened or endangered species. By working to conserve biodiversity at this level, scientists are also conserving genetic diversity. Although biologists have been using a species-level approach to conserve biodiversity for a longer time compared with other approaches focused on other scales, the understanding of what it takes to prevent species from going extinct is far from complete.

Populations of organisms become in danger of extinction when the number of individuals in a population is reduced below a certain count. A reduction in population size caused by humans typically might result from a loss of habitat, habitat degradation, habitat fragmentation, or overharvesting. A number of chance events can also force small populations to extinction. Although these events may also occur in larger populations, they do not then have as deleterious an effect as they do when populations are small. Shaffer (1981) has described these chance events as: (1)demographic stochasticity, (2)environmental stochasticity, (3)natural catastrophes, and (4)genetic stochasticity.

Demographic stochasticity refers to things such as the random variation in the birth and death rates of individuals within the population, deviations from a normal sex ratio, the dysfunction of social structures, and the inability of individuals in a population to find mates (Primack 1993). All of these chance events become amplified when populations are small. Environmental stochasticity includes chance events in the environment that all members of a population experience (Gilpin and Soulé 1986). These might include effects of competitors, predators, parasites, and diseases on the population (Shaffer 1981). Natural catastrophes such as fires, storms, and droughts also may occur randomly and may have impacts on small populations. Genetic stochasticity results from a loss of genetic variability in a population and may be caused by genetic drift, random fixation, the founder effect, or inbreeding (Shaffer 1981, Primack 1993).

The loss of genetic variability is of concern for several reasons. Species with greater genetic diversity are able to adapt to changes in the environment better. Greater levels of heterozygosity within species also have been correlated with greater disease resistance, better reproductive performance, and increased survival rates (Packer et al. 1991, Groombridge 1992). If heterozygosity is lost, deleterious recessive alleles are more likely to be expressed (Packer et al. 1991), and mortality rates are likely to increase (Soulé and Simberloff 1986). High levels of inbreeding have been correlated with increased levels of abnormal sperm as well as lower sperm counts (O'Brien et al. 1983, Packer et al. 1991). Inbreeding may also lead to fewer offspring or offspring that are weak or sterile (Wayne at al. 1991, Primack 1993). Conserving existing genetic diversity is essentially ensuring a species' future evolution; once genetic diversity is lost, it is regained only through the slow process of evolution, which is dependent on random mutations.

Conservationists have attempted to counteract the effects of chance events and human activities on populations by calculating and managing for MVPs. A population's MVP is the smallest number of individuals it can contain and still continue to survive natural stochastic and human-induced events. However, in the calculation of MVPs, neither the length nor the probability of survival are known for most species. Estimates of the percent chance of survival and duration of existence for a population are determined arbitrarily by scientists before calculating an MVP. These estimates commonly are not agreed upon, and typically they vary from 95 percent chance of survival for 100 years (Groombridge 1992) to 99 percent chance of survival for 1,000 years (Shaffer 1981).

Presently, conservation biologists attempt to calculate MVPs using what is known as a population viability analysis (PVA). This analysis attempts to take into consideration the chance events that may affect populations; however, efforts to design models that consider all four stochastic events have been slow to develop (Gilpin and Soulé 1986). So far, PVAs developed have dealt with only one or two of these factors (Shaffer 1991). This is due partly to the fact that the interactions between the various factors (genetic, demographic, environmental) are not understood well (Woodruff 1989). Consequently, current estimates of MVPs have been criticized because they are based on very few data and lack feedback among demographic and genetic processes. Estimates of MVPs also are criticized because they are based on the assumption that loss of genetic diversity within populations affects all species equally (Lacy 1992); however, few data are available that can be used to verify this assumption. Estimating MVP size also is difficult because there is no reliable way of predicting how severe future or human-induced environmental fluctuations will be or of assessing their consequences on population growth rates.

Some things may help biologists be able to predict MVP sizes in the future better. More data on the genetic variation within a species are needed. Although genetic information is currently known about many captive populations, few wild populations have been studied in depth. One of the best-studied mammal populations is the lions of Ngorongora crater in Tanzania. Here, scientists have combined data from genetic surveys with genealogical data in order to estimate the rate of genetic change within a population (Packer et al. 1991). Biologists compared the genetics of this small lion population with that of a much larger population from the Serengeti; such a comparison may help them decide when inbreeding may become a problem for wild populations (Woodruff 1989). If biologists know when a population is small enough to be in danger of inbreeding, they may then be able to manage the population by establishing corridors between existing populations or by transporting individuals between populations in an attempt to prevent the problems of inbreeding. However, knowing when a population has reached the size where inbreeding is a problem has been difficult to predict. Extensive studies of the genetics and responses to inbreeding of the population of concern may be necessary before estimates of MVPs can be made (Lacy 1992).

The existence of metapopulations also may be helpful in terms of the long-term survival of threatened species, because gene flow may occur between subpopulations. However, estimating MVP size for metapopulations is much more complex than for single populations (Shaffer 1991). More information is also needed about life histories of species as well as about the temporal and spatial distribution of resources (Gilpin and Soulé 1986). Detailed demographic studies of populations will be necessary in order to have data adequate to predict MVP sizes.

It is important to understand that estimates of MVPs must still be performed on a case-by-case basis and that models should be built for particular species of concern (Soulé

1987, Shaffer 1991). No single MVP model is applicable to all species, and scientists do not see this as a possibility in the future (Gilpin and Soulé 1986, Soulé 1987). The genetic, demographic, environmental, and human-induced factors that influence the survival of one species are not the same for another species, so developing a universally acceptable protocol for determining MVP size is unlikely. "Comprehensive, realistic, and reliable methods for applications to all situations or precise prescriptions that can be applied uncritically will take a long time to develop, if they are even possible" (Shaffer 1991).

Despite these problems, it is important to note that even with limited knowledge, biologists have been able to estimate MVPs for several populations, including Florida panthers (*Felis concolor coryi*), Bali starlings (*Leucospar rothschildi*), and the Sumatran rhinoceros (*Didermocerus sumatrensis*) (Groombridge 1992). What has become evident from this work is that in order to ensure survival of some of these species, especially mammals, populations must contain a substantial number of individuals. Using both genetic and demographic extinction models, scientists have estimated the MVP size for the Tana River crested mangebey, an endangered forest primate, to be 8,000 individuals. It is hoped that this will ensure a 95 percent chance of survival of these primates for 100 years (Kinnaird and O'Brien 1991). In order to ensure similar survival success for the grizzly bear (*Ursus arctos*), researchers estimate that a population of 50 to 90 individuals must be maintained. This MVP was calculated using models that incorporated estimates of environmental and demographic stochasticity (Shaffer and Sampson 1985). Although the MVP is smaller for grizzlies than for mangebeys, conservation biologists also must consider the amount of habitat needed by MVPs. Large mammals such as grizzlies are often wide-ranging and have a large habitat requirement per individual. It has become apparent, therefore, that most national parks and other protected areas are not of sufficient size to maintain MVPs of these types of species (Newmark 1985). *In situ* conservation will require that these threatened species have habitats extensive enough for their continued survival (Soulé and Simberloff 1986).

To summarize, scientists can do certain things in order to get a better idea of the MVP size of a population, including the gathering of more genetic and demographic data for species of concern. The use of models to predict MVPs will continue to be difficult due to the inherent unknown factors affecting a population's survival, such as the occurrence of natural environmental fluctuations and human activities. Also, the chance that a single model will be applicable to all populations is highly unlikely, and studies therefore will have to be performed on a case-by-case basis. However, the work of conservation biologists in this area has pointed out three important issues: (1)the effect of various chance events on a population's continued survival, (2)the time frame to use in conservation planning, and (3)the degree of security desired for populations of concern (Shaffer 1987). The second and third issues relate to the criteria that should be used in defining MVPs. Should we plan for continued survival of a species for 100 or 1,000 years or longer? Do we plan for an 80 percent chance of survival or a 99 percent chance? Answering these questions will involve disciplines not traditionally considered a part of science, such as ethics, economics, and politics.

In addition to the variety of species, biodiversity also includes diversity at levels such as the community, ecosystem, and landscape. Although conservation efforts traditionally have been focused at the population or species level, scientists are now assessing the conservation of biodiversity at larger scales (Scott et al. 1987, McNaughton 1989, Pickett et al. 1992). Proponents of this wider view claim that the traditional reductionist approach of science wherein biological systems are reduced to their component parts in order to try to understand them does not necessarily foster successful conservation of biodiversity. By employing a more holistic approach to the study of biodiversity and by enlarging the scale of reference, scientists can obtain a more accurate assessment of biodiversity, because each level has unique properties that can only be understood by studying that level. For example, if healthy ecosystems depend on specific interactions among organisms functioning within food webs and responding to the abiotic environment, then storing species *ex situ* (in zoos or botanical gardens) and later reintroducing them to the wild will not recreate a functional ecosystem (NcNaughton 1989). This view is reflected by the fact that studies at the ecosystem level emphasize the processes essential to healthy ecosystem functioning rather than end products (Walker 1989, Pickett et al. 1992). Ultimately, however, the question of what scale is appropriate will be determined by the particular research question being asked.

Conserving biodiversity at the ecosystem level means striving to protect ecosystems' basic trophic structure and the energy flow and nutrient cycling patterns that result from that structure. If an ecosystem is protected adequately, then the assumption is that all of its resident species also are protected. Whereas traditionally species that have economic or instrumental value have been the targets of conservation, here other species, such as bacteria, fungi, invertebrates, and even those species that are not yet known are protected as well. These species may in fact be more important to the healthy functioning of an ecosystem than large, charismatic species (McNaughton 1989). Further, by protecting large areas, species that move between habitats or live where two habitats meet will have a better chance of survival. Another important objective of conservation at this scale is to protect a representative sample of ecosystems on a worldwide basis. Most of the large areas that have some sort of conservation status today are in national parks, protected reserves, or forests. However, these areas were not originally set aside with the goal of protecting biodiversity, and their boundaries do not make sense ecologically. Grumbine (1990b) points out the inability of U.S. national parks to provide habitats of adequate size for large vertebrates and some long-lived plant species.

Although the importance of the concept of protecting ecosystem diversity is now agreed upon by most conservation biologists, the question of how best to achieve conservation at this level is still being debated. To begin with, the classification of ecosystems into a manageable system has been a major problem. The range of classification systems is great, with some systems classifying terrestrial ecosystems, for example, according to their plant communities and other systems taking a more general approach based on an area's physical characteristics and appearance. Part of the problem is that these systems are based upon the assumption that ecosystems are discrete units that can be delineated and distinguished from each other instead of a series of interacting parts of a greater and highly variable continuum. It is extremely difficult to determine exact areas for ecosystems and even more difficult to estimate rates of habitat loss.

Some research on spatial and temporal scales of ecosystems has focused on the biogeographic consequences of fragmentation. Ecosystem fragmentation causes large changes in the physical environment as well as biogeographic changes. A result is that landscapes consist of remnant areas of native flora and fauna surrounded by land modified by human influences. Consequently, fluxes of physical and biotic inputs across ecosystem boundaries are altered, thereby affecting native species in natural remnant areas. Biological consequences of the isolation of protected areas due to the modification of adjacent lands is significant and varies as a result of the time since isolation and distance from other natural remnant areas. Research also has indicated that the consequences of fragmentation are influenced by size, shape, and distance of remnant areas from each other. Controversy exists regarding whether one large reserve will protect more species than several smaller areas with a total area equivalent to that of the larger reserve. Generally, larger reserves are more buffered from adverse consequences of fragmentation and are therefore thought to be better than smaller reserves. Unfortunately, most research on size and shape of reserves has provided little of practical value to resource managers for the reason that managers of protected lands are dealing with ecological conditions that are a *fait accompli*. In other words, with few exceptions, protection of biodiversity must occur on lands already set aside for particular management goals but whose boundaries were based upon political or cultural as opposed to biotic considerations. Consequently, a critical need exists to develop integrated approaches to land management that place conservation of biodiversity in the context of factors that influence overall use of landscapes.

One strategy being employed by conservation biologists to help determine what areas are in need of protection is gap analysis. Although the term "gap analysis" is relatively new, the process has been used for many years (Burley 1988). The idea is to compare the locations of habitats or ecosystems with those of existing reserves in order to find the gaps in the system (Allen 1992). In the area to be considered, the biodiversity is identified and classified, often in several different ways (e.g., by ecosystems, vegetation types, habitat types, species). Existing and proposed protected areas are then identified, and by comparing these with the biodiversity data, protection of biodiversity can be enhanced (Burley 1988). Gap analysis can be done on a large international scale or at more local levels. At the largest scale, biogeographers designated eight terrestrial biogeographical regions worldwide which were then further subdivided into 227 provinces and have evaluated whether these regions are protected in existing reserves, national parks, national monuments, wildlife reserves, or protected landscapes. These provinces are also associated with 14 biomes. This information can be used to identify high-priority ecosystems and to guide recommendations for the establishment of future parks and reserves (Burley 1988, Primack 1993). For example, after assessing what percent of each of the 14 biomes is currently protected, it was discovered that least protected are temperate grasslands (0.78 percent) and lake systems (1.28 percent), whereas the biomes protected most are subtropical/temperate rain forests/woodlands (9.32 percent) and mixed mountain systems (7.71percent) (Groombridge 1992).

Gap analyses are also being used in all U.S. states to determine if existing preserves, parks and refuges are protecting biodiversity adequately. This effort is primarily being orchestrated by the Nature Conservancy, a private nonprofit organization, along with Defenders of Wildlife and the U.S. Fish and Wildlife Service (Allen 1992). Even though there is no agreed-upon system of vegetation classification, the Nature Conservancy, working with state government agencies, has established Natural Heritage Data Centers in most states (Burley 1988). Using several vegetation classifications as well as available data on species distribution, data needed for conservation efforts are gathered and organized. One application of this information is in preserve selection, so that limited resources can be better used to protect biodiversity in priority areas (Jenkins 1988). Scientists have used gap analysis in several states by overlaying maps displaying current land ownership patterns, the location of threatened and endangered plants and animals, and vegetation types in order to look for gaps in the protected area network. Because data were unavailable for locations of all species, vegetation maps were used to predict where animals might be found, and these areas were then field-checked to test these predictions (Allen 1992).

Much of the recent work in gap analysis has been aided by a technology known as Geographical Information Systems, a computer-based process that allows maps containing diverse data such as vegetation types, climate, soils, species distribution, and current land ownership to be overlaid (Scott et al. 1987). After converting traditional maps to digital (computer-compatible) maps, scientists can combine these data with information from other sources, such as Landsat Thematic Mapper Imagery, to create overview maps that provide useful information for specific conservation questions. Geographical Information Systems are becoming a powerful tool in the conservation of biodiversity, but it should be noted that some sources of error are associated with this technology. Errors can result from mistakes in data input (including using data from inappropriate or out-of-date sources), processing and analyzing data, and the output and presentation of data (DeGloria 1991).

In recent years, efforts to conserve biodiversity at the ecosystem level have yielded new information. However, some areas of uncertainty and several limitations in this approach must be recognized. As mentioned previously, defining, classifying, and delimiting an ecosystem is still problematic. In addition, many ecologists traditionally have thought that ecosystems and their communities would eventually reach a steady state, and many management activities in conservation areas have attempted to stabilize the ecosystems being managed (Walker 1989). Accordingly, concepts of ecosystem stability have been a focus of study for many conservation biologists. However, the concept of stability has stimulated considerable debate among scientists. For example, it is not known whether stability is due to species diversity or the cause of it. Further, concepts of stability can variously emphasize the resistance to disturbance of an ecosystem, the time an ecosystem requires to recover from damage, the zone from which an ecosystem will return to a stable state, the degree to which the pattern of secondary succession is not an exact reversal of the retrogression following environmental impact, and the degree to which a stable ecosystem established after disturbance differs from the original steady state. More problematically, some ecologists question whether concepts of stability have any real ecological meaning (Westman 1990). Consequently, management decisions based upon concepts of stability must recognize the uncertainty surrounding the different concepts and the practical implications of managing for one concept as opposed to another.

Further, the goal of managing ecosystems in relatively stable or equilibrium states may conflict with the goal of maintaining high levels of biodiversity, because some systems need to be unstable in order for natural biodiversity to remain high (Huntley 1988, Walker 1989). Management of ecosystems for stable states also conflicts with recent views on the nonequilibrium nature of ecosystems. Pickett et al. (1992) describe the classical paradigm in ecology as the "equilibrium paradigm," and the new paradigm as the "nonequilibrium paradigm." The equilibrium paradigm emphasized the stable endpoints of ecological systems and the idea that ecosystems were functionally and structurally complete and self-regulating. This implies that ecosystems, once set aside in parks or preserves, will maintain themselves as they were at the time of protection and that if disturbed they will return to their original state. In contrast, the contemporary nonequilibrium paradigm includes the following ideas: (1)natural systems are open, (2)processes rather than endpoints are emphasized, (3)a variety of scales are considered, and (4)episodic disturbances are recognized. In order to practice conservation under the modern paradigm, conservationists must focus on the processes of communities and ecosystems and work to maintain the dynamics of the system while recognizing that change and disturbance are important to the continued health or integrity of ecosystems (Kay 1991). These processes might include the effect of herbivores on vegetation, fire, rainfall, and other natural disturbances. The nonequilibrium nature of ecosystems has important consequences for decisions about sustainable development and the conservation of biodiversity. If decisions are made to attempt to maintain static ecosystems so that they may provide specific biological resources or services to humans on a sustainable basis, then significant and perhaps unrealistic levels of intervention in biological and ecological systems may be required of resource managers. Because of these concerns, Angermeier and Karr (1994) have proposed that concepts of ecological integrity rather than elements of biological diversity be used as a basis for policies to protect biological resources because the former emphasize the organizational processes of ecosystems that generate and maintain all of the elements of biodiversity instead of only the presence or absence of particular elements.

By practicing conservation at larger scales, scientists have the opportunity to see how various processes operate at the ecosystem level, and it then becomes important for those involved in the management of protected areas to enable these dynamic processes to occur. This approach has been adopted by those in the field of landscape ecology who realize that ecosystems are not isolated units but are interacting systems which exchange nutrients, energy, and species. However, despite the apparent advantages of managing biodiversity at the landscape level, there is a concern that such a scale is so large that not all species will be able to be conserved. Communities that are small but rich in species (such as streams, wetlands, and coastal habitats) may not be apparent at larger or coarser scales (Huntley 1988). For example, much of the diversity of African ecosystems is found in areas that are too patchy or narrow to be picked up by large-scale approaches using vegetation maps. Endangered or threatened species in these smaller areas may not be adequately protected by ecosystem-level strategies, and endemic species are often not identified at such large scales. Consequently, the use of finer-scale resolution may be necessary to discern additional areas of conservation concern. Further, maps displaying vegetation types are often used to predict the locations of animal species in landscape analysis because of the assumption that if all vegetation types are included in protected areas, then all animal species also will be protected (Burley 1988). However, it is not known whether this assumption should be accepted.

4.2. THE STATUS OF ECOLOGY AS A BASIS FOR MANAGEMENT

The conservation of biodiversity requires not only scientific information regarding numbers of species, extinction rates, and MVPs but also the successful management of biological resources and ecosystems so that they are protected from adverse impacts of human activities. In this sense, conservation of biodiversity on a sustainable basis requires knowledge of the impacts of human activities upon biological and ecosystem attributes.

Cognizant of the aforementioned theoretical and practical problems which constrain the scientific understanding of biodiversity as well the ability to assess the ecological impacts of human activities, a number of researchers have critically analyzed the extent to which the methods and techniques of science are capable of yielding reasonably certain information appropriate to serve as a basis for management and public policy decisions (Lemons in press). Westman (1990) concludes that many current policies to conserve biodiversity are based upon concepts no longer accepted in the ecological community, including notions that all species in a community are interdependent. He suggests that policies should recognize the individualistic distribution of species over a landscape and that research should be directed toward better understanding of the relative abundance of coevolved relationships in different biomes and whether and to what extent one species can substitute functionally for another in an ecosystem. Sagoff (1988) argues that the role of ecology should be to identify ecological indicators that might allow scientists to diagnose perturbations in species or ecosystems early enough so that mitigation measures could be implemented. This type of diagnosis does not depend on knowing generalizable laws and basing predictions upon them; rather, it involves the integration of diverse information to make a general argument for one rather than another interpretation of the causes or consequences of ecological impacts. Recently, Bella et al. (1994) also have argued that the role of the ecological sciences in problems of global environmental change ought to be in the identification of indicators of ecological change rather than in the prediction of the consequences of human activities with reasonable certainty. Lemons (1986) has analyzed the different meanings of "stress" as applied to species and ecosystems and has concluded that the theoretical differences between the different meanings are so great that when combined with informational uncertainty concerning the assessment and evaluation of the causes of stress and their effects, little basis for reasonably certain predictions exists. Lubchenco et al. (1991) have identified numerous scientific uncertainties regarding sustainable development and protection of biodiversity and have proposed a research agenda to obtain more information about biodiversity. In their report, they acknowledge the limited role scientists can play in making reasonably accurate predictions about the effects of human interventions in ecological systems. Cairns and Niederlehner (1993) note that in theory, both structural and functional attributes of ecosystems can be used as a basis for ecological predictions but that practically speaking, there is a significant lack of knowledge about them. Based upon a review of the role of science in the protection of biodiversity in national parks, Lemons (1994b) concluded that scientific information should be used in the planning process in order to help decisionmakers make more environmentally sound decisions, but that this information is not adequate to serve as a basis for firm predictions. Shrader-Frechette and McCoy (1994) maintain that science may have some role in informing public policy decisions in matters of environmental quality and biodiversity, but that its role should be considered to be heuristic. More specifically, they argue that site-specific case studies may yield information useful for decisionmaking but that such information should not be used to make more generalizable laws for predictive purposes. Based upon these types of analyses, our view is that both informational and theoretical uncertainty exists and is so pervasive that the science of ecology and conservation biology should be considered as having heuristic value but not predictive capabilities suitable as a firm basis for decisions about intensive management of biological resources.

The pervasive uncertainty inherent in the methods of ecology and the understanding of biodiversity combined with science's emphasis on minimizing type I error (thereby increasing the chances of accepting a false conclusion that no harm will be done to biological resources) means that those promoting the enhanced protection of biodiversity will have difficulty in meeting burden of proof requirements imposed by law, science, and business (Lemons and Junker in press). Allowing uncertainty to delay decisions to protect biodiversity is to make a tacit decision to allow and thereby promote the status quo; no decision is in fact still a decision. A precautionary approach to protection of biodiversity might very will be to shift the burden of proof to those seeking to undertake activities that potentially threaten biodiversity to demonstrate that their activities will not cause harm.

4.3. IMPLICATIONS OF SCIENTIFIC UNCERTAINTY AND COST-BENEFIT ANALYSIS

Agenda 21 also calls for the use of cost-benefit analysis as a means for assessing the values of species and ecosystems. A variety of methods are used to assess the values of species, and it is beyond the scope of this chapter to discuss such methods in detail. Extensive discussions of the use of cost-benefit analysis in natural resources problems can be found in Norton (1987) and Costanza (1991).

Generally speaking, cost-benefit analysis considers the following values associated with species and ecosystems: (1)use values, (2)option values, (3)quasi-values, and (4)existence values. The first three types of values are considered to be instrumental values. Use values include all of the direct and indirect ways in which people use species and ecosystems, including amenities such as aesthetic enjoyment. Option values include those values assigned to species or ecosystems that are not currently being used by humans but that could be used. Quasi-values represent values that might be enhanced due to expected growth in knowledge or aspirations which would enable new or as yet unknown uses for species or ecosystems. Existence values are those that are independent of the use of species or ecosystems. Existence values infrequently are included in cost-benefit analysis because it is questionable whether and how they can be amenable to quantification.

The use of cost-benefit analysis is prevalent in decisionmaking about natural resources. Proponents of cost-benefit analysis assume that its techniques can provide quantitative and objective information to decisionmakers on the present and future values of species or ecosystems. However, many people concerned with conservation of biodiversity object to the use of cost-benefit analysis for several reasons: (1)it may systematically bias decisions by ignoring, discounting, or miscalculating values; (2)different methods of assigning dollar or other quantitative values to species or ecosystems can lead to different approximations of benefits; (3)pervasive scientific uncertainties imply that it is not possible accurately to compute present or future values for most species or ecosystems; (4)it ignores distributional problems such as when the future worth of species or ecosystems is discounted and therefore greater benefits are distributed to present generations and greater potential costs or harm are distributed to future generations; (5)a decision not to proceed with a development project based on cost-benefit analysis is reversible, whereas a decision to let a species become extinct due to development is not; and (6)the use of cost-benefit analysis ignores so-called intrinsic values which are independent of market or instrumental values. Norton (1987) offers a more detailed discussion of these types of problems pertaining to the use of cost-benefit analysis in preserving species.

Despite the aforementioned concerns about cost-benefit analysis, other fundamental problems also exist. Normally, in cost-benefit analysis, the instrumental value of a single species of interest or the aggregate values of multiple species of interest are included in calculations. However, from the standpoint of conserving biodiversity, what matters is a species' role in maintaining within-habitat, between-habitat, or total diversity. This would imply that the attributes of species in maintaining all of the various forms of diversity be assessed, and this would involve considerations of their genetic and population characteristics as well as other characteristics that determine their role in landscape ecology. In other words, cost-benefit analysts attempt to assess the values of particular species of instrumental interest, wherein conservation biologists attempt to assess the role of a species as a unit of biological diversity. These are two very different types of assessments. Because of pervasive scientific uncertainty, it is highly problematic to assign accurate values for species or ecosystems for use in the first type of assessment. The problem is even worse for the second type of assessment, and we therefore posit that for most species it is simply not possible to assign any reasonably accurate quantitative values for biodiversity assessment purposes.

In addition, the treatment of species and ecosystem attributes as traditional commodities has been identified as a significant cause of loss of biodiversity for several reasons (Goodland et al. 1993, Cairns and Meganck 1994): (1)those who receive the benefits of exploiting biological resources usually do not pay the full costs of the exploitation; (2)the benefits of utilizing biotic resources are easier to quantify than the benefits of preserving them; (3)many biological resources are publicly owned and treated as free or inexpensive commodities; (4)discount rates in cost-benefit analysis often are set too high compared with biological growth rates, thereby enabling more efficient depletion of biotic resources; and (5)gross national product measurements normally do not consider the depletion of biological resources as a reduction of net natural capital but rather treat it as net income, so that gross national product rises while biological resources decline.

4.4. RECOMMENDATIONS TO IMPROVE SCIENTIFIC CAPABILITIES

Agenda 21 includes recommendations to improve the capabilities of science for use in sustainable development and conservation of biodiversity, both in terms of yielding more accurate information for use in making predictions about the consequences of human activities on species and ecosystems and in terms of providing more accurate information for use in cost-benefit analysis. Of course, Agenda 21 is not alone in recommending improvements in scientific capabilities to make more accurate predictions of human impacts on biodiversity; similar recommendations can be found in Franklin et al. (1990), Magnuson (1990), Swanson and Sparks (1990), Lubchenco et al. (1991), and CNIE (1994). Murphy (1990) and Drew (1994) specifically suggest that the science of conservation biology be redirected away from descriptive studies and toward controlled experimentation sufficient to meet the qualifications of testable scientific questions. In this manner, they argue that scientific results will be less speculative and more powerful as a basis for management decisions.

While these recommendations are worthwhile, they also contain an inherent danger. Such recommendations seem to imply that existing problems of conserving biodiversity are primarily or significantly a function of a lack of scientific knowledge and management techniques and that as such knowledge increases through improved scientific capacity, species and ecosystems can then be managed intensively and successfully on a long-term sustainable basis. Further, they seem to imply that increases in knowledge can be obtained rapidly enough to be of real assistance in redirecting the world toward sustainable development and conservation of biodiversity in the time required. If our interpretation of Agenda 21 recommendations is correct, then they conflict with the conclusions of many ecologists and philosophers of science who have argued extensively that the ecological sciences are descriptive and inherently limited in their ability to provide information suitable for long-term predictions. Consequently, the danger of which we speak is twofold. First, the implementation of recommendations calling for the improved capacity of science to yield more predictive information requires a significant allocation of financial and other resources to accomplish the task. Such allocations will conflict with other significant financial needs to promote sustainable development and environmental protection and likely will not result in the intended improvements in the predictive capabilities of science for the reasons already mentioned. Second, implementation of the recommendations runs the risk of contributing to a "business as usual" approach, wherein society continues to use and manage ecosystems in existing (i.e., ecologically harmful) ways until such time as more scientific information becomes available.

An alternative role for science in problems of sustainable development and conservation of biodiversity would be to increase its capacity to assist in a more adequate formulation of public policies and goals by directing research toward useful indicators of change rather than precise predictions. Consistent with this approach, Noss (1990) has identified compositional, structural, and functional indicators for assessing biodiversity at the genetic, population/species, community/ecosystem, and regional landscape levels. In this manner, science used in problems of sustainable development and the conservation of biodiversity would contribute to a more broad and integrated view of problems and would place more emphasis on professional judgment and intuition and be less bound by analytically derived empirical facts; it would seek to assist in the management of human interactions with ecosystems rather than attempt to manage them toward predetermined ends. In addition, the role of science could be reformulated to study the question of how much area of different types of ecosystems is required to protect them from human intrusion in order to ensure the conservation of total biodiversity upon which both within-habitat and between-habitat diversity depends. Consequently, this alternative role for science would be more amenable for practical public policy purposes compared with the claims of a predictive science approach with its inherent limitations of predictive capabilities.

5. Linkages Among Sustainability Problems

If sustainable development and conservation of biodiversity is going to be achieved, it is important that linkages with other problems be understood and resolved. Myers (1993) has described various types of linkages important to sustainability problems, including: (1)those between one environmental problem and another, such as between biodiversity and climate change; (2)those between different spheres of human activity, such as between environmental protection and development generally; (3)those between the developed and developing world; (4)those between present and future generations; (5)those between protection of natural resources and basic human needs; (6)those between ecology and economics; and

(7)those between economic efficiency and social equity. If the linkages among different problems are not understood and resolved, then long-term solutions likely will not be found for particular problems.

For example, local or regional efforts to conserve biodiversity will fail if a solution is not found to the problem of global warming, because the latter problem threatens biodiversity on a global scale. Typically, problems of managing and protecting terrestrial and marine biodiversity are dealt with separately. However, they need to be linked, because deforestation and other land practices can lead to increased pollution and degradation of important coastal habitats for marine species. Use of well-managed forest plantations may contribute to social and economic sustainability while providing for acceptable levels of biodiversity, but they contain less carbon per unit area than more mature forests and therefore contribute to a buildup of atmospheric greenhouse gases. Increased population growth will lead to increased use of both mature ecosystems and marginal lands and will lead to additional losses of their biodiversity; hence, population growth must be limited in order to conserve biodiversity. At the same time, increasing affluence by people of developed nations will lead to increases in pollution and resource use, thereby threatening biodiversity. Because the loss of stratospheric ozone threatens UV-B-sensitive species and potentially will alter food chain relationships, the problem of ozone depletion must be resolved in order to conserve biodiversity. Stresses on forests from slash-and-burn agriculture increases the sensitivity of forests to acid precipitation, and acid precipitation on undisturbed forests increases their vulnerability to slash-and-burn agriculture, because increased numbers of dead and dying trees make the forests easier to clear for agriculture. Consequently, conditions leading to an increase in slash-and-burn agriculture and acid precipitation must be dealt with simultaneously. The conservation of biodiversity is linked to financial policies of governments and corporations, whereas many developing nations feel forced to sell off their national resources as a means to generate revenue to pay off foreign debts. The problem of the debt burden of developing nations is exacerbated by the military buildup of the lending developed nations, because such a buildup contributes to increases in the floating interest rates for foreign debts (George 1992). Finally, local and regional warfare has both direct and indirect effects on biodiversity, and therefore conditions that create conflicts between peoples and nations must be alleviated to protect biodiversity better.

Even if we assume that the world adopts the goal of sustainable development and conservation of biodiversity, the fact that problems of conserving biodiversity are linked with other problems creates potential conflicts. How these linkages are understood and dealt with will have implications for biodiversity as well as for the solution of other environmental problems relevant to sustainable development. Most ecologists and others concerned about sustainability and biodiversity will favor the adoption of integrated and interdisciplinary approaches to solving the problems to ensure that the linkages between the problems are understood and dealt with adequately. Although this approach makes the most sense theoretically speaking, it is problematic from a practical standpoint for several reasons: (1)people can be expected to disagree about the importance of some problems compared with others because the distribution of benefits and harms/costs varies spatially and temporally for different problems, (2)some problems are understood better than others, and (3)more substantial financial and technical resources exist to deal with some problems compared with others. Consequently, the rate of the world's progress toward sustainable development and environmental protection can be expected to vary for each of the types of environmental problems. However, success in solving any one type of problem may require that one or more other problems be solved simultaneously because of the nature of the linkages that exist among them.

6. Value-Laden Issues of Science and Decisionmaking

In part, actions taken to conserve biodiversity inevitably will conflict with other economic or developmental interests for which agencies have responsibilities or with which they have to deal. Consequently, even though many of the laws that might be applied in support of conserving biological diversity contain normative language to promote or encourage biodiversity values, such language generally does not offer firm legal or ethical prescriptions regarding how or where an administrator should balance conservation of biodiversity with conflicting uses of resources. Consequently, many decisions about use of natural resources and biodiversity reflect the values of the decisionmakers themselves (Lemons 1993). Unless or until the value-laden dimensions of such decisions are made more explicit, most people will assume erroneously that decisions about biodiversity are made on sound scientific information and legislative mandates that prescribe particular agency decisions.

The U.S. Endangered Species Act can be used to demonstrate the fact that many decisions about biodiversity are value-laden. Specifically, the law permits the secretary of the interior to list a plant or animal as endangered for any one of five reasons: (1)present or threatened destruction of habitat; (2)overutilization for commercial, recreational, scientific, or educational purposes; (3)losses due to disease or predation; (4)the inadequacy of existing laws and regulations to protect the organism in question; and (5)other natural or human factors affecting the continued existence of a species (including subspecies and populations). The law also mandates that listing decisions be based on the best scientific and commercial data available.

Briefly, decisions reflect the values of decisionmakers to a large extent in several key areas. First, significant uncertainty exists about the ecological status of most species and their habitat requirements. Decisionmakers must make the value-laden decision of how conservative to be given conditions of scientific uncertainty. Second, although the Endangered Species Act directs federal departments and agencies to utilize their authorities in furtherance of the purposes of the act, most agencies have the authority to use administrative discretion so long as it is not arbitrary or capricious. Consequently, many agency decisionmakers attempt to balance decisions about listing of endangered species with the economic and social costs associated with protecting such a species, as well as with the goals of their own particular agencies. Neither the Endangered Species Act nor the legislative mandates for federal agencies prescribe how or where concerns about protection versus concerns about social and economic costs should be balanced. Third, although the Endangered Species Act permits the listing of subspecies and populations, it does not mandate it. Hence, decisionmakers have discretion concerning the taxonomic basis for listings. Further, there is no firm scientific or practical definition of a subspecies or population. Consequently, the judgments of scientists themselves regarding how a particular group of organisms should be classified are not based on scientific information solely. Fourth, the Endangered Species Act requires that conservation measures include all methods and procedures necessary to bring any endangered or threatened species to the point at which the measures pursuant to the act are no longer necessary. Because of the scientific uncertainties regarding what constitutes an MVP, decisionmakers have considerable discretion and can therefore base their decisions on numbers that are so low that prospects for recovery are low. This is a likely outcome if the decisionmakers hold the social and economic costs of listing to be higher than its benefits. Fifth, there often is bias toward such taxonomic groups such as mammals and birds compared with others. Sixth, lists of endangered species may carry their own bias which is not recognized by decisionmakers. For instance, rare or restricted species are not necessarily the most endangered but may receive more attention. Seventh, the Endangered Species Act is biased toward the protection of recognized species but cannot effectively target unrecognized species or taxa for action. Eighth, cultural and political bias exist insofar as nations are more willing to protect species within their own boundaries. Similar types of value-laden issues are inherent in other laws that attempt to protect species and ecosystems.

Many so-called scientific methods and tools upon which scientific information is derived also are embedded with the subjective values of scientists. Mayo and Hollander (1991), Shrader-Frechette and McCoy (1993), and Westra and Lemons (in press) have presented critical analyses of how and why numerous value-laden judgments, evaluations, assumptions, and inferences are embedded in scientific methods of ecosystem and human health risk identification, assessment, evaluation, and management, as well as in more basic research methods of ecology. For example, scientists often have to make judgments about which species or ecosystem attributes to study without having a firm scientific knowledge base to inform their choice. Often, ecologists use simplified models with many built-in assumptions that cannot be validated or verified. Interdisciplinary studies used in ecology require the synthesis of information and methods from different disciplines, which introduces subjectivity into the studies. Many studies are by necessity limited to small spatial and temporal scales, yet scientists often make long-term predictions extrapolated from them, even though such predictions cannot be verified or validated. Scientists also have to make decisions about whether to minimize type I or type II errors in their evaluation of acceptance or rejection of testable hypotheses. In other words, they must decide whether it is better to have a higher probability of accepting false positive or false negative results. Finally, scientists often have personal interests in certain attributes of biodiversity and attach their own values to them. Consequently, they often acquire knowledge and define scientific problems based, in part, upon their interests and values. Each of these types of decisions and judgments reflects a combination of the professional expertise of scientists as well as some of their values.

Our mention of the fact that scientific methods and tools are value-laden is not a criticism of science. Rather, we raise this because a failure to recognize the existence of the value-laden dimensions of science casts serious doubts about even the best and most thorough of so-called scientific and technical studies used to inform decisions about sustainable development and the conservation of biodiversity. In other words, unless the value-laden dimensions of scientific and technical studies and information are disclosed, the positions of decisionmakers will appear to be justified on objective or value-neutral scientific reasoning, when in fact they will be based, in part, on often controversial and conflicting values of scientists and decisionmakers themselves.

7. Ethical Principles to Guide Decisionmakers

Problems of sustainable development and conserving biodiversity involve scientific, social, and economic considerations. They also fundamentally concern matters of ethics, because appropriate criteria must be identified and utilized in decisions regarding whether and to what extent humans have obligations to members of nonhuman species. Almost all of the language of Agenda 21 pertaining to sustainable development and conservation of biodiversity suggests that the reasons for conserving biodiversity are derived from the instrumental values and uses that biological resources provide humans. Consequently, Agenda 21 recommendations reflect anthropocentric ethical theories which hold that decisions about sustainable development and conservation of biodiversity should be based upon the rights, interests, or welfare of humans.

Anthropocentric ethical theories and the use of instrumental values of resources prevail in most economic analyses and public policy decisions, especially when cost-benefit analysis is used. When such theories are used, the values of nonhuman species—which can include use values, option values, quasi-values, and possibly but not usually existence values—must be quantified for use in decisionmaking. Examples of such values include life-support (i.e., ecosystem services and products to maintain life) values, recreational values, scientific values, genetic-diversity values, aesthetic values, cultural values, historical values, character-building values, therapeutic values, sacramental values, and commercial or market values. When these types of values are used to assess the utility of nonhuman species, there is the question of whether all values should be accorded the same importance. When Agenda 21 speaks of conserving biodiversity, does it do so in order to promote some or all of these types of values? When conflicts arise between different values, on what basis should it be decided whether one has precedence over the other? Historically, market or commercial values have been accorded greater worth in cost-benefit analysis and public policy decisions. The prevalence of market or commercial values in public policy decisions is tantamount to a judgment that anthropocentric values should have precedence over nonanthropocentric values and that market or commercial values should have precedence over other types of instrumental values; such a judgment is one of ethics, and it has ethical implications. The view that anthropocentric values and, in particular, market or commercial values should have precedence over other values has been criticized extensively on ethical grounds as well as for the pragmatic reason that the use of market or commercial values as a basis for decisions about biodiversity inevitably leads to a loss of biodiversity. Rolston (1985) has argued that insofar as the protection of biodiversity is concerned, some values should have precedence over others. Importantly, he has argued that where market or commercial values conflict with the other types of values in matters of biodiversity, the latter should have precedence over the former.

While the ethical stance adopted by Agenda 21 may reflect prevailing public policy practices and ethical theory, substantial alternative theories regarding duties and obligations of humans to nonhuman species also have been formulated and may be suitable as a basis of public policy regarding sustainable development and conservation of biodiversity. Nowhere do Agenda 21 recommendations seem to reflect a consideration of other such ethical views. Efforts to extend moral consideration to nonhumans have given rise to two general types of nonanthropocentric ethical theories (Norton 1987): (2)inherentism, where all value in nonhuman nature is dependent on human consciousness, but some of this value is not derivative from human values; and (2)intrinsicalism, where some value in nature is independent of human values and human consciousness. In other words, nonhuman organisms are said by many to have value and a right to existence independent of their value to humans.

In addition to deciding whether decisions about biodiversity should be based on anthropocentric or nonanthropocentric ethical theories, the question also arises whether moral consideration should be applied to individuals or to higher levels of biological organization such as species or biotic communities. Traditional systems of ethics have included individuals in moral considerability but have not included species or communities. However, increasingly these traditional theories are being criticized by those who view them as being too atomistic and not according proper worth to biotic communities. Rolston (1988) contends that moral consideration must extend beyond individuals to biotic communities or ecosystems as well. Leopold (1949) was one of the first advocates of this view when he proposed a land ethic in which the scope of ethics is enlarged to include soils, waters, plants, and animals, or collectively, the land. A rationale for the enlargement of our ethics was based upon the recognition that whenever biotic or abiotic components are interdependent, they bear mutual interactions and dependencies within their communities. More recently, deep ecologists have proposed ethical theories supporting the idea that nature relatively undisturbed by humans has a right to existence and that humans have an ethical obligation not to interfere with such a right (Devall and Sessions 1985). Westra (1994) has argued that ethical approaches to conserving biodiversity be predicated upon concepts of ecological integrity;

this would imply that large tracts of relatively undisturbed ecosystems be established and maintained.

Achieving the goals of sustainable development and conservation of biodiversity requires successful resolution of conflicts between humans and nonhumans. If, for the sake of argument, in addition to the human species at least some or all nonhuman species are said to be due moral consideration, on what bases should conflicts between species be decided? Naess (1973) has argued that, insofar as possible, species should be treated with equality. In contrast, Devine (1978) posits that humans must be preferred to nonhumans when conflicts between species exist. Singer (1990) maintains the utilitarian view that the balance of good over evil must be maximized and that the interests of both human and nonhuman beings should be taken into account and given the same weight. Others have argued that sentient beings or beings with interests should be accorded moral consideration and that these attributes should be used as a basis for resolving conflicts between species. Modifications of egalitarianism also have been proposed in an attempt to incorporate a greater variety of species into moral considerations (VanDeVeer 1979). For example, any member of a species has some interests that are essential and others that are peripheral to its well-being without threatening its survival. It is reasoned that essential interests are to be preferred to serious and peripheral interests, and that serious interests are to be accorded greater weight than peripheral ones. This proposal would require that equal consideration be given to the basic interests of members of different species. Attfield (1991) suggests that theories calling for consideration of interests as a means of resolving conflicts between different species are confronted with the objection that humans with interests and capacities like those of nonhumans may be deemed to have priority. Finally, Callicott (1980) and Rolston (1988) argue that any system of ethics that favors nonhumans over humans is simply an extension of the privileged class with moral standing—i.e., humans—which is permitted to exploit all members of other species. These scholars are more concerned with extending moral consideration to species and to the biosphere than with individuals.

Because of inequities that exist between the developed and developing nations, policies must be identified and employed that can meet the needs of all people in an ethical manner. For example, members of developed nations might have an obligation to reduce consumption of resources and utilize less harmful technologies in order to contribute to the conservation of biodiversity. The task to conserve biodiversity has been criticized recently as being predicated upon elitist and Western cultural attitudes and traditions. Gómez-Pompa and Kaus (1992) maintain that conservation policies are predicated upon Western beliefs about nature and that they ignore perspectives of Third World rural peoples. They argue that rural peoples have long maintained a relationship with nature and that their views and practices in terms of both utilizing the land and caring for it must be taken into account in conservation plans. According to this view, conservation must reflect the values and practices of rural Third World people who depend upon the land for their physical and cultural subsistence. Theoretically, this argument may have merit. However, given the fact that there are literally millions of impoverished people with high fertility rates living in, for example, the tropics and subtropics, there is simply no reason to believe that an effective conservation policy in these areas can be developed that will accommodate the needs of these people while at the same time protecting nonhuman species and ecosystems.

Problems of sustainable development and conserving biodiversity are compounded because of the lack of consensus on whether or to what extent present generations have obligations to those of the future. In other words, on what basis should we evaluate the needs of the present versus those of the future? Philosophical viewpoints regarding obligations to the future include: (1)no moral obligations beyond the immediate future exist; (2)rights and interests of members of future generations are the same as those of contemporary generations; and (3)moral obligations to the future exist, but the future is assigned less weight than the present. Agenda 21 recommendations do not offer a prescription on how to resolve conflicts between present and future generations with respect to the problem of conserving biodiversity. Although the question of the nature of our ethical obligations to future generations is relevant to problems of sustainable development and conservation of biodiversity, a more detailed discussion is beyond the scope of this chapter. Extensive discussions of responsibilities to future generations can be found in Partridge (1981).

8. Conclusion

Agenda 21 recognizes that biodiversity should be conserved in order to promote the goals of social, economic, and environmental sustainability, and it calls on science to develop methods and technologies for the long-term conservation of biodiversity and the sustainable use of biological resources at the genetic to the ecosystem levels. Such methods will have to be capable of ascertaining the MVPs of species and the attributes to use in measuring and monitoring the status of species in changing environmental conditions. The development of such methods is problematic, because little or no scientific information is available for most species. In addition, science is limited in its ability to provide reasonably certain predictions for decisionmakers. Consequently, decisionmakers should look to science to identify ecological indicators of change and criteria to assist in the understanding and application of normative concepts to conserve biodiversity rather than in the manipulation and control of nature to achieve predetermined ends. This also implies that scientific information used in determining the values of species for use in cost-benefit analysis may not be adequate and that alternative methods of valuation be developed.

The twin problems of conserving biodiversity and achieving sustainable development are embedded with many value-laden public policy and economic questions. Although Agenda 21 seems to promote the conservation of biodiversity to satisfy human needs primarily, biodiversity nevertheless has many values ranging from instrumental to intrinsic. Because some of these values may be in conflict with one another, appropriate decisionmaking procedures to decide what value(s) should have precedence over others will need to be developed. In addition, conservation of biodiversity can conflict with the needs of humans and with the goals of economic and social sustainability, depending on the levels and types of resource use. Decisions to conserve biodiversity also are value-laden because many of the methods of science and economics involve value-laden judgments, assumptions, evaluations, and inferences. Ways to resolve these issues also will have to be developed. Practically speaking, in order for conflicts between biodiversity and sustainable development to be resolved, a reconciliation of traditional legislative mandates and management goals of most governmental agencies that do not have an exclusive focus on conserving biodiversity or promoting sustainable development also will have to occur.

9. References

Allen, L. 1992. Ecology Forum: Plugging the Gaps. *Nature Conservancy* (Sept/Oct): 8-9.

Angermeier, P.L. and J.R. Karr. 1994. Biological Integrity versus Biological Diversity as Policy Directives. *BioScience* 44: 690-697.

Attfield, R. 1991. *The Ethics of Environmental Concern*. The University of Georgia Press, Athens, GA. 249 pp.

Bella, D.A., R. Jacobs, and L. Hiram. 1994. Ecological Indicators of Global Climate Change: A Research Framework. *Environmental Management* 18: 489-500.

Berger, J. 1991. The Federal Mandate to Restore: Laws and Policies on Environmental Restoration. *The Environmental Professional* 13: 195-206.

Bonnicksen, T.M., and E.C. Stone. 1985. Restoring Naturalness to National Parks. *Environmental Management* 9: 479-486.

Burkardt, N., B.L. Lamb, and R. Bartlett. 1990. Self and Organizational Efficacy of Front-Line Managers in the U.S. Fish and Wildlife Service. *The Environmental Professional* 12: 241-248.

Burley, F.W. 1988. Monitoring Biological Diversity for Setting Priorities in Conservation. In *Biodiversity*, E.O. Wilson (ed.). National Academy Press, Washington, DC, pp. 227-230.

Cairns, J., Jr. 1988. Can the Global Loss of Species be Stopped? *Speculations in Science and Technology* 11: 189-196.

Cairns, J., Jr., and B.R. Niederlehner. 1993. Ecological Function and Resilience: Neglected Criteria for Environmental Impact Assessment and Ecological Risk Analysis. *The Environmental Professional* 15: 116-124.

Cairns, M.A., and R.A. Meganck. 1994. Carbon Sequestration, Biological Diversity, and Sustainable Development: Integrated Forest Management. *Environmental Management* 18: 13-22.

Callicott, J.B. 1980. Animal Liberation: A Triangle Affair. *Environmental Ethics* 2: 311-338.

(CNIE) Committee for the National Institute for the Environment. 1994. *The Environmental Professional* 16: 93-191.

Cole, C.F. 1992. New Dimensions to the Education of Fish and Wildlife Professionals. *The Environmental Professional* 14: 325-332.

Costanza, R. (ed.). 1991. *Ecological Economics: The Science and Management of Sustainability*. Columbia University Press, New York. 525 pp.

Daily, G., and P.R. Ehrlich. 1992. Population, Sustainability, and Earth's Carrying Capacity. *BioScience* 42: 761-771.

DeGloria, S.D. 1991. Elements of Geographic Information Systems for Resource Conservation. In *Challenges in the Conservation of Biological Resources: A Practitioner's Guide*. D.J. Decker, M.E. Krasny, G.R. Goff, C.R. Smith, and D. W. Gross (eds.). Westview Press, Boulder, CO, pp. 153-166.

Devall, B. and G. Sessions. 1985. *Deep Ecology*. Peregrine Smith Books, Layton, UT. 266 pp.

Devine, P. 1978. The Moral Basis of Vegetarianism. *Philosophy* 53: 481-505.

Drew, G.S. 1994. The Scientific Method Revisited. *Conservation Biology* 8: 596-597.

Ehrlich, P.R. and E.O. Wilson. 1991. Biodiversity Studies: Science and Policy. *Science* 253: 758-762.

Erwin, T.L. 1983. Beetles and Other Insects of Tropical Forest Canopies at Manaus, Brazil, Sampled by Insecticidal Fogging. In *Tropical Rainforest: Ecology and Management*, S.L. Sutton, T.C. Whitmore, and A.C. Chadwick (eds.). Blackwell Scientific Publications, Edinburgh, pp. 59-75.

Franklin, J.F., C.S. Bledsoe, and J.T. Callahan. 1990. Contributions of the Long-Term Ecological Research Program. *BioScience* 40: 509-523.

George, S. 1992. *The Debt Boomerang: How Third World Debt Harms Us All*. Pluto Press, London.

Gilpin, M.E., and M.E. Soulé. 1986. Minimum Viable Populations: Processes of Species Extinction. In *Conservation Biology: The Science of Scarcity and Diversity*, M.E. Soulé (ed.). Sinauer Associates, Sunderland, MA, pp. 13-18.

Gómez-Pompa, A., and A. Kaus. 1992. Taming the Wilderness Myth. *BioScience* 42: 271-279.

Goodland, R.J.A., H.E. Daly, and S. El Serafy. 1993. The Urgent Need for Rapid Transition to Global Environmental Sustainability. *Environmental Conservation* 20: 297-309.

Groombridge, B. 1992. *Global Biodiversity, Status of the Earth's Living Resources.* Compiled by the World Conservation Monitoring Centre, Chapman and Hall, London. 585 pp.

Grumbine, E. 1990a. Protecting Biodiversity Through the Greater Ecosystem Concept. *Natural Areas* 10: 114-120.

Grumbine, E. 1990b. Viable Populations, Reserve Size, and Federal Lands Management: A Critique. *Conservation Biology* 4(2): 127-134.

Grumbine, E.R. 1994. What Is Ecosystem Management? *Conservation Biology* 8: 27-38.

Hargrove, E.C. 1989. *Foundations of Environmental Ethics.* Prentice Hall, Englewood Cliffs, NJ. 227 pp.

Hirsch, A. 1993. Improving Consideration of Biodiversity in NEPA Assessments. *The Environmental Professional* 15: 103-115.

Huntley, B.J. 1988. Conserving and Monitoring Biotic Diversity: Some African Examples. In *Biodiversity*, E.O.Wilson (ed.). National Academy Press, Washington, DC, pp. 248-262.

Jenkins, R.E. 1988. Information Management for the Conservation of Biodiversity. In *Biodiversity*, E.O.Wilson (ed.). National Academy Press, Washington, DC, pp. 231-239.

Johnson, S.P. (ed.). 1993. *The Earth Summit. The United Nations Conference on Environment and Development (UNCED).* Graham & Trotman/Martinus Nijhoff, London. 532 pp.

Kay, J. 1991. A Nonequilibrium Thermodynamic Framework for Discussing Ecosystem Integrity. *Environmental Management* 15: 483-495.

Keiter, R.B., and M.S. Boyce (eds.). 1991. *The Greater Yellowstone Ecosystem.* Yale University Press, New Haven, CT. 428 pp.

Kinnaird, M.F., and T.G. O'Brien. 1991. Viable Populations for an Endangered Forest Primate, the Tana River Mangabey (*Cercocebus galeritus galeritus*). *Conservation Biology* 5(2): 203-213.

Lacy, R.C. 1992. The Effects of Inbreeding of Isolated Populations: Are Minimum Viable Population Sizes Predictable? In *Conservation Biology: The Theory and Practice of Nature Conservation, Preservation and Management*, P.L. Fiedler and K.J. Subodh (eds.). Routledge, Chapman and Hall, New York, pp. 277-296.

Larkin, P.A. 1977. An Epitaph to the Concept of Maximum Sustained Yield. *Transcripts of the American Fisheries Society* 106: 1-11.

Lemons, J. 1986. Research in the National Parks. *The Environmental Professional* 8: 127-137.

Lemons, J. 1987. United States' National Park Management: Values, Policy, and Possible Hints for Others. *Environmental Conservation* 14: 329-340, 328.

Lemons, J. 1993. Conservation Biology: The Role of Science, Values, and Ethics. In *Conservation and Resource Management*, S.K. Majumdar, E.W. Miller, D.E. Baker, E.K. Brown, J.R. Pratt, and R.F. Schmalz (eds.). The Pennsylvania Academy of Sciences, Easton, PA, pp. 333-348.

Lemons, J. 1994a. The Use of Science in Environmental Impact Assessment. *International Journal of Ecology and Environment* 20: 303-315.

Lemons, J. 1994b. The Role of Science in Preserving Biodiversity in National Parks. In *Biological Diversity: Problems and Challenges*, S.K. Majumdar, F.J. Brenner, J.E. Lovich, J.F. Schalles, and E.W. Miller (eds.). The Pennsylvania Academy of Science, Easton, PA, pp. 184-198.

Lemons, J. (ed.). In press. *Scientific Uncertainty and Environmental Problem-Solving.* Blackwell Science, Cambridge, MA.

Lemons, J. and K. Junker. In press. The Role of Science and Law in the Protection of National Park Resources. In *National Parks and Protected Areas: Their Role in Environmental Protection*, R.G. Wright (ed.). Blackwell Science, Cambridge, MA.

Leopold, A. 1949. *A Sand County Almanac*. Oxford University Press, New York. 226 pp.

Lonergan, S.C. 1993. Impoverishment, Population, and Environmental Degradation: The Case for Equity. *Environmental Conservation* 20: 328-334.

Lovejoy, T.E., R.O. Bierregaard, Jr., A.B. Rylands, J.R. Malcolm, C.E. Quintela, L.H. Harper, K.S. Brown, Jr., A.H. Powell, G.V.N. Powell, H.O.R. Schubart, and M.B. Hays. 1986. Edge and Other Effects of Isolation on Amazon Forest Fragments. In *Conservation Biology: The Science of Scarcity and Diversity*, M.E. Soulé (ed.). Sinauer Associates, Sunderland, MA, pp. 257-285.

Lubchenco, J., A.M. Olson, L.B. Brubaker, S.R. Carpenter, M.M. Holland, S.P. Hubbell, S.A. Levin, J.A. MacMahon, P.A. Matson, J.M. Melillo, H.A. Mooney, C.H. Peterson, H.R. Pulliam, L.A. Real, P.J. Regal, and P.G. Risser. 1991. The Sustainable Biosphere Initiative: An Ecological Research Agenda. *Ecology* 72: 371-412.

MacArthur, R.H., and E.O. Wilson. 1967. *The Theory of Island Biogeography*. Princeton University Press, Princeton, NJ. 203 pp.

Magnuson, J.J. 1990. Long-Term Ecological Research and the Invisible Present. *BioScience* 40: 495-501.

Mann, C.C. 1991. Extinction: Are Ecologists Crying Wolf? *Science* 253: 736-738.

May, R.M. 1992. How Many Species Inhabit the Earth? *Scientific American* (4): 42-48.

Mayo, D. G. and R.D. Hollander (eds.). 1991. *Acceptable Evidence: Science and Values in Risk Management*. Oxford University Press, Oxford.

McNaughton, S.J. 1989. Ecosystems and Conservation in the Twenty-first Century. In *Conservation for the Twenty-first Century*, D. Western and M. Pearl (eds.). Oxford University Press, Oxford, pp. 109-120.

Murphy, D. 1990. Conservation Biology and Scientific Method. *Conservation Biology* 4: 203-204.

Myers, N. 1993. The Question of Linkages in Environment and Development. *BioScience* 43: 302-310.

Naess, A. 1973. The Shallow and the Deep, Long-Range Ecology Movement. *Inquiry* 16: 95-100.

Newmark, W.D. 1985. Legal and Biotic Boundaries of Western North American National Parks: A Problem of Congruence. *Biological Conservation* 33:197-208.

Norton, B.G. 1987. *Why Preserve Natural Variety*. Princeton University Press, Princeton, NJ. 271 pp.

Noss, R.F. 1990. Indicators for Monitoring Biodiversity: A Hierarchical Approach. *Conservation Biology* 4: 355-364.

Noss, R.F. 1992. Issues of Scale in Conservation Biology. In *Conservation Biology: The Theory and Practice of Nature Conservation, Preservation and Management*, P.L. Fiedler and K.J. Subodh (eds.). Routledge, Chapman and Hall, New York, pp. 239-252.

O'Brien, S.J., D.E. Wildt, D. Goldman, C.R. Merril, and M. Bush. 1983. The Cheetah Is Depauperate in Genetic Variation. *Science* 221: 459-462.

Odum, E.P. 1994. Conservation of Biodiversity. In *Biological Diversity: Problems and Challenges*, S.K. Majumdar, F.J. Brenner, J.E. Lovich, J.F. Schalles, and E.W. Miller (eds.). The Pennsylvania Academy of Sciences Press, Easton, PA, pp. 18-25.

Packer, C., A.E. Pusey, H. Rowley, D.A. Gilbert, J. Matenson, and S.J. O'Brien. 1991. Case Study of a Population Bottleneck: Lions of the Ngorongoro Crater. *Conservation Biology* 5(2): 219-230.

Partridge, E. (ed.). 1981. *Responsibilities to Future Generations*. Prometheus Books, Buffalo, NY. 319 pp.

Pickett, S.T.A., V.T. Parker, and P.L. Fiedler. 1992. The New Paradigm in Ecology: Implications for Conservation Biology Above the Species Level. In *Conservation Biology: The Theory and Practice of Nature Conservation, Preservation and Management*, P.L. Fiedler and K.J. Subodh (eds.). Routledge, Chapman and Hall, Inc. New York. pp. 65-90.

Primack, R.B. 1993. *Essentials of Conservation Biology*. Sinauer Associates, Sunderland, MA. 564 pp.

Redford, K.H. 1992. The Empty Forest. *BioScience* 42: 412-422.

Robinson, J.G. 1993. The Limits to Caring: Sustainable Living and the Loss of Biodiversity. *Conservation Biology* 7: 20-28.

Rohlf, D.J. 1991. Six Biological Reasons Why the Endangered Species Act Doesn't Work-And What to Do About It. *Conservation Biology* 5: 273-282.

Rolston, H., III. 1985. Valuing Wildlands. *Environmental Ethics* 7: 23-48.

Rolston, H., III. 1988. *Environmental Ethics, Duties to and Values in the Natural World*. Temple University Press, Philadelphia, PA. 391 pp.

Sagoff, M. 1988. Ethics, Ecology, and the Environment: Integrating Science and Law. *Tennessee Law Review* 56: 78-229.

Scott, J.M., B. Csuti, J.D. Jacobi, and J.E. Estes. 1987. Species Richness: A Geographic Approach to Protecting Future Biodiversity. *Bioscience* 37(11): 782-788.

Shaffer, M.L. 1981. Minimum Population Sizes for Species Conservation. *Bioscience* 31(2): 131-134.

Shaffer, M. 1987. Minimum Viable Populations: Coping with Uncertainty. In *Viable Populations for Conservation*, M.E. Soulé (ed.). Cambridge University Press, Cambridge, pp. 69-86.

Shaffer, M.L. 1991. Population Viability Analysis. In *Challenges in the Conservation of Biological Resources: A Practitioner's Guide*, D.J. Decker, M.E. Krasny, G.R. Goff, C.R. Smith, and D. W. Gross (eds.). Westview Press, Boulder, CO, pp. 107-118.

Shaffer, M.L., and F.B. Samson. 1985. Population Size and Extinction: A Note on Determining Critical Population Size. *American Naturalist* 125: 144-152.

Shrader-Frechette, K.S., and E.D. McCoy. 1993. *Method in Ecology*. Cambridge University Press, Cambridge. 328 pp.

Singer, P. 1990. *Animal Liberation, A New Ethics for Our Treatment of Animals*, Jonathan Cape, London. 320 pp.

Slocombe, D.S. 1993a. Environmental Planning, Ecosystem Science, and Ecosystem Approaches for Integrating Environment and Development. *Environmental Management* 17: 289-303.

Slocombe, D.S. 1993b. Implementing Ecosystem-Based Management. *BioScience* 43: 612-622.

Soulé, M.E. 1985. What Is Conservation Biology? *BioScience* 35: 727-734.

Soulé, M.E. (ed.). 1986. *Conservation Biology: The Science of Scarcity and Diversity*. Sinauer Associates, Sunderland, MA. 584 pp.

Soulé, M.E. 1987. Where Do We Go from Here? In *Viable Populations for Conservation*, M.E. Soulé (ed.). Cambridge University Press, Cambridge, pp. 175-183.

Soulé, M.E., and D. Simberloff. 1986. What Do Genetics and Ecology Tell Us About the Design of Nature Reserves? *Biological Conservation* 35: 19-40.

Swanson, F.J., and R.E. Sparks. 1990. Long-Term Ecological Research and the Invisible Place *BioScience* 40: 502-508.

VanDeVeer, D. 1979. Interspecific Justice. *Inquiry* 22: 55-79.

Walker, B. 1989. Diversity and Stability in Ecosystem Conservation. In *Conservation for the Twenty-first Century*, D. Western and M. Pearl (eds.). Oxford University Press, Oxford, pp. 121-132.

Wayne, R.K., N. Lehman, D. Girman, P.J.P. Gogan, D.A. Gilbert, K. Hansen, R.O. Peterson, U.S. Seal, A. Eisenhawer, L.D. Mech, and R.J. Krumenaker. 1991. Conservation Genetics of the Endangered Isle Royale Grey Wolf. *Conservation Biology* 5(1): 41-51.

(WCED) World Commission on Environment and Development. 1987. *Our Common Future*. Oxford University Press, New York and London.

Wellman, J.D., and T.J. Tipple. 1990. Public Forestry and Direct Democracy. *The Environmental Professional* 12: 77-86.

Westman, W.E. 1990. Managing for Biodiversity. *BioScience* 40: 26-33.

Westra, L. 1994. *An Environmental Proposal for Ethics: The Principle of Integrity*. Rowman and Littlefield, Lanham, MD. 235 pp.

Westra, L., and J. Lemons (eds.). In press. *Scientific and Ethical Perspectives on Ecological Integrity*. Kluwer Academic Publishers, Dordrecht, The Netherlands.

Woodruff, D.S. 1989. The Problems of Conserving Genes and Species. In *Conservation for the Twenty-First Century*, D. Western and M. Pearl (eds.). Oxford University Press, Oxford, pp. 76-88.

Chapter 7
CLIMATE CHANGE AND SUSTAINABLE DEVELOPMENT

John Lemons[1]
Rudolf Heredia[2]
Dale Jamieson[3]
Clive Spash[4]

1. Introduction

One of the environmental problems Agenda 21 focused on is protection of the atmosphere. Programs to protect the atmosphere include: (1)improving the scientific basis for decisionmaking, including addressing scientific uncertainties; (2)promoting sustainable development by better use of energy and consumption of materials, transportation, industrial development, and terrestrial and marine resources; (3)preventing stratospheric ozone depletion; and (4)mitigating transboundary air pollution.

The objective of the program to improve the scientific basis for decisionmaking is to facilitate understanding of physical, chemical, and biological processes that influence and are influenced by the earth's atmosphere on global, regional, and local scales and to improve understanding required for mitigation of threats to the atmosphere. The basis for action is the increased concern about the effects of climate change and atmospheric pollution that has created new demands for scientific knowledge to reduce uncertainties.

The objective of the program to promote sustainable development is to reduce adverse effects on the atmosphere from the energy sector through less polluting and more efficient energy production and use, particularly by the development of renewable energy sources. Importantly, this program recognizes the need for equity, adequate energy supplies, and increasing energy consumption in developing countries. It also suggests a consideration for the situations of countries that are dependent on the income generated from the production and consumption of fossil fuels and associated energy-intensive products for which countries have difficulties in switching to alternatives, and of countries that are highly vulnerable to the adverse effects of climate change.

The program objectives to prevent stratospheric ozone depletion are based on concern about the increasing concentrations of reactive chlorine, bromine, chloroflurocarbons (CFCs), halons, and other substances. While this program recognizes that the 1985 Vienna Convention for the Protection of the Ozone Layer and the 1987 Montreal Protocol on Substances That Deplete the Ozone Layer (amended in London in 1990) were important steps to protect the ozone layer, the total chlorine loading of the atmosphere of ozone-depleting substances has continued to rise. Consequently, further measures to reduce this loading in

[1]Department of Life Sciences, University of New England, Biddeford, ME 04005, U.S.A.; [2]Social Science Centre, St. Xavier's College, 5, Mahapalika Marg, Bombay 400 001, India; [3]Department of Philosophy, University of Colorado, Boulder, CO 80309, U.S.A.; [4]Department of Economics, University of Stirling, Stirling FK9 4LA, Scotland.

J. Lemons and D. A. Brown (eds.), Sustainable Development: Science, Ethics, and Public Policy, 110–157.

the atmosphere through compliance with control measures identified in the Montreal Protocol are needed.

The objectives to mitigate the effects of transboundary air pollution on human health and ecosystems focus on ways to improve the lack of reliable emissions data outside of Europe and North America and on the need to acquire better information on the environmental and human health effects of air pollution.

Clearly, the language contained in the aforementioned programs' objectives mandates the use of science, economics, and ethics. In addition, Agenda 21 includes other recommendations calling for the use of these disciplines in protection of the atmosphere. For example, it calls on the sciences for better understanding and prediction of the various properties of the atmosphere and of affected ecosystems, as well as health impacts and their interactions with social and economic factors. Further, scientific knowledge is required to identify threshold levels of atmospheric pollutants and greenhouse gases that would cause dangerous levels of anthropogenic interference with the climate system and the environment and to identify the associated rates of changes that would not allow ecosystems to adapt naturally.

Many recommendations to protect the atmosphere refer to the necessity to base decisions on economic methods of analysis and information. Recommendations refer to energy as being essential to economic and social development and improved quality of life, and they refer to the need to develop at the national level appropriate methodologies for making integrated energy, environment, and economic policy decisions for sustainable development through environmental impact assessments. Many recommendations discuss the need to achieve environmental protection by the use of cost-effective policies.

Ethical considerations also are apparent in recommendations to protect the atmosphere. For example, energy sources need to be used in ways that respect the atmosphere, human health, and the environment as a whole. Recommendations call for taking into full account the legitimate priority needs of developing countries for the achievement of sustained economic growth and the eradication of poverty. Many recommendations stress the need to develop equitable solutions to problems of protecting the atmosphere and achieving sustainable development.

Although application of science, economics, and ethics is required for implementation of Agenda 21 recommendations to protect the atmosphere, many problems regarding their application need to be resolved. For example, the status of scientific knowledge about the state of the atmosphere needs to be understood, particularly with reference to the determination of how certain we are of such knowledge and what its predictive capabilities are. In addition, because scientific uncertainty about the state of the atmosphere is prevalent, conclusions about the atmosphere often are more value-laden than is commonly thought. Economic tools and methods are required to assess problems of the atmosphere and how to resolve them by the application of cost-benefit analysis and/or alternative methods of valuation; application of such methods often is controversial. Although sustainable development and environmental protection fundamentally is an ethical problem, the language of Agenda 21 is not prescriptive with respect to the ethical criteria that should be used to resolve intergenerational or intragenerational conflicts among humans, how to resolve conflicts between humans and the nonhuman environment, or who should decide and on what basis how conservative or precautionary decisions should be given scientific uncertainty.

In this chapter we: (1)summarize the scientific basis for climate change and its projected environmental consequences, including areas of scientific uncertainty; (2)analyze the ethical implications posed by problems of climate change; (3)analyze the adequacy of traditional and alternative methods of economic analyses used to assess climate change problems; and (4)present a representative perspective of southern nations' views on problems of protecting the atmosphere.

Protection of the earth's atmosphere requires consideration of problems due to increasing concentrations of greenhouse gases, acid precipitation and other air pollutants, and global ozone depletion. While these problems have many common features, all are complex and controversial (IGBP 1990). An adequate treatment of all of these problems is not possible in a single chapter. Consequently, we focus mostly (but not exclusively) on global climate change due to increased concentrations of greenhouse gases, primarily carbon dioxide. We do this for several reasons. First, global climate change is likely to have the most significant impacts on humans and the environment. Second, the effects of other principal air pollutants are known with more scientific certainty and are regulated to a greater extent by laws of many nations. Third, conventions and voluntary measures have been established to begin the regulation of ozone-depleting chemicals. Fourth, the United Nations Environmental Programme recommends that climate change studies focus on carbon dioxide (Hogan et al. 1991).

2. Scientific Assessment of Climate Change

2.1. WARMING OF THE EARTH-ATMOSPHERE SYSTEM

Equilibrium of the temperature of the earth-atmosphere system is maintained by a balance between the amount of incoming solar energy absorbed by the system and the amount of outgoing radiant energy. Most of the outgoing radiant energy is in the long-wave or infrared region, in the wavelengths of 4 to 100μm. Numerous human activities have the potential to cause significant climate change by altering the factors responsible for maintaining the temperature equilibrium of the earth-atmosphere system. Such activities include: (1)release of carbon dioxide by burning of fossil fuels; (2)release of methane, chlorofluoromethanes, nitrous oxide, carbon tetrachloride, and carbon disulfide; (3)release of particles and aerosols from industrial and agricultural practices; (4)release of heat; (5)upward transport of chlorofluoromethanes and nitrous oxide into the stratosphere; (6)release of trace gases such as nitrogen oxides, carbon monoxide, or methane that increase tropospheric ozone by photochemical reactions; and (7)patterns of land use and deforestation. The primary reason that the listed chemicals (so-called greenhouse gases) potentially can cause warming of the atmosphere is because they absorb radiant energy in the infrared region and because they have long residence times in the atmosphere.

Greenhouse gases have increased since preindustrial times (c. 1750-1800). Carbon dioxide has increased from about 280 ppmv to 354 ppmv, methane from 0.8 ppmv to 1.74 ppmv, CFC-11 from 0 pptv to 280 pptv, CFC-12 from 0 pptv to 485 pptv, and nitrous oxide from 288 ppbv to 312 ppbv. Annual rates of increase are approximately 0.5 percent for carbon dioxide, 0.9 percent for methane, 4 percent for CFC-11, 4 percent for CFC-12, and 0.25 percent for nitrous oxide. Residence times are estimated to be 50-200 years for carbon dioxide, 10 years for methane, 65 years for CFC-11, 130 years for CFC-12, and 150 years for nitrous oxide. Between 1980 and 1990, carbon dioxide is estimated to have accounted for about 55 percent of the change in radiative forcing, methane 15 percent, CFC-11 and CFC-12 (combined) 17 percent, and nitrous oxide 6 percent (IPCC 1990). However, Hansen et al. (1988) suggest that the total greenhouse effect is now due slightly more to other gases collectively than to carbon dioxide alone.

Data for the quantities of carbon found in the climate system provide an example of how humans have modified the amount of chemicals found there. Presently, the atmosphere contains about 750 Gt of carbon, compared with about 575 in the preindustrial atmosphere. The annual release of carbon to the earth's atmosphere is more than 5 Gt from the burning of fossil fuels and is about 2 Gt from deforestation. The amount of carbon stored in all of the earth's phytomass is approximately 560 Gt, compared with 4,000 Gt stored in recoverable coal and oil and 5,000-10,000 stored in potentially recoverable fossil fuels. Because a large amount of carbon is stored in recoverable and potentially recoverable fossil fuels relative to the amount in the atmosphere or phytomass, there is considerable potential for the amount of atmospheric carbon to increase if humans burn fossil fuels in large amounts or at rapid rates. Most of the carbon added to the earth-atmosphere system since 1860 has come from the burning of fossil fuels (Clark 1982).

Based on apparent correlations between atmospheric carbon dioxide concentration and temperature change over the past 160,000 years and the past 100 years, respectively, there is presumptive evidence that an increase of atmospheric carbon dioxide concentration has resulted in an increase of the earth's atmospheric temperature (Hansen and Lebedeff 1987). However, other factors such as variations in the energy output of the sun, levels of sulphur-oxides, land use changes, and volcanic eruptions also can contribute to temperature increases. Consequently, the relationship between past increases in atmospheric carbon dioxide and temperature is not conclusive.

2.2. METHODS TO MODEL CLIMATE

Ideally, decisions on whether and how to attempt to prevent or mitigate climate change must be predicated on reasonably accurate scientific information regarding the causes of change, the magnitude and rate of atmospheric temperature increase, and the ecological and human health impacts of change (Lemons 1991).

In recent years, scientists have developed various general circulation models (GCMs) to predict future climate (Trenberth 1992). All GCMs are limited in the physical, chemical, and biological detail they can handle, as well as in the spatial detail they can resolve. Most GCMs focus on the physical climate system and ignore or use oversimplified information and assumptions about chemical processes, land surface processes, and biological or ecological processes. In addition, the feedbacks that exist in climate change, such as processes involving deep ocean circulation, oceanic biogeochemical cycling, water vapor, clouds, snow, sea ice, vegetation distribution, ultraviolet radiation and phytoplankton, and soil carbon storage, are understood poorly and infrequently included in GCMs (IPCC 1990). Because GCMs are built with different assumptions and include different factors and levels of detail and certainty, large uncertainties exist in our ability to project future climate change.

Generally speaking, a GCM is a mathematical model composed of systems of partial differential equations based on laws of physics. The equations describe basic atmospheric processes such as large-scale wind, temperature, and distribution in the atmosphere and surface climate. The GCMs also incorporate with varying degrees of success interactions with oceans, clouds, land surfaces, and sea ice. Equations used in GCMs are too complex to be solved analytically; they must be converted to arithmetic form suitable for computations by digital computers. The GCMs are run with the current carbon dioxide concentration until it reaches a steady state; this represents an experimental control. Typically, subsequent runs are made using two or three times the current concentration of atmospheric carbon dioxide. After these types of runs are completed, they are compared to determine the changes caused by increased atmospheric carbon dioxide.

Most GCMs represent the earth-atmosphere system in a three-dimensional grid system. Depending on the model, horizontal spacing of grid points ranges between 4° to 8° latitude and 5° to 10° longitude, with 2 to 12 vertical atmospheric layers extending to 30 km above the surface (Schneider 1991). A few models have higher resolution grids of approximately 2.5° by 2.5°. No models are likely to be developed within the foreseeable future with grids less than 100 km by 100 km, which is far larger than the scale of most ecological research and microclimate processes (Root and Schneider 1993).

Once models are developed, attempts need to be made to verify their predictive capabilities. Generally, three methods of verification exist; none by itself is sufficient (Schneider 1992). The first method checks model simulation against present-day seasonal cycles of surface air temperature. Such a method provides verification of rapid processes, such as cloud formations. But it does not provide verification of slow changes that occur over long time periods for variables such as ice cover, soil organic matter, or deep-ocean temperatures. A second method of verification tests individual components of a model directly against real data. For example, upward infrared radiation emitted from the earth can be measured from satellites and compared with predictions made by GCMs. This method of verification does not, however, guarantee that the net effect of the interacting components of a model has been defined or accounted for properly. A third method of verification is an a priori one, in that some researchers have more confidence in models that include a maximum amount of spatial resolution and physical, chemical, and biological data. The problems of verifying GCMs introduces additional uncertainty into our confidence in the predictive capabilities of GCMs.

Various attempts have been made to verify predictions made by GCMs. For example, the IPCC (1990) compared observed mean global temperature changes from 1861 through 1989 with values predicted by GCMs. The typical prediction of 0.5 to 1°C warming over this century is consistent generally with, but larger than, that observed. Schneider (1992) provides the following possible explanations for the discrepancy: (1)the sensitivity of the models to greenhouse gases has been overestimated by a factor of two or so; (2)modelers have not accounted properly for external factors such as volcanic dust, changes in solar output, or regional tropospheric aerosols from biological, agricultural, and industrial activities; (3)modelers have not accounted for the large capacity of the oceans to absorb increased heat from the atmosphere; (4)both present models and observed climatic trends could be correct, but models typically are run for equivalent doubling of carbon dioxide, whereas the world has experienced only a quarter of this increase; (5)the incomplete and inhomogeneous network of thermometers has underestimated warming; and (6)there may have been a natural cooling trend of up to 0.5°C during this century. Although global temperature trends and those anticipated by GCMs disagree somewhat, the difference may not be fundamental. Depending on what assumptions one makes about the above explanations for the discrepancy, the observed temperature trend could be consistent with an equivalent doubling of carbon dioxide and an equilibrium temperature response of 0.5 to 5.0°C.

To be valid, climate models also must be able to differentiate atmospheric temperature increases and changes of other climate variables from the natural variation of climate that occurs over both short and long time periods. Characteristically, this includes periods of several days, periods ranging from about 10 days to a season, periods of several years, and periods of decades or longer. Only limited observational and modeling efforts have been devoted to climate variability on time scales of decades and longer. The natural variability of climate makes the detection of changes due to human activity difficult, especially given the fact that greenhouse gas-induced mean atmospheric temperature increases are expected to occur at a rate of between 0.1 and 0.8°C per decade (IPCC 1990). This rate of increase is within the magnitude of natural mean atmospheric temperature fluctuations.

Climate models also have been used to estimate the immediate reduction in emissions for stabilization of greenhouse gases at present atmospheric levels (Lashof and Tirpak 1989, IPCC 1990). In order to stabilize greenhouse gases at present atmospheric levels, estimates indicate that carbon dioxide, nitrous oxide, and CFCs would need to be reduced by approximately 60 percent or more; methane would require a reduction of about 15 percent. It is important to note that results from climate modeling indicate that the longer emissions continue at present rates, the greater will be the reductions in future emissions that will have to be made to stabilize atmospheric carbon dioxide concentrations at a given level.

2.3. PROJECTED CLIMATE SCENARIOS

Because of limited knowledge, various projections of future mean annual atmospheric temperature increases have been made. A number of modeling studies have yielded projections of future mean atmospheric temperature increases in the range of 2.8 to 5.2°C by the end of the next century for a doubling of atmospheric carbon dioxide concentration (see, e.g., Washington and Meehl 1984, Wetherald and Manabe 1986, Wilson and Mitchell 1987, Schlesinger and Zhao 1989, IPCC 1992, Trenberth 1992). However, some estimates of temperature increases based on models that attempt to take into account full ocean processes are in the range of 1-2°C (Washington 1992). The IPCC provided a best estimate of 1 to 2°C warming by the year 2030 and 3°C warming by the end of the next century (Tolba 1991). Projected rates of temperature increase are 0.2 to 0.8°C per decade. This rate of warming is greater than that ever experienced in human history. The differences between various models are difficult to understand because their construction varies and the feedback mechanisms may be substantial. Models also have projected the geographical distribution of temperature increases from a doubling of atmospheric carbon dioxide concentration (Washington 1992); warming of about 2-3°C in the tropics and up to 20°C in the winter poleward regions is projected. However, geographical regions at the same latitude are projected to experience different amounts of warming.

Two recent studies indicate that long-term climate warming may be more serious than has been projected by earlier studies. Manabe and Stouffer (1993) ran their climate model to 500 years into the future. The eventual quadrupling of carbon dioxide during the next 140 years implied by current trends would increase temperature by 7°C or more. During the first 50 years of this period, there would be a drastic reduction in the ocean currents that flush the deep sea with oxygen-rich waters, lift nutrient-rich deep waters to the surface, and carry heat toward the polar regions. Projected consequences include a decrease in the oxygen levels of the ocean and a nearly stagnant deep circulation, eventually killing off much ocean life. Walker and Kasting (1992) took into account the rate at which the ocean and vegetation remove carbon dioxide from the atmosphere and assumed that conservation of fossil fuels would slow the emission of atmospheric carbon dioxide but would not stop it eventually. They then ran their climate model for different conservation scenarios. If the rates of fossil fuel use and deforestation continue as they have over the past few decades, it was found that the atmospheric concentration of carbon dioxide would be more than seven times preindustrial levels by the 23rd century. If fossil fuel use remains at today's level, the concentration of atmospheric carbon dioxide reaches seven times the preindustrial in the year 2700. Ending deforestation would lower the peak carbon dioxide concentration to four times preindustrial levels. Projected mean atmospheric temperature increases are on the order of 10°C. According to this model, the only way to limit the rise in carbon dioxide to a doubling of preindustrial levels is to reduce present emissions by a factor of approximately 25—something neither the developing nations nor the developed nations are likely to accomplish.

2.4. PROBLEMS OF DETECTION

Detection of climate warming due to increased emission of greenhouse gases requires careful evaluation of signal-to-noise ratios to be sure apparent change is not due to random fluctuations. The approximate 0.5°C or so mean atmospheric temperature increase observed during the past few decades has been attributed to natural fluctuations by some researchers (Klein 1982) and to statistically significant warming by others (Hansen and Lebedeff 1987). Consequently, it is not possible to conclude with confidence that atmospheric warming has occurred.

Detection of climate warming due to increased concentrations of greenhouse gases will require evidence that warming is due to such emissions and not due to some other factor(s). Researchers have proposed measuring various physical factors in order to discern whether any observed atmospheric warming is attributable to greenhouse gas emissions. These factors include surface temperature, temperature in the stratosphere, temperature in the troposphere, infrared radiation, the cryosphere, the oceans, and hydrologic variables. Measurements of these factors are problematic in that they have statistical or random fluctuations. Further, researchers disagree on the priority of their importance in detection of greenhouse gas-induced warming.

2.5. ENVIRONMENTAL IMPACTS

Climate change may result in many impacts to ecosystems, species, and humans. Ideally, scientific studies should be able to provide knowledge required for making informed decisions regarding mitigation or prevention of adverse impacts of climate change. Following, we provide a brief descriptive summary of the status of knowledge regarding some of the impacts of climate change on ecosystems, species, and humans. We provide a more detailed analysis of the status of knowledge for global ecological impacts in order to show some of the approaches used to acquire knowledge about climate change impacts.

2.5.1. *Assessing Greenhouse Gas Emissions and a Greenhouse Gas Index*

The patterns of greenhouse gas emissions vary between different countries (WRI 1992). Cumulative emissions of carbon dioxide for 1950-1989 range from approximately 155 billion metric tons for the United States to 90 billion metric tons for the European Community to less than 10 billion tons for most developing nations. In 1989, per capita emissions for carbon dioxide were approximately 20 metric tons for the United States, 10 metric tons for the United Kingdom, and a little over 1 metric ton for India. As would be expected, the cumulative as well as the per capita emissions for carbon dioxide and other greenhouse gases for developed countries greatly exceed those of developing nations.

The IPCC (1990) has adopted a conceptual unit called the "global warming potential" (GWP) for comparing the impact of gases that have different lifetimes in the atmosphere and different capacities for absorbing heat. Based on its use of the GWP, countries have been ranked on the basis of their total annual greenhouse gas index and on their relative per capita greenhouse gas index. According to GWP rankings for 1989, the United States contributed approximately 18 percent of global greenhouse gas emissions, the former U.S.S.R. 14 percent, the European Community 11 percent, China 9 percent, Japan 5 percent, and India 4.5 percent. These six countries were responsible for about 50 percent of the total atmospheric impact of current emissions. Most other nations contributed less than 1 percent. On a per capita basis, the average person from the United States and other developed countries contributed significantly more to atmospheric impact than the average person from a developing nation. For example, the per capita impact of a person in the United States was about 8.7 times that of a person from China and about 14.3 times that of a person from India.

The GWP for each greenhouse gas is determined by integrating an expression for the removal rate of the gas from the atmosphere and multiplying it by an expression for the infrared absorption potency of the gas. Consequently, the GWP for a particular gas depends on the period of years over which the integration is performed, which by necessity must be somewhat arbitrary. Integration periods of 20, 100, and 500 years have been used. The GWP values are normalized so that the value for carbon dioxide is 1; corresponding values for methane and CFCs are 21 and 5,873, respectively. To calculate a greenhouse gas index, national emissions from each country are weighted by the appropriate GWP, and the result is summed to provide an estimate of the impact of a country's total emissions in carbon dioxide equivalents. Despite its use, there is not universal acceptance of the GWP approach for several reasons. First, the warming potency of a greenhouse gas depends on its concentration in the atmosphere, which in turn is dependent on assumptions about future emissions. Second, atmospheric residence times for most greenhouse gases are not known with precision; this is especially true for carbon dioxide. Residence times are determined primarily by estimates of greenhouse gas removal rates based on models of atmospheric, oceanic, and ecosystem processes that are controversial. Some scientists believe that a more reliable method to calculate a greenhouse gas index would be to use observational data rather than models of how the atmosphere behaves to determine atmospheric residence times. Third, considerable debate exists about the use of arbitrary integration periods.

Beyond these problems, no scientific consensus has emerged regarding how to develop a greenhouse gas index that is appropriate for use in public policy decisionmaking. Smith (1991) argues that economic development in most developed countries has been fostered by energy use that has resulted in a so-called "natural debt." Because of their earlier and more extensive use of fossil fuels, industrial countries have significantly larger cumulative emissions of greenhouse gases than developing nations. A natural debt occurs when greenhouse gases are emitted into the atmosphere faster than they can be removed. Consequently, the natural debt is the cumulative portion of anthropogenic greenhouse gases on a per capita basis, allowing for the different warming potencies and atmospheric residence times for each gas. Although there are many uncertainties in this approach, as an approximation, the estimated total carbon released into the atmosphere and still present as carbon dioxide is about 260 metric tons per living person in the United States, compared with about 6 metric tons for the average citizen of India.

Fujii (1990) attempts to calculate a greenhouse gas index based on concerns for intergenerational and intragenerational equity. Somewhat arbitrarily, he assumes that all persons born between 1800 and 2100 have equal rights to equal quotas of carbon dioxide emissions. His method establishes regional quotas designed to equalize per capita carbon dioxide emissions in each region for this 300-year period, with the assumption that carbon dioxide levels and world population double from present levels. On a regional basis, future generations can inherit unspent quotas. According to Fujii's method, the North American carbon dioxide quota is about one tenth of the region's present emission levels due to its longer history of intensive energy use. Agarwal and Narain (1991) developed an index that allocates the natural sinks for carbon dioxide proportional to a nation's population, and they calculate each country's excess emissions beyond what its share of the global sink can absorb. They also propose that nations that exceed their emission quotas could buy emission rights from other nations. Other alternative methods to calculate greenhouse gas indices have been developed to overcome the problem of selecting an arbitrary integration period in calculating global warming potentials by choosing a period based on discount rates employed by economists to make estimates about the future (Lashof and Ahuja 1990).

2.5.2. *Global Ecology*

The distribution of the world's biomes depends primarily on climate, particularly temperature and precipitation. If warming of the climate lasts for decades, biomes may adjust to the new climatic conditions by modifying structural and functional attributes and changing their boundaries, thereby approaching a new equilibrium. If significant climatic change lasts for a century or more, succession to new biomes may occur.

There are several general approaches for assessing global ecological changes; none singly or in combination are sufficient to forecast such changes. Site-specific studies focus on understanding responses of different species to climate change. These studies are based on the recognition that each species has its own unique ecological and physiological needs, and as a result, each will exhibit different responses to the rate, magnitude, and duration of environmental perturbations (Cohn 1989). Most site-specific studies are based on inductive reasoning, and while they might make use of models and make generalizations about the ecological behavior of species, their results should be interpreted as having heuristic as opposed to predictive value. In other words, the results of site-specific studies are best viewed as being relevant to the particular study areas and their conditions. Although ecologists often make generalizations based on the results of site-specific studies, these studies often contain internal inconsistencies and assumptions and are accurate, at best, in only a probabilistic sense (Cairns and Niederlehner 1993).

Based on site-specific studies, Clark (1991) has summarized ecosystem sensitivities to climate change. For example, broad-scale processes such as net primary productivity may be sensitive to small changes in temperatures and water balance in deserts, grasslands, and temperate and conifer forests. Net primary productivity seems more sensitive to changes in temperature than to precipitation changes resulting from climate change, because the magnitude of temperature changes is relatively larger. Decomposition rates and the accumulation of detritus may be more sensitive in temperate hardwood forests, because rates of decomposition slow with increased latitude to a greater degree than do production rates. Decomposition rates in hardwood forests may be more sensitive to small climate shifts compared with those in conifer forests because of the higher litter quality in hardwood forests. However, protracted climate change or large-magnitude changes potentially could have greater effects in boreal conifer forests because of their greater accumulation of organic matter. Nutrient cycles respond differently to macroclimate, microclimate, seasonality, local vegetation cover, and disturbance. Consequently, it is difficult to predict their response to climate change. Fire frequency and magnitude also are sensitive to climate change, and it is likely that drier conifer forests will display greater sensitivity to climate change than will mesic forests. Finally, existing patterns of species composition would be expected to be altered as a function of climate change and the consequent fragmentation of ecosystems that is expected to occur.

Long-term climatic changes would be significant for the tropics and the Arctic tundra. In semiarid regions, trees are susceptible to decreases in precipitation. In wet forests, trees are vulnerable to insect pests, and infestations are influenced by temperature and precipitation. In the Arctic tundra, a warming trend would cause a reduction in the permafrost; consequently, trees would grow poleward farther, the upper layers of the tundra peat would dry out, and oxidation and decay of organic matter would increase. The additional carbon dioxide and methane that would be released would enhance warming, thereby creating a positive feedback.

Other approaches to predict the responses of species or ecosystems to global shifts such as climate change have been developed. Statistical models have been used to test hypotheses or to generate descriptions of the responses of species or ecosystems to perturbations. Based on an examination of studies focusing on species invasions and deletions in ecosystems, Ehrlich (1989) and Lodge (1993) conclude that ecologists can make some powerful and wide-ranging statements about invasions. For example, they can state that the addition or deletion of one species in an ecosystem can have profound impacts on community structure and function. However, they cannot accurately predict the results of a single (particular) invasion or deletion of a species in an ecosystem.

Mechanistic models also have been used by ecologists to predict the ecological consequences of environmental perturbations. These models normally are built on the assumption that the underlying causes of ecosystem structure and function are known, along with detailed knowledge of a species' individual physiological requirements or of a population's demographic characteristics. However, Pace (1993) points out that almost all mechanistic models fail to serve as a basis for reasonable predictions because they cannot capture all of the complexities involved in determining even a single species' response to perturbations, especially if ecosystem structure or function is to be regarded as fluid and not fixed.

As a partial remedy to problems posed by use of mechanistic models, Peters et al. (1991) propose the use of comparative studies which consider the responses of many populations, communities, or ecosystems to environmental perturbations across a specified gradient, region, or larger geographic area . Comparative studies attempt to describe and answer questions about general ecosystem patterns or responses by acquiring data and making statistical inferences. The advantages of such studies are that: (1)by sampling numerous populations or ecosystems, one can develop baseline data against which to evaluate future change; (2)studies involving many species or ecosystems are more likely to document large-scale human impacts than studies focused on a few systems; and (3)they provide a means for developing probabilistic models that can forecast large-scale changes. The disadvantages of comparative studies are that: (1)statistical inferences based on regression and correlation do not lead directly to mechanistic understanding; (2)the studies may fail when changing environmental conditions extend beyond the range of a model's prediction; (3)the studies often do not detect subtle ecological interactions.

Ecosystem simulation modeling is another tool that is used to examine potential ecological responses to global climate change. Studies using GCM scenarios generally use the output of equilibrium climate experiments as their starting point to forecast ecological responses to climate change. Most models focus on either structural or functional attributes of ecosystems. Structural models focus on processes that control vegetation structure and distribution, whereas functional models focus on biogeochemical processes and cycling, nutrient dynamics, soil carbon storage, and plant production. Several aspects of ecosystem modeling determine its suitability in forecasting the consequences of climate change (Root and Schneider 1993).

There is a mismatch of scales between GCM models and ecological studies, wherein the scale of the former normally is orders of magnitude larger than the latter. Consequently, knowledge from GCMs is not able to be applied to local or even regional scales. To overcome this problem, attempts have been made to develop regional forecasts from GCMs, but these are more uncertain than those at larger scales (IPCC 1992). Another problem with linking GCM output with multiscale ecological processes is that estimates of climatic variability during the transition to a new climatic equilibrium at the local or regional scale are important determinants of a species or ecosystem response to climate change. However, such variability estimates are not able to be included in GCMs because they are not capable of including such regional information.

Linking GCMs with multiscale ecological studies is problematic because there is an unpredictability of time-evolving transient climates in regional areas (Root and Schneider 1993). Although there is a fairly uniform increase and distribution of greenhouse gases in the atmosphere, a uniform or global ecological response is not likely. The timing of responses will vary among regions, and some will be more transitory than others. Further, the character of transitory responses will be different from that of a long-term climatic equilibrium. This means that not only will ecological consequences of climate change be difficult to predict at the local or regional level but also that transitory responses are likely to increase extinction rates in local environments because the vast majority of habitats cannot be protected from transient effects through prevention or mitigation efforts (Watt 1992). Presently, climate change scenarios as used in GCMs and ecological response models apply to equilibrium conditions, whereas actual climatic and ecological changes will be transient in character until such time as equilibrium conditions are achieved.

Most climate scenario studies do not provide for linkages among plants, animals, and climate on a large scale. Assessing the effects of climate change on animals by linking GCM output with multiscale ecological studies also is complicated by the fact that while the ranges of many animals have been found to be linked to vegetation, others are more directly linked to temperature or to competition with other species (Root 1988).

Paleoecological studies also are used to assess ecological responses to climate change. Past climate changes have caused large-scale shifts in species' ranges, the species composition of biological communities, and species extinctions. A 3°C atmospheric temperature increase would result in a climate warmer than experienced in the past 100,000 years. A 4°C increase would make the earth warmer than anytime since the Eocene, 40 million years ago (Webb 1992). In addition, the projected rate of human-induced climate warming is up to 100 times faster than past natural fluctuations. Based on paleoecological data, both the rate and magnitude of projected climate warming and associated changes exceed the ability of many species to adapt. Problems posed by climate change might be more acute in poleward temperate regions, since temperature changes there are projected to be larger than the mean global increase.

Recent data have suggested that even slow temperature changes have been linked to rapid periods of species' extinction and evolution (Kerr 1993). However, the causal explanation for how climate change affects rapid extinction and evolution of species is not clear. Any increase in species extinctions would be superimposed on current extinction rates. It is estimated that from 4,000 to 6,000 species become extinct annually due to the activities of humans; this rate is approximately 10,000 times so-called natural rates. However, there are many uncertainties in the number of species that exist presently, and the actual number of species becoming extinct may be two orders of magnitude higher than thought previously (Ehrlich and Wilson 1991). While paleoecological studies may provide useful information in understanding the effects of climate change, they should be used with caution in predicting ecological responses to future climate change if the rates and magnitudes of the latter exceed the paleoclimatic data base.

Regardless of the approach used to assess ecological effects of climate change, all projected ecosystem changes may have to be evaluated in the context of increasing human interventions in natural ecosystems. Few of the world's ecosystems are free from human influence, and in many parts of the world, human intervention probably will have an equal or greater ecological impact than that of climate change in the next 50 to 100 years. Significant attempts to link ecological effects of climate change with other effects of human activities have not occurred.

2.5.3. *Human Health and Disease*

There have been few studies of the effects of global change on human mortality (Longstreth 1990). The effects of climate on specific diseases are difficult to assess, owing to the many different geographical conditions and controls, together with the uncertainties regarding projected magnitudes and rates of climatic change. An increase in the incidence of certain diseases and change in their geographical ranges has been postulated by some investigators as possible consequences of global warming. Examples include schistosomiasis (Weihe 1979), bacillary dysentery (WHO 1977), hookworm (CCTA/WHO 1963), malaria, dengue fever, and yaws and cholera (Brown 1977). Because the complex natural histories of diseases are influenced also by such conditions as water quality, dietary conditions, food sanitation, refuse disposal, and level of economic development and education, they must not be linked solely to climate factors.

Some of the effects of climate change on human health are observable directly. For example, statistical relationships are known to exist between temperature and mortality from heart disease, stroke, acute bronchitis, asthma, and pneumonia (Rogot and Padgett 1976). Generally speaking, there may be increases in summertime deaths for areas that experience warming trends. Although these areas might experience a reduction in winter deaths for the same diseases, the increase in summer deaths is expected to exceed the reduction in winter deaths. Sudden changes in temperature also are correlated with increases in deaths. Consequently, if climatic variability increases, morbidity and mortality also are expected to increase.

It is important to remember that adverse consequences to health from climate warming might occur in tandem with increasing exposure to UV-B radiation due to depletion of stratospheric ozone. Recent data suggest that for every 1 percent decrease of ozone there is up to a 2 percent increase in cutaneous melanoma incidence and between 0.3 and 2 percent increase in mortality due to the melanoma (EPA 1987). The role of UV-B also has been confirmed in inducing cutaneous melanoma in animal models. Data also indicate that between a 0.3 and 6 percent increase in cataracts can be expected for every 1 percent decrease in stratospheric ozone.

2.5.4. *Population Settlements*

There is good evidence that alterations in human settlement patterns result from global warming and consequent shifts of rainfall patterns and deserts (Lemons 1985). Examples include: (1)the dispersal of the ancient Mycenaeans circa 1230 B.C.; (2)abandonment of agricultural areas and villages in Europe circa 1450 due to severe winters and variable summers; (3)the Irish potato famine, which resulted from warmer and wetter than usual summers between 1845 and 1851; (4)the displacement of hundreds of thousands of farmers and settlers from the western North American plains due to drought in the 1890s and 1930s; and (5)recent deaths and resettlement of nomadic populations during the Sahelian drought of 1968-73.

The adverse effects of climate change will disproportionately affect the people of developing countries, since it is estimated that they will comprise about 78 percent of the world population in the year 2000, and because they utilize marginal lands that are more susceptible to climate change for their livelihood (UNPF 1991). Of course, one of the most catastrophic impacts would result from the disintegration of the unstable West Antarctic ice sheet, should this occur. For example, estimates are that in the United States alone 11 to 16 million people would be displaced (Bentley 1980); 8 to 12 million people could be displaced along the Nile River delta (El-Sayed 1991).

Although the historical evidence indicates the significant impact of increased regional warming upon human settlements, it is not feasible to make detailed predictions of the effect of future warming for many regions because changes in regional temperature and precipitation variability cannot be ascertained at this time.

2.5.5. *Agriculture, Livestock, and Fisheries*

The stability and distribution of food production could be affected greatly by climate warming, in terms of both agricultural productivity and trade (Dudek 1991). Changes in the world food system will be due to: (1)direct biological effects of increased carbon dioxide concentrations, which would tend to increase productivity; (2)interactions of temperature and precipitation in rainfed agriculture; (3)changes in water demand and availability for irrigation; (4)longer growing seasons in temperate latitudes; (5)changes in soil nutrients and salinity; (6)increased infestations of agricultural pests and diseases; (7)stress on livestock production; and (8)the exacerbation of water and air pollution problems by climate warming. Although the causal mechanisms resulting in changes in the world agricultural system due to climate change are known with some confidence, the directions and magnitudes of some of the changes are not known well for at least four reasons.

First, as discussed previously, the likely changes in variability of temperature and precipitation for specific geographical regions are not known. Second, every crop responds to climatic factors differently, and effects must be examined individually. For example, estimates are that a 1°C. temperature increase results in a 2 percent reduction in U.S. corn yield or a 12 percent reduction if combined with a 10 percent reduction in precipitation; other projections indicate that wheat yields would increase (Waggoner 1983). Third, changing climate affects the frequency and severity of food pest infestations. Pimentel (1989) indicates that warmer and longer growing seasons induced by climate change could enable many insect pests to pass through an additional one to three generations. The exponential increase of some pest populations under new favorable environments could increase losses due to insects and make their control more difficult. Fourth, economic dislocations due to climate change limit food availability and distribution. Most food traded is surplus, and yearly weather fluctuations affect the amount of surplus and demand for it. These fluctuations create wide price swings, which influence local supplies and the ability of people to afford them.

Chameides et al. (1994) have examined possible implications of regional ozone pollution for the three most agriculturally productive regions of the world. These regions cover 23 percent of the earth's continents but account for most of the world's energy consumption, fertilizer use, food-crop production, and food exports. They also account for more than half of the world's atmospheric nitrogen oxide emissions. As a result, they are prone to high levels of ground-level ozone during summer months. Approximately 10-35 percent of these agriculturally productive regions currently may be exposed to levels of ozone that may reduce crop yields. If abatement of anthropogenic nitrogen oxide does not occur, by the year 2025 approximately 30-75 percent of the world's cereals may be grown in ozone-damaging regions. This could result in a 5-10 percent reduction in crop yields.

Recent models have been used to calculate population size, food production and consumption, and storage of grain under different climate scenarios over a 20-year period (Daily and Ehrlich 1990). According to results of this modeling, it is possible that there will be a 10 percent reduction in global grain harvest an average of three times a decade. This could result in the starvation of between 50 and 400 million people. Further, global warming could reduce cropland by 10 to 50 percent due to increased temperatures, increased rainfall in certain areas, and coastal flooding. Developing regions whose agriculture appears to be at most risk from climate warming include the Sahel, Egypt, southern Africa, India, eastern Brazil, and Mexico (IPCC 1990, Parry 1990).

The IPCC (1990) recognizes three primary areas of uncertainty that need to be resolved to understand the responses of agricultural systems to climate change: (1)understanding the effects of climatic and atmospheric changes, singly and interactively, on major crop, forest, and livestock species; (2)understanding how pests and diseases will change in impact, spatial and temporal distribution, and variability, and to model these changes so that they may be incorporated into change and management scenarios; and (3)development of the capacity to predict the effects of changes in climate and atmospheric composition on the quality of land through changes of *in situ* soil processes and in soil erosion.

Supplies of fish are important to many countries for economic reasons and as a protein source, and some studies have noted the effect of temperature changes on fisheries resources. For example, the periodic reduction of oceanic upwelling due to coastal and surface water warming and the consequent nonreplenishment of nutrients to surface waters has caused declines of fish catches off the coasts of Peru, California, Namibia, Somalia, and Mauritania (Ryther 1969). These regions contribute a significant fraction of the world's fish supply. Although there is some understanding of how temperature fluctuations can affect net fish productivity, other climatic and oceanic variables such as prevailing winds and ocean currents, cloud cover and rainfall patterns, and availability of nutrients also influence production. Because it is not known how all of these might change as a result of climate warming, it is not yet possible to predict impacts to fisheries accurately.

2.5.6. *Water Resources*

Climate warming will alter precipitation, both globally and regionally. Potential sensitivity of water resource issues to global warming determined from GCM sensitivity analysis includes inadequate surface water supply/storage, groundwater mining, flooding, conflicts in water use, salinity problems, drought, water for vegetation, surface water contamination, waterborne diseases, inefficient irrigation management, availability of potable water, and reservoir sedimentation. However, researchers disagree considerably about the levels of confidence that should exist regarding predictions of climate change on water resources. Recent estimates suggest that 10-50 years are needed before predictions can be made with confidence (Schneider et al. 1990).

Typically, climate models project that the largest changes in precipitation will occur in the vicinity of 30°S and 30°N (Sulzman et al. in press). Increased precipitation at higher latitudes is expected throughout the year, and at midlatitudes during winter months. Many models project little change in precipitation for the dry subtropics. Other models that project geographic distributions of hydrological changes show different responses at different latitudes. Decreases in precipitation are predicted to occur between latitudes 40°N and 10°S, while increased precipitation is expected to occur between 10°N and 20°S in regions north of 50°N and south of 30°S (Washington 1992). Such changes would have profound effects on the distribution of the world's water resources. The combination of increased evaporation and decreased rainfall in the Colorado River system of the United States would diminish the flow of the river by 50 percent or more. Other river systems that provide needed water for prime agricultural areas and that would experience greatly increased flows include: the Hwang Ho in China; the Amu Darya and Syr Darya in Asia; the Tigris-Euphrates system in Turkey, Syria, and Iraq; the Zambezi in Zimbabwe and Zambia; and the São Francisco in Brazil. Increased water flows resulting from increased precipitation could occur in the northern Africa rivers of the Niger, Chari, Senegal, Volta, and Blue Nile. Projected increased flows in the Mekong and Brahmaptura rivers could lead to widespread and destructive flooding in Thailand, Laos, Cambodia, Vietnam, India, and Bangladesh.

Even if effects of climate change on water resources were not catastrophic, significant changes in water supply systems could still result from decreases in mean stream flow or increases in variance of stream flow. Such attributes include water quality and yield from unregulated streams, reservoirs, and groundwater. In addition, changes in storm frequency and drought are likely to be brought about by climate change. It is estimated that the destructive potential of hurricanes might increase by 40 to 50 percent with a doubling of atmospheric carbon dioxide (Hansen et al. 1989).

2.5.7. *Sea Level Rise*

The general effect of sea level rise is to increase beach erosion, the loss of marshes, storm damage, and salt water intrusion and to threaten the lives and well-being of people and their buildings. Because a large fraction of the world's people live in coastal zones, they are prone to the adverse consequences of even a small increase in sea level. Countries such as the Netherlands, Egypt, and Bangladesh particularly are at risk.

Average surface temperatures have risen approximately 1°C in the last century, and sea levels have increased at an average rate of approximately 1 to 1.5 mm per year during this time (Gornitz and Lebedeff 1987). Typical projections of future sea level rise range from about 0.3 m to 3.5 m by the year 2100, although some projections are greater (Hoffman et al. 1983, Meier 1990). However, it is not clear exactly how accelerated atmospheric warming will affect sea level rise because the interactions between the atmosphere and the ocean are not understood well.

Sea level rise also is a function of whether and to what extent a reduction of the amount of sea ice drifting in the Arctic Ocean occurs and whether partial melting of the West Antarctic ice sheet occurs. The effects of the former would be to increase sea surface temperature and consequently shift major climatic zones 200 km or more northward. These effects might occur over a period of a few decades if atmospheric temperature increases approach 4 to 5°C (Flohn 1982). The effects of a partial melting of the West Antarctic ice sheet might be a 5 m or more elevation rise of the world's sea level, with consequential flooding of many coastal and lowland areas. Considerable debate exists concerning the likelihood of the Antarctic ice sheet melting. Based upon paleoclimatic evidence, Flohn argues that an atmospheric warming of 4 or 5°C would result in an ice-free Arctic Ocean but would not cause significant melting of the Antarctic ice sheet. Based on data from the geologic record, Leatherman (1991) argues that melting of most of the ice in ice caps and glaciers could result in a 70-m rise in sea level, but that such melting would likely occur over a time span of millions of years.

Mitchell (1982) postulates that the paleoclimatic data are tenuous and suggests that some climate models project melting of the Antarctic ice sheet occurring over a period of 1,000 years. In theory, the advantage of using paleoclimatic analogues is that they represent realistic solutions for sets of equations that only nature can solve; main disadvantages are that changes in boundary conditions (e.g., atmospheric composition, sea level, land surface changes) over time are not known well and data resolution allowing for mapping of past climates is insufficient. On the other hand, current climate models, while projecting melting of the Antarctic ice sheet, are not sufficient to allow for reasonably accurate predictions.

2.6. CLIMATE LINKAGES

In order to understand global climate changes, linkages among different phenomena need to be understood. In some modeling experiments, the influence of other greenhouse gases in addition to carbon dioxide has been considered. Results from these experiments indicate that projected mean annual atmospheric temperature increases should be approximately 20 percent higher than those based on carbon dioxide concentration only (Wang et al. 1991). These results suggest that more definitive studies of climate change should use other greenhouse gases in addition to carbon dioxide. Inclusion of other greenhouse gases into climate change studies is difficult because fundamental aspects of factors that regulate their atmospheric concentrations are not well understood.

For example, although the atmospheric concentrations of greenhouse gases have been increasing, beginning in 1991 the buildup of carbon dioxide, methane, and nitrous oxide slowed. Only recently have those gases resumed their historical rates of increase. However, the buildup of carbon monoxide continues to slow (Novelli et al. 1994). Presently, researchers have not developed causal mechanisms to explain these observations. Some researchers believe that the chemistry and recent buildup of the atmospheric concentrations are related, perhaps by the eruption of Mt. Pinatubo and atmospheric cooling it may have caused. However, they have been unable to develop a coherent explanation of the role of the eruption of Mt. Pinatubo as it might have affected the recent anomalies for carbon dioxide, methane, nitrous oxide, and carbon monoxide (Kerr 1994). Other factors that have been implicated (but not proved) in causing a temporary slowdown in these greenhouse gases include a reduction in methane sources (e.g., biomass burning, rice paddies, and natural wetlands that might have slowed in a cooler climate due to the eruption of Mt. Pinatubo) and the patching of natural gas pipeline leaks in the former Soviet Union. Recent dry spells in the tropics might have affected levels of carbon monoxide due to less biomass burning because of a reduction in the amount of agricultural waste needing to be burned and by a slowed expansion of slash-and-burn agriculture brought about by dryer conditions. Researchers do not really know how to explain the rise of nitrous oxide, because it has many sources in the soil and water, and it is not known how human activity has been affecting them.

Recently, the impact of stratospheric ozone depletion on global warming has been assessed. Although CFCs are effective at trapping heat in the lower atmosphere, they are the primary source of chlorine, which degrades stratospheric ozone. There are indications that decreased stratospheric ozone may exert a cooling effect on the lower atmosphere that might offset (partially) the warming attributed to CFCs (WMO 1993). Although there are many uncertainties surrounding these estimates, if accurate, they might partially explain why observed atmospheric temperature increases lagged behind those predicted by GCMs. In essence, until these types of linkages are established, the links between climate and atmospheric composition that might amplify global warming in the future will not be understood.

In order to make more accurate projections of temperature change, the linkages of climate models with such factors as future levels and rates of fossil fuel use, emission rates of other greenhouse gases, deforestation, and population growth rates need to be established (Lemons 1985, UNPF 1991). Projections of future fossil fuel use are based on numerous assumptions concerning future rates and levels of population growth, gross national product growth rates, informational inputs for energy models such as governmental energy policies and mix of energy sources, and whether significant energy conservation is realized. Various projections of world energy use for the year 2000 have ranged between 384 and 646 quads, and between 334 and 847 quads for the year 2020. Generally speaking, high consumption scenarios are characterized by low or moderate use of coal and little or no use of oil shale. Obviously, the mix of energy fuels actually utilized will be of paramount importance to future climate change. Unfortunately, the uncertainties regarding energy use are so large that they have precluded projections of "most likely" scenarios.

The net effect of changing land use on future concentrations of greenhouse gases can be significant (Houghton and Skole 1990). Forests can serve as a larger sink for atmospheric carbon dioxide if they are increased, or they can serve as a source of additional atmospheric carbon dioxide if they are cleared. Projections of future rates of deforestation vary due to uncertainties regarding the need for agricultural land and fuel wood, a sustained demand for wood operating in the absence of effective programs for forest conservation, population growth rates, and increases in standards of living. Typical rates of deforestation have been projected to range between 4 and 20 million hectares per year through the end of the century (WRI 1992). Estimates of the release of carbon from deforestation over the past decade have ranged from between 20 and 100 percent of the annual emission of carbon from fossil fuels (Woodwell et al. 1978), although a figure of about 20 to 30 percent commonly is accepted (Houghton 1991).

The emission of greenhouse gases is linked to population growth (Harrison 1990). During the period 1950-1985, worldwide emissions of carbon dioxide increased an average of 3.1 percent per year. During the same period, population growth grew by 1.9 percent per year, and per capita production of carbon dioxide increased 1.2 percent per year. Presumably, this latter increase was due to the higher per capita consumption of goods that involved the production of carbon dioxide. According to this type of analysis, population growth was responsible for approximately two thirds of the increase in carbon dioxide emissions during this 35-year period.

If carbon dioxide emissions in developing countries increase at the same rate that they have during the past 40 years, they will more than double from the 1985 level of 0.8 to 1.7 metric tons on a per capita basis by the year 2025. During this time, the populations of these countries are projected to almost double from 3.7 to 7.2 billion people. The increase in population would produce an additional 5.8 billion metric tons of carbon dioxide compared with the present worldwide total of about 6.9 metric tons. To be sure, this type of analysis is subject to uncertainty for a variety of reasons. For example, it implies a linear progression of consumption patterns and trends and does not reflect the fact that economic processes are subject to nonlinear changes.

The linkage between population growth and greenhouse gas emissions can, perhaps, best be exemplified by India. During recent years, the government of India has developed a number of initiatives to promote economic development and improve living standards. It is projected that this development will induce a doubling of India's carbon dioxide emissions (Dave 1988, Oppenheimer and Boyle 1990). More to the point, consider that India's 1990 population of about 850 million people is growing at the rate of about 2.1 percent per year and is projected to increase by 1.4 billion people by 2024. If we assume that India manages to reduce its fertility rate to replacement level within the next three or four decades and if it only doubled its per capita use of energy by use of fossil fuels, given the multiplier effect of India's present population and its rate of growth, the annual per capita emission of carbon dioxide in 2024 would be about one metric ton, which is the 1990 world average. Because of the multiplier effect of population, this increased amount of carbon dioxide emitted would more than cancel stringent reductions of carbon dioxide emissions made elsewhere. For example, it would exceed the reduction in carbon dioxide emissions if the United States stopped all coal burning without replacing it with any other fossil fuels (Ehrlich and Ehrlich 1990).

Assessing the linkages between population growth and greenhouse gas emissions becomes more problematic when other gases such as methane are considered. Methane is a potent greenhouse gas, and half of all anthropogenic emissions come from rice paddies, irrigated lands, and ruminant livestock. These sources have expanded in recent years to meet the needs of increasing populations and because of the demand for improved diets. About 14 percent of greenhouse gas emissions are from agricultural sources, and the overall amount can be expected to increase as more food is required to feed an expanding human population. While the argument can be made that some carbon dioxide emissions should be reduced because they result from inefficient patterns of production and consumption, this argument is not made easily in the case of methane, because some of its production is tied to food to support an expanding human population.

Some researchers have made projections of future atmospheric temperature increases by taking into account the scientific as well as other uncertainties such as future energy use. Using carbon dioxide emissions from fossil fuel combustion as an example, traditional thought is that the future growth of atmospheric carbon dioxide should depend primarily on the rate of fossil fuel combustion and the manner in which the carbon cycle responds to the increased carbon dioxide (Baes et al. 1976). Assuming a high-energy-use scenario whereby the initial growth rate of fossil fuels is 4.3 percent per year (and which is reduced in proportion to the fraction of the supply of fossil fuel that has been used), various models predict more than half of the 7,000 Gt of recoverable fossil carbon will be released in less than 100 years. This represents a predicted range of temperature increase of between 2°C to 10°C and 2.5°C to 12.5°C. (Two minimum and maximum temperatures are given, which reflect uncertainties in the behavior of the carbon cycle, extent of deforestation, etc.) If a low-energy-use scenario is assumed, where the fossil fuel growth rate is only 2 percent per year until the year 2025 (followed by a symmetrical decrease as solar energy becomes more available and fossil fuel use is discouraged), the models predict that the total carbon released will be about 25 percent of the high-energy-use scenario and that the carbon dioxide content of the atmosphere will be approximately 1.5 times preindustrial levels. Projections for this scenario indicate a temperature increase between 0.5°C to 2.5°C and 1.0°C to 5.0°C. If present fossil fuel emission levels continue unchanged, doubling of carbon dioxide does not occur until into the 23rd century. In contrast, an annual fossil fuel growth rate of 4.3 percent would double atmospheric carbon dioxide within the lifetime of today's children (Clark 1982). Significant environmental impacts have been predicted if the globally averaged temperature increases by approximately 3°C to 4°C; such an increase is projected if atmospheric carbon dioxide concentration increases to approximately 500-620 ppm. Assuming even moderate world economic growth and world fossil fuel energy increases of 2 to 3 percent per year, the estimated atmospheric carbon dioxide concentration would be approximately 580-650 ppm by the middle of the next century or before (Washington 1992).

When considering the effects of climate change, it also is important to remember that extremes, variability, and means of climate conditions will affect the severity of other environmental problems such as acid deposition, stratospheric ozone depletion, and attainment of ambient air pollution standards (White 1989). For example, the amounts of sulphur dioxide and nitrogen oxides are influenced by climate in their source regions. Cold winters increase demands for heating oil in some regions and therefore the production of nitrogen oxides. Warmer summers increase demands for electricity and therefore the production of sulphur dioxide if coal is used as an energy source. Emissions of carbon monoxide, nitrogen oxides, and volatile organic compounds stemming from transportation use are influenced by the effect of weather on combustion efficiency. Warming of the atmosphere also will increase the transformation rates of primary acidifying gases and the production of ozone. This change in transformation rates will lead to a change in the relative amounts of acidifying materials deposited and in their concentrations.

Land-use and resource policies will affect and be affected by changes in the atmosphere. Policies that affect the quality of terrestrial and marine resources can decrease greenhouse gas sinks and increase atmospheric emissions. Loss of biological diversity may reduce the resilience of ecosystems to climatic variations and to air pollution damage. Climate change and agriculture may affect the natural environment due to regional changes in crop and livestock production. Changes in agriculture may increase soil erosion, intensify the demand for water for irrigation, degrade water quality, reduce forested land, and impact wildlife habitat. The problems of population growth and the human demands on natural resources also will be exacerbated by consequences of climate change. For example, within another few decades Bangladesh may lose a sizable portion of its land to sea-level rise, but by that time its population is projected to increase to twice its present level of 116 million people (Meyers 1993).

All of these types of linkages introduce many more uncertainties into the assessment of future impacts of climate change in addition to those described already.

3. Ethics and Climate Change

The possibility of climate change poses many ethical issues. These include questions about global environmental justice, duties to future generations, duties to nonhumans, our obligations as individuals, and what constitutes ethical national policies. In addition, questions in these areas interact with science and economics. How do we make morally responsible decisions under conditions of ignorance or scientific uncertainty or when facts are indeterminate? How do economic considerations relate to moral reasons? We cannot hope to answer these questions here. Instead, we will provide an introduction to some of the most important ethical issues posed by anthropogenic climate change.

3.1. GLOBAL ENVIRONMENTAL JUSTICE

Questions about global environmental justice take on their meaning and significance against an empirical background. Primarily it is the industrial or materially rich countries that have loaded the atmosphere with greenhouse gases. The fact that they are rich is closely related to their use of fossil fuels in key stages of their development. While the rich countries have reaped private benefits from emitting greenhouse gases, the negative effects of these emissions will be felt by people all over the world, including those who did not benefit from the economic development that the massive use of fossil fuels made possible. For example, the people of Bangladesh have benefited very little from the large-scale use of fossil fuels. Yet some projections suggest that climate-change-induced floods may kill or harm hundreds of thousands of Bengalis (IPCC 1990).

Inequities in the emissions of greenhouse gases are not only historical facts. A handful of industrial countries still emit between one half and three quarters of all greenhouse gases (Brown et al. 1994). Furthermore, despite frequently stated worries about potential increases of emissions in the developing world, the annual increases of greenhouse gas emissions is greater in the United States than it is in India (IPCC 1990).

Inequities in greenhouse gas emissions are part of an international system that is characterized by increasing inequality. Tickell (1992) states that in 1880 the real per capita income between Europe on the one hand and India and China was 2 to 1; by 1975 it was 40 to 1, and now it is 70 to 1. According to the World Bank (1992), poverty is increasing; there are now 1.1 billion people living in poverty, more than 20 percent of the world's population. Yet resources continue to be transferred from poor to rich countries. George (1992) says that between 1982 and 1990, rich countries sent about $900 billion to poor countries in the form of loans, credits, and grants, while during the same period, poor countries paid more than $1.3 trillion to rich countries in interest and principal payments on loans.

In the background are problems of overpopulation and overconsumption. Despite efforts such as the 1994 Cairo conference on population, global population is expected at least to double from what it is at present before stabilizing, and even this may prove to be an optimistic expectation. Most people in the rich countries show little inclination to stabilize consumption, and many people in the poor countries would like to increase their rate of consumption. One way of understanding the possible impacts of the conjunction of growth in population and per capita consumption is to consider the following example. Sweden enjoys a high quality of life, yet its greenhouse gas emissions are only 40 percent of those in the United States on a per capita basis. If global per capita greenhouse gas emissions were the same as Sweden's, global emissions would more than triple, reflecting the large populations and low emissions of some underdeveloped countries (Streets 1990). This tripling of emissions is beyond even the worst scenarios that have been contemplated in most studies.

In the face of these profound problems, philosophers have had little influence. Indeed, there has been some question about whether questions of justice (as opposed to obligation) arise at all in international relations. Even if we assume (as we should) that these questions involve matters of justice, it is difficult to see how traditional theories of justice apply to them. The most influential theories of justice in the contemporary literature are those that center on equality and those that focus on entitlements. These theories provide precise formulations of two of our deepest intuitions about justice. The egalitarian intuition is that everyone should have the same. The entitlement intuition is that everyone should have what they deserve.

The most influential egalitarian theory is Rawls's (1971) theory of justice. According to Rawls, fair principles of justice are those that would be chosen by agents in the "original position," in which they do not know their particular tastes, preferences, or place in society. Rawls thinks that these agents would reject the idea of absolute equality (in itself a very difficult idea to formulate) and choose instead two lexically ordered principles, the first concerning liberty and the second concerning distribution. The second principle (the "Difference Principle") requires social and economic inequalities to be attached to positions and offices that are open to all under conditions of fair equality of opportunity and that they be to the greatest benefit of the least advantaged members of society.

The most influential entitlement theory is that of Nozick (1974). He argues that the justice of a distribution depends entirely on how it was arrived at, no matter how equal or unequal it may be. According to Nozick, a complete theory of justice is comprised of three principles: a principle of just acquisition, a principle of justice in transfer, and a principle for rectifying past injustices.

On the face of it, it would appear that both theories imply that the present international order is unjust. Clearly, the inequalities that exist do not benefit the disadvantaged, and in part the present distribution reflects the global history of domination, imperialism, and exploitation. Yet Rawls and Nozick have little to say about international and environmental justice. Moreover, many environmental goods appear to resist treatment as distributable benefits and burdens (the stuff of distributive justice). We are situated in an environment that conditions everything we do and that in part constitutes our identities. Furthermore, on any reasonable human time scale, a stable climate, unlike standard commodities, is irreplaceable (Jamieson 1994).

While it is plausible to suppose that both historical and present patterns of greenhouse gas emissions are part of an unjust international order, philosophical theories of justice thus far have provided only limited conceptual resources for dealing with these problems.

3.2. FUTURE GENERATIONS

Our contemporaries are often victims of injustice, but there are mechanisms for representing their interests. These range from systems of justice in individual countries to the United Nations. Those who come after us are likely to live in a very different world from what they would have due to the climate change that we may be bringing about. Yet future people have no representation in the deliberations of today.

That future people have no political representation is an obvious fact. They cannot vote, and there are presently no trustees who are charged to defend their interests (Weiss 1988). A more complicated question concerns the representation of future interests in present economic decisions.

It is common practice in economic decisionmaking to discount future costs and benefits. There are both good and bad reasons for such an approach. The good reason is that if we undertake a project now that will entail a cost of N in 10 years, the project will be worth doing if the present benefits invested for 10 years will equal or exceed the cost. But this approach becomes problematical in cases in which there is uncertainty, irreversibility, or uncompensable harms.

Suppose that our present climate change activities will result in damages of N for our descendents living in a century. If we can obtain a 5 percent return (compounded monthly) on present benefits, our climate change activities would only have to be worth $.0068N$ in order for them to be justified. For example, a present benefit of \$100,000 would justify inflicting a cost of \$14.68 billion on those living a century hence. Not only does this specific result seem suspect, it seems ludicrous to suppose that we can do the calculation at all, for that would require assigning meaningful economic values to the loss of many wild species, the destruction of societies and cultures, and the unknown health effects of climate change. But even if these problems could be overcome, we would still be faced with the moral issue of climate change depriving future people of significant choice. They might prefer to live in a world characterized by our current stable climate regime rather than to enjoy a higher standard of living.

One line of argument suggests that present people owe nothing to future generations (Schwartz 1978). Since the actions we undertake now will determine which future individual people will come to exist, nothing we do now will make future individuals worse off than they otherwise would have been. Thus no future person can complain that he or she would have been better off had present people made different choices; had our choices been different, the person with the complaint would not have existed at all.

In our view, the moral of this argument is not that present people are ethically absolved of the effects of their actions on the future, but rather that actions can be wrong even if no individual is made worse off. This is an important result, for it compels us to reject some otherwise plausible, person-affecting moralities, and perhaps the view that intergenerational morality is a matter of justice or rights (MacLean 1983).

The utilitarian tradition has claimed the other extreme. Sidgwick (1907) argued that impartial morality requires that the time someone exists has no relevance to the urgency of that person's interests. To value the interests of future people less because they are remote in time from us is as morally arbitrary as discounting the interests of people who are remote from us in space, ethnicity, or psychological constitution. While this is a powerful argument, it seems to have the consequence that the interests of present people will always be swamped by those of future people. If large human populations continue to exist for, say, a million years, the interests of those living now will inevitably lose to those who will come after. There are vastly more of them than there are of us.

Once again we must conclude that an important moral problem has not been solved, even in theory. Most of us believe that we owe something to the future but not as much as to the present. This intuition may be correct, but as of yet it suffers from a lack of secure philosophical grounding.

3.3. NONHUMANS

Future people are not adequately represented in present decision processes, but at least they will be represented when they become decisionmakers; nonhumans are not represented at all, yet the effects of climate change on the nonhuman environment may be even greater than on humans. Climate change is likely to be much too rapid for many plants and animals to migrate or adapt. Even when migration would in principle be possible, few migration routes will be available in an environment that has become highly fragmented due to widespread and densely populated areas. Despite the intensity of these impacts, nonhuman nature is completely without representation in our decision processes. It must depend entirely on the preferences of human sympathizers for support.

Of course, some people would say that nonhuman nature is not entitled to moral consideration because, they say, it has only instrumental value and therefore serves as "raw material" for us to use as we please. This view has been criticized in recent years (see, e.g., Gruen and Jamieson 1994). Singer (1975/1990) has argued that we have moral obligations to all sentient creatures. This would include many nonhuman animals, such as other mammals. Goodpaster (1994) and Taylor (1986) have argued that we have moral obligations to every living thing. Rolston (1988) has argued that we have obligations to virtually every element of the natural order, including whole species and ecosystems.

If any of these views are correct, then climate change poses serious moral problems with respect to our obligations to nonhuman nature. Our usual approach, to consider the value of nature to be the value that humans place on nature, simply will not do. If nature is entitled to direct moral consideration, then it would be as wrong to think that the value of nature is exhausted by "contingent evaluation" as to think that this approach exhausts the value of children, the aged, or any other human.

3.4. ETHICS AND ECONOMICS

Thinking about ethical issues relating to climate change is difficult for many reasons. One complexity concerns the relations between ethics and economics.

Economic analyses and evaluation often work in two distinct ways. In one way they are hypothetical. They tell us what the economic implications are of various courses of action. Such analyses and evaluation provide one important piece of information, but in themselves they do not tell us what to do. Economic values are not the only values, and often we think that it is right for someone to do something that makes little economic sense. For example, most of us would say that someone who chose her friends or lovers strictly on the basis of economic considerations has an inadequate, one-dimensional value system. However, while economic analyses often begin as hypothetical, they often quickly turn to the categorical. That something makes economic sense is too often regarded as a decisive reason for action. The appropriation of such words as *rational* and *efficient* (as well as *good* and *bad*) by economists has contributed to the conflation of hypothetical and categorical evaluations.

However, once economic and moral reasons are clearly distinguished, there is a tendency to veer to the other extreme and to suppose that they have nothing to do with each other. One tradition in moral philosophy, deontology, often seems to suppose that right actions are those that are in conformity with moral rules, regardless of the consequences, economic or otherwise (see e.g., Bennett 1981). But surely this cannot be entirely right. Certainly we need to construe the consequences of actions or policies in a way that is much broader than is typically done in economic evaluation, and perhaps even then the consequences may not in themselves be a decisive reason for undertaking the action or policy. However, it is quite implausible to deny that consequences should play an important role in the evaluation of actions and policies. It may generally be wrong to lie, but if the entire fate of the world hangs on someone lying, then surely she should lie.

Economic results are an important consequence of many decisions, and therefore it is often important to know what they are. The possibility of climate change poses many important moral questions, but they are not completely separable from economic considerations. What we need to understand clearly is that moral considerations are not exhausted by economic concerns. What this means in the case of abating emissions of greenhouse gases is that while the costs and benefits of doing so are important to assess, the policy decision about whether or not to abate should not be decided solely on economic grounds.

3.5. SCIENTIFIC UNCERTAINTY

Policy decisions about climate change are made even more difficult by the problem of scientific uncertainty. Uncertainty often provokes people to divide into two camps. One camp insists that no action be taken until more research is done. The other camp claims that enough is known to take some action now. These arguments often have the effect of delegitimizing science in the eyes of the public, which sees science being brought in to provide justifications for policy decisions that are really being made on other grounds. In order to understand better the problem of trying to determine how much knowledge is enough for action to be warranted, it is important to make some distinctions and to appreciate the social context in which questions of uncertainty arise (Wynne 1992).

First, consider the distinction between uncertainty and ignorance. When we say that we are uncertain of something, this suggests that we know what it would take to make us certain. Ignorance, on the other hand, relates to the fact that we could be wrong about almost any proposition to which we give our assent and in many cases have no reasonable way of assessing this probability. However, from the fact that we could be wrong about almost anything, it doesn't follow that we are uncertain about almost everything. We can say crudely that uncertainty arises from ignoring ignorance. We take various features of a problem as given and focus on other dimensions. For example, it is widely agreed that the case for climate change is weakened by the fact that we are uncertain about the effects of clouds on the climate system. To identify clouds as an area of uncertainty is to presuppose that our general knowledge of the climate system is largely correct. This general background knowledge is "black-boxed"; it is taken as a set of fixed assumptions from which we proceed. This process of black-boxing is part of what makes science possible, for not every proposition can be interrogated simultaneously.

Uncertainty also needs to be distinguished from indeterminacy. Often what appears to be uncertainty cannot be reduced because there is no fact of the matter that can be learned that will reduce the apparent uncertainty. There is a great deal of indeterminism in the climate change debate because we do not know how people will behave in the future—what policy decisions governments will undertake, what firms will do, how individuals will change their lifestyles, and so on. There is no uncertainty about these matters because there are not now any facts about which we can become more certain. A second source of indeterminacy flows from the fact that any piece of data is evidence for a multiplicity of distinct hypotheses (Quine 1960). This is why different people with varying worldviews can feel vindicated by one and the same experience.

Once these distinctions have been made, we can see that regarding some proposition as uncertain is already to make some very large assumptions. Various problems about ignorance and indeterminacy have been pushed aside. Large social forces as well as small scientific ones can be involved in this pushing aside.

When we say that something is uncertain, we are relating it to a purpose. Some people claim that it is uncertain whether emitting greenhouse gases into the atmosphere will change climate; others seem to deny this. But in some cases they are not really disagreeing. Both parties may agree that for the purposes of scientific knowledge more research is needed. But those who deny that there is significant uncertainty may be claiming that there is no uncertainty for the purposes of policy formation, that what we ought to do is clear.

We should recognize the rhetorical role of claims of uncertainty. Often such claims are a way of arguing that no action should be taken. Those who want to take action then feel compelled to claim that there is no significant uncertainty. A debate that is really about values is disguised as a discussion of epistemology. In our view it would be better to discuss our ethical differences explicitly and directly rather than to mask them in the language of science (Jamieson 1992).

3.6. ETHICAL NATIONAL POLICY

While we have raised more questions than we have provided answers, still something can and should be said about what constitutes ethical national policy with respect to climate change.

The first point to make is that just as a policy should be based on good science and in some sense be economically reasonable, so a national policy should be responsive to the ethical concerns that we have identified. Second, it should be clear that ethical national policies will be different for different countries. For the United States to continue to increase dramatically its emissions of greenhouse gases is unethical in a way in which it is not for India or Haiti. Indeed, an ethical policy for the United States will be different from such a policy for other industrial countries, given their different histories, access to resources, alternatives, and so on. The framework for an ethical policy is established in part by the network of agreements to which a nation is party. The United States, along with more than 160 other countries, is committed to stabilizing greenhouse gas emissions. Exactly what this means remains a matter of negotiation and commitment, but the direction of change is clear.

What more an ethical policy would require depends on how one weights the various considerations that we have identified: global environmental justice, duties to future generations, obligations to nonhuman nature, and so on. However, it does seem clear that an ethical policy probably requires more than the United States is currently doing to reduce greenhouse gas emissions. Even a reasonable regard for the nation's long-term self-interest would seem to require more. While it is often argued that the developing countries are more likely to suffer serious adverse impacts from climate change than the developed countries, still the developed countries should be more risk-averse because they also will experience significant losses. A prudent government will protect a rich nation from what may be a serious risk. Moreover, it seems clear that the United States would benefit from reducing to some degree its use of fossil fuels. Energy in the United States is currently being used inefficiently compared with most European countries, and in addition to the direct economic effects of such inefficient uses, it also indirectly results in air and water pollution, land disturbance, and congestion and also makes the United States strategically dependent on the Middle East.

While it is unclear exactly what range of policies would constitute an ethical national policy for the United States with respect to climate change, it is clear that this range of policies would more effectively reduce the use of fossil fuels than those currently in place.

3.7. INDIVIDUAL RESPONSIBILITY

Whatever policy a nation adopts with respect to climate change, individuals are not thereby freed from acting in a morally responsible way in their everyday lives. Part of what it means to act in a morally responsible way is to work for political and other collective solutions to public problems. But it also means adopting lifestyles that are themselves ethically responsible. With respect to climate stabilization, this means reducing both the use of fossil fuels and engagement in other activities that promote the release of greenhouse gases. Changing lifestyles can be effective both in their cumulative effects and as one way of trying to bring about political and social change. If decisionmakers see that people are willing to change their behavior in the absence of coercion or legal mandates, this can help give the decisionmakers the courage to adopt ethical national policies. But even if societies do not change their behavior and a destructive climate change occurs, morally responsible individuals will at least have the satisfaction of knowing that they did what they could in a time of decision. They were a part of the solution, not just a part of the problem.

4. Greenhouse Economics

The contribution of economists to the greenhouse debate can be broadly divided into determining how seriously the threat needs to be taken and what action is most efficient to achieve agreed-upon policies. The first area is the realm of cost-benefit analysis and modeling intergenerational welfare. The second concentrates upon alternative policy instruments such as carbon taxes versus tradeable pollution permits and the impacts of different tax structures on various industrial sectors. The majority of economists are far more comfortable with this latter role, because the tools of conventional economics can be applied and many of the existing models developed for other purposes can be used—for example, merely by changing one sector to represent energy production or increasing the price of fossil fuel inputs. Hence trade, optimal control, and game theory models have been repeatedly applied in the economic literature on global warming. However, this second area of research also works within the framework set up by the first and must accept the theoretical approach that is common to both, in particular the utilitarian philosophy and trade-off assumptions. Thus, while the following sections concentrate on cost-benefit analysis and intergenerational issues, the constraints to economic techniques that are identified have broader implications. In the next section, the cost-benefit analysis approach is outlined and critically analyzed. A comprehensive treatment of cost-benefit analysis in the environmental context is given by

Hanley and Spash (1993). The discussion here raises the issues of uncertainty, individual preference formation, and intergenerational ethics, each of which is dealt with in turn.

4.1. COST-BENEFIT ANALYSIS OF GREENHOUSE GAS CONTROL

The movement toward the adoption of a cost-benefit analysis approach to this issue can be seen on at least two fronts. First, legislation concerning public projects has become increasingly environmentally concerned because of a publicly recognized need to conserve scarce resources. Current legislation in Europe requires the use of environmental impact assessment (where impacts are measured in physical units) for certain projects, under Directive 85/337. While cost-benefit analysis is an alternative paradigm for measuring environmental impacts, in the United States, environmental impact assessment was followed chronologically by Reagan's Executive Order 12291, mandating the use of cost-benefit analysis for public projects and policies. Hence, cost-benefit analysis has been more commonly applied in the United States, so influencing the economic literature and the policy debate on global warming. Second, the imposition of greenhouse gas constraints and/or alternative technologies in developing countries will need some justification. Preventing development projects because of their adverse impacts on global climate may disproportionately affect the economies of less developed countries, who can rightly point out that developing countries increased their own greenhouse gas emissions levels during early industrialization.

Faced with the threat of global warming, society has three options: do nothing, prepare to adapt, or reduce emissions of greenhouse gases. The first implies that the greenhouse effect is either unimportant or beneficial. The second and third options take the problem seriously enough to warrant action and could be carried out simultaneously. Adaptation would include measures such as strengthening sea defenses, changing cropping patterns, organizing population migration, and increasing irrigation. A policy solely relying on adaptation implies that humans have the ability to adapt to all future consequences and to offset undesirable physical effects and that this option is less costly than control. Irreversible damages, uncertainty, and ignorance of future consequences argue in favour of controlling greenhouse gases. However, to the extent that global warming is already irreversibly underway, society has no choice but to adapt. The third option is the one most commonly studied by economists and is the one we concentrate on here.

The economic approach to deciding how serious the problem is and what action to take involves weighing the costs of control against the benefits of preventing damages. Global warming could be reduced by cutting greenhouse gas emissions and/or by increasing sinks for the gases (e.g., reforestation). A stream of costs and a stream of benefits are associated with such actions. Optimal levels of greenhouse gas reductions could, in principle, be deduced from an examination of how costs and benefits of control vary with the level of reduction. Control costs will be higher the greater the reductions in emissions are and the faster a given reduction is attempted. The marginal benefits of reducing greenhouse gases will fall with the level of control, since fewer damages are avoided per unit of greenhouse gas reduced. The optimal level of control will occur when the marginal benefits of greenhouse gas reductions, in present value terms, are just equal to marginal control costs. If the assumptions concerning control costs and benefits are correct (e.g., there are no discontinuities in the functions), this analysis implies that the optimal reduction in greenhouse gases will be less than 100 percent, since the output associated with greenhouse gas production is valued more highly the scarcer it becomes.

The earliest example of a cost-benefit analysis of greenhouse gas control is d'Arge (1975), with little work since then until the early 1990s (a notable exception is Cumberland et al. 1982). Recent approaches range from the country-specific (Ingham and Ulph 1991) to world models (Manne and Richels 1991) and from partial equilibrium (IEA 1989) to general equilibrium studies (Bergman 1991). Surveys of this work may be found in Hoeller et al. (1991) and Ayres and Walter (1991). The almost exclusive focus of these studies is the control cost of carbon dioxide reductions with exceptions such as Nordhaus (cited below) and Cline (1992).

The work of Nordhaus (1982, 1991a, 1991b) is well known and worth analyzing more closely to convey the general cost-benefit analysis approach and some of its flaws. In his most recent studies, Nordhaus divides the U.S. into three sectors by susceptibility to climate change: (1)very susceptible, such as agriculture; (2)medium susceptibility, such as construction; and (3)unsusceptible, such as finance. These sectors accounted for 3 percent, 10 percent, and 87 percent respectively of U.S. Gross National Income (GNI) in 1981. The economic benefits of emissions reductions in the high and medium sensitivity sectors is slight (only 0.25 percent of GNI, or $6.23 billion for double carbon dioxide-equivalent), because these account for a low proportion of total GNI. Marginal damage costs under three scenarios are $1.83/ton carbon dioxide for low damages (0.25 percent of GNI), $7.33/ton for medium damages (1 percent of GNI), and $66/ton for high damages (2 percent of GNI). Nordhaus excludes undesirable effects of global warming on nonmarketed resources (such as wildlife), viewing such impacts as too difficult to value. However, he states, "My hunch is that the overall impact upon human activity is unlikely to be bigger than 2 percent of total world output" (Nordhaus 1991a). In calculating control costs, he assumes greenhouse gas reductions will be achieved by methods offering the lowest control cost. He argues that control costs will depend on how fast reductions in greenhouse gases are required and that marginal control costs will increase steeply beyond a 10 percent reduction. Thus, Nordhaus calculates the optimal control policy for the greenhouse effect as being to cut CFCs by 9 percent and carbon dioxide by 2 percent under the medium damages scenario (assuming a 1 percent discount rate).

Such minimalist recommendations have been criticized as misleading, for example by Daily et al. (1991) and Ayres and Walter (1991). The latter make three main points. First, up to a certain point, the costs of reducing greenhouse gases are negative. In other words, society would be better off reducing its use of substances generating greenhouse gases. This principally means cutting energy demand, since energy production and consumption comprise the single largest source of greenhouse gases. There are two reasons for this conclusion: (1)due to market distortions, energy is currently overused, and (2)profitable opportunities for energy conservation exist but are currently ignored. Ayres and Walter provide case-study evidence for Italy and the United States, while Fitzroy (1992) cites similar evidence produced by Flavin and Lenssen (1990). Thus, some greenhouse gas emissions can be cut at no net cost. This implies, *ceteris paribus*, a higher optimal level of emission reduction than the case where control costs are always positive.

Second, cutting greenhouse gas emissions has environmentally beneficial side-effects in addition to reducing global warming. CFC reductions will help reduce stratospheric ozone depletion. If a carbon tax were imposed, coal consumption would be cut, since coal would face a higher tax rate than either oil or natural gas due to its relatively high carbon content by weight. Reduced coal use would reduce sulphur dioxide emissions and so lower acid deposition. Substitution of renewable energy sources for fossil fuels would reduce pollution externalities. In general, fossil fuels are associated with dispersed temporal and spatial chemical impacts, while renewable energy sources tend to have local physical ones, i.e., lower external costs (Spash and Young in press). Afforestation would generate a stream of nonmarket amenity benefits, depending on the type of forestry planted. In fact, the UK Forestry Commission now includes carbon absorption benefits when appraising new tree planting (Whiteman 1991).

Finally, Nordhaus extended his estimates for the U.S. economy to the world level (as does Cline 1992), and Ayres and Walter target their criticism at these world figures. As d'Arge and Spash (1991) have pointed out, developing countries are more susceptible to global warming, with extensive dependence on climate-sensitive production, a limited ability to adapt, and a sizable population of subsistence farmers. In criticizing Nordhaus, Fitzroy (1992) points out that climate change combined with soil erosion in food-producing regions would reduce world food supplies at a time when the world population will have doubled. Declining levels in major world aquifers would aggravate this situation. Ayres and Walter revise Nordhaus's estimates of the area of land lost upwards by a factor of ten and increase the value of land lost in less developed countries, such as Bangladesh. They also add an amount to cover the cost of resettling refugees forced to move as a result of sea-level rise. Even without attempting to include nonmarket effects, these revisions result in benefits of reducing global warming ten times greater than the medium damage scenario estimates given by Nordhaus.

An obvious next step would be to include the economic value of nonmarket benefits related to actions that reduce global warming. While much work in environmental economics during the last 20 years has focused on such nonmarket valuation, the application of benefit measurement techniques to the greenhouse effect confronts two key problems. First, many individuals may be unsure as to the meaning of the greenhouse effect and its related damages and the implications to them of preventing an increase in emission of greenhouse gases. While the valuation of benefits under uncertainty has been the subject of much attention in the environmental economics literature (e.g., Meier and Randall 1991), others have expressed concerns that poorly informed consumers cannot be relied upon to make sensible decisions about complex environmental phenomena (e.g., Sagoff 1988). Second, individuals may be unwilling to trade off increases/decreases in global warming against losses/gains in income. If a certain proportion of the population hold rights-based beliefs, this would prevent them from agreeing to such trade-offs. For example, environmental campaigners might believe that future generations have the right to live in their own homeland regardless of the utility this gives or of the costs to society. Such noncompensatory decision rules are referred to by neoclassical economists as representing "lexicographic preferences." These two issues are now considered in more detail.

4.2. UNCERTAIN FUTURES

Introducing uncertainty has lead some economists to argue that reducing greenhouse gas emissions is desirable even if the expected costs of doing so are known to exceed the expected benefits (e.g., Cline 1992, Spash and Hanley 1994a). The reasoning is based upon society being risk averse. Thus, the costs of reducing greenhouse gas emissions by 75 percent might be known to be $1 trillion. The benefits of reducing greenhouse gas emissions might range from $0.25 trillion to $10 trillion, with an expected value of $0.8 trillion. If society is risk averse, it can prefer to incur the certain loss of $1 trillion (the "certainty equivalent") rather than the expected loss of $0.8 trillion with the potential for higher losses. Thus, greenhouse gas control could be regarded as an insurance premium against known but uncertain future states of the world, where the probability of those states occurring is known or knowable. This would be consistent with an expected utility framework and could justify a safe minimum standard approach. Once a threshold with a safe margin has been chosen, the economy could be "safely" allowed to emit greenhouse gases.

However, in a fragmented world, risk aversion leads to a risk externality; that is, the risk is placed upon "other" societies (e.g., future generations), rather than leading to greenhouse gas control. Thus, (world?) government intervention would then be required to correct both a pollution and a risk externality. More seriously, this economic approach to an uncertain world requires that potential future states be reduced to probabilistic events. As a result,

Spash and Clayton (in press) note, several questionable, implicit assumptions are being made by the analyst:

1. A cause-and-effect relationship can be established to determine the outcomes to be included in the set of possible future states; this is a difficult task for global warming.
2. Probabilities can be associated with all future states of the world. The problem is that an action leading to an event may be recognized as a possible state but without a probability being attached to the outcome. Thus, an event can be expressed as uncertain yet have no associated probability of occurrence. The probability itself may be unknown or nonexistent. (Such a division of risk and uncertainty can be found in Keynes [(1921) 1973].)
3. The type of missing knowledge being analyzed concerns the risk associated with the occurrence of outcomes. However, all the models of the behavior of complex systems, such as environmental and economic systems or their interactions, are imprecise and limited in their scope. These limitations arise for a number of reasons: ignorance about a particular system, ignorance about the behavior of a class of systems, and the indeterminate nature of some complex systems (which can become chaotic at various points). This means the behavior of such systems can only be modeled in probabilistic terms, for limited domains, or for a limited time.
4. The distribution of risk over space and time is unimportant when judging appropriate action. Yet many decisions involve choosing among options that have different risks for different people at different times. Part of the issue here concerns the perception of risk. The general public has been observed to reject very low-probability, high-loss risks which experts judge to be acceptable (Freeman 1993). Thus, the experts could vastly underestimate the potential welfare costs that these risks impose upon people.

In addition to these problems, there are areas of ignorance related to sources of utility. First, some elements, substances, and organisms on the planet have yet to be utilized directly by humans. This can be viewed as uncertainty and ignorance over future use patterns. For example, losses in biodiversity due to global warming can cause future losses of which present humans are ignorant. Second, many of the features of nature that are directly utilized in economic processes are dependent on features of nature that are indirectly utilized. Current biomass depends on an ecological infrastructure that enables flows into human systems but is ignored itself. Thus, stratospheric ozone can be depleted by CFCs, allowing higher levels of UV-B radiation to reach the surface of the planet; this would in turn affect the marine biota at the base of the food chain on which harvested species of fish depend. In this way, uncertainty and ignorance pertain to ecosystems functions in addition to risk.

Once the above arguments are accepted, an optimal level of the insurance premium would be undefinable. Thus, while greenhouse gas control can be viewed as an insurance premium, this definition tends to reject the wider concepts of uncertainty and of ignorance. Society needs to accept that some areas of ignorance cannot be easily placed into the framework of knowledge about systems (Faber et al. 1992). In general, where altering the potentialities of systems causes changes that are, in principle, unpredictable, the appropriate response is to maintain options. This implies accepting the importance of different views on the same problem, questioning current knowledge, and emphasizing criteria of flexibility and reversibility (Spash and Clayton in press).

4.3. NONCOMPENSATORY CHOICES

The typical approach to the valuation of nonmarket environmental assets (such as wildlife) in environmental economics has been to treat such assets identically to marketed goods and services (e.g., Braden and Kolstad 1991). A standard theoretical assumption is the existence of the direct utility function which includes all items of value. The willingness-to-pay (WTP) of an individual to prevent a loss of an item relates to the impact on its utility function. An individual would therefore be prepared to give up some consumption of other goods to maintain a constant utility level if reducing greenhouse gases made him/her better off. The WTP amounts are typically summed across all affected individuals to obtain an aggregate WTP figure. Similarly, the minimum compensation demanded to accept an increase in greenhouse gases can be calculated (WTAC). In this case, expenditure on other goods needs to rise to compensate for the damages caused by global warming, keeping the agent at their initial level of welfare. The welfare measures of WTP and WTAC are expected to diverge, due to the potential for loss aversion (Knetsch 1990), income effects (Willig 1976), and substitution effects (Adamowicz et al. 1993).

However, besides the information problems outlined above, some individuals may treat certain environmental goods differently from the manner suggested by this theoretical framework. If an individual believes that aspects of the environment, such as wildlife, have an absolute right to be protected, then that individual will refuse all money trade-offs that decrease what is regarded as an environmental commodity in the neoclassical framework (Spash and Hanley 1994b). Thus, WTAC would be infinite, since the respondent believes that greenhouse gas damages should remain at or below their current level (i.e., no increases in greenhouse gases should be allowed). Simultaneously, WTP to reduce greenhouse gases can be positive or zero depending upon the income constraint. In fact, individuals may express a zero WTP as a protest against the implication that such things as the rights of future generations could be traded for other goods or money.

Such a noncompensatory stance can be viewed as evidence of a lexicographic preference. Lexicographic preferences mean that utility functions including greenhouse gas reductions are undefinable for an individual (since the axiom of continuity is violated) and that indifference surfaces are single points (Gravelle and Rees 1992). The implication is that one good is immeasurably more important than another, which leads to lexicographic preferences being regarded as unrealistic and unlikely to occur in economics (Malinvaud 1972). However, some evidence for the existence of lexicographic preferences has been put forward (Stevens et al. 1991, Spash and Hanley 1994b).

A belief system that denies trade-offs drives at the heart of modern welfare economics, which has been built around the Kaldor-Hicks potential compensation test. This test allows for projects to be approved where there is the potential to make at least one person better off and none worse off—i.e., some potential resource distribution after the project could achieve a Pareto improvement. Thus, knowledge of the required potential compensation is necessary and, in the neoclassical framework, would be based upon individual preferences. This criterion becomes inoperable once compensatory amounts become infinite. Furthermore, cost-benefit analysis itself is meaningless under noncompensatory preferences. The extent to which this issue is relevant to greenhouse gas control depends, at least partially, upon how far future generations can be compensated for damages they suffer as a result.

4.4. RESPONSIBILITIES TO FUTURE GENERATIONS

Spash (1994) has argued that the greenhouse effect could have serious impacts upon future generations while actually benefiting their predecessors. The standard application of cost-benefit analysis to the greenhouse effect, even if all costs and benefits could be calculated from individual preferences, would give the impression that the future is almost valueless, largely due to discounting. As Nordhaus (1991a: 936) has stated:

> The efficient degree of control of greenhouse gases would be essentially zero in the case of high costs, low damages, and high discount-

> ing; by contrast, in the case of no discounting and high damages, the efficient degree of control is close to one third of greenhouse gas emissions.

The distribution of net costs in the future, and net benefits now, makes the emission of greenhouse gases appear falsely attractive. Spash (1993) has criticized four common reasons for giving less weight to the expected future damages of long-term environmental pollution than if they were to occur now. These concern who constitutes the electorate, uncertainty over future preferences, the extinction of the human race, and uncertainty over future events. Without these justifications, discounting loses its moral imperative. Cost-benefit analysis as commonly applied would use an arbitrary but positive social discount rate. Thus, implicitly, some concern for the future effects of global warming would be shown, but the extent of this concern would depend upon the discount rate chosen. The problem that faces economists, in falling back on the use of a positive rate, is that their policy conclusions still have serious long-term implications which raise the need for a moral justification for the procedure.

However, there is a persistent view that the current generation should be unconcerned over the loss or injury caused to future generations because they will benefit from advances in technology, investments in both human-made and natural capital, and direct bequests. Adams (1989) has raised this exact issue in terms of alleviating our responsibilities for global warming. While fossil fuel combustion implies foregone opportunities for future generations, they "typically benefit (in the form of higher material standards of living) from current investments in technology, capital stocks, and other infrastructure." However, this line of reasoning confuses actions taken for two separate reasons. That future generations may be better off has nothing to do with societies consciously deciding to compensate the future.

If society has in fact been undertaking investments with the express purpose of compensating future generations for global warming, the lack of publicity has been conspicuous. More importantly, this would imply that the extent to which the future will be better off has in some sense been balanced against all the long-term environmental problems. That is, society cannot take global warming and see the future as better off, and then ignore global warming and take ozone depletion as compensated, and then ignore ozone and balance nuclear waste against supposed future well-being. Each case of long-term damage implies compensation which is distinct from catering to the general needs of future individuals.

This distinct nature of such compensatory transfers has been neglected (Spash and d'Arge 1989; Spash 1993, 1994). The greenhouse effect as characterized earlier creates an asymmetric distribution of losses and gains over time. Intergenerational compensation would counterbalance the negative outcomes of global warming by positive transfers, while not interfering with basic transfers. For example, assuming egalitarianism, the maintenance of the same welfare level fails to compensate for global warming. Yet the suggestion has been made that spreading the costs of global warming equitably across generations is an acceptable solution (Crosson 1989).

The problem with the latter approach arises from the economic view that changes in units of welfare are equivalent regardless of their direction. The standard approach of economists can be traced at least as far back as Bentham ([1843] 1954: 438):

> ...To the individual in question, an evil is reparable, and exactly repaired, when after having sustained the evil and received the compensation, it would be a matter of indifference whether to receive the like evil, coupled with the like compensation, or not.

Unfortunately, this approach treats harm as reversible by good. In general, doing harm is not canceled out by doing good. If an individual pays to have a road straightened and saves two lives a year, that person cannot shoot one motorist a year and simply calculate an improvement (Barry 1983). This argument is most apparent where the right to life is involved, but it can be extended to other areas where rights are accepted to exist. For example, assume individuals of a nation are accepted to have a right to live in their own homeland. Sea level rise due to global warming floods the Maldives and violates this right. Of course the Maldavians can be relocated and compensated, but this approach is unacceptable given the previously stated right.

The objection free-market economists might raise to the imposition of such rights is that freely contracting parties are prevented from entering into agreements of their own free will. As Bentham went on to point out:

> What is manifest is—that to no person, other than the individual himself, can it be known whether, in this instance, between an evil sustained, and a benefit received on account of it, any compensation have place or not.

That is, the individual is her or his own best judge of welfare changes. If the Maldavians believe they are better off in their new homeland, then who is to deny the acceptability of this exchange? The difficulty in the intergenerational context is that the individuals who will be impacted are unavailable for comment. In order to protect these individuals from unjustified harm, rights could be used, so that what appeared to be a problem for the use of rights can be viewed as an argument in their favor. In fact, this approach would define harm as a violation of the rights adopted by society.

The appeal to the "safe minimum standard" can be viewed as an example of constraining economic trade-offs by introducing rights. This standard advocates the protection of species, habitats, and ecosystems unless the costs of doing so are unacceptably large. In the case of global warming, Batie and Shugart (1989) argue that the safe minimum standard would support emission reductions despite apparently high costs. However, the withdrawal of the right of, say, a species to exist at some cost implies a basis of the right within utilitarian morality. This view contrasts with rights in the context of a deontological philosophy.

More generally, the economic process of exchange can be viewed as the transfer of goods and services within a framework of established rights. In this case, rights are only valid in as far as the institutional setting allows them to exist. Yet the question being probed here is one of the existence of a right of future generations in the sense of a natural right, not merely the recognition by a piece of legislation in a particular society at a particular time that such a right is valid. A natural right can be defined as a right based on intrinsic value (Nash 1989). The United Nations charter of human rights represents an internationally accepted set of goals to which the world aspires. The fact that these rights are violated does not reduce their importance. Yet within these rules, there is little comfort for future generations. A generous reading would only protect the future indirectly under articles intended to protect the current generation. Public concern is starting to be expressed regarding this oversight, and this has reached the extent of a global petition to the United Nations (Cousteau Society 1991).

If rights that protect future individuals from the results of our greenhouse gas emissions are accepted to exist, the scope for trade-offs commonly assumed in economics will be drastically reduced. Compensation payments are no longer licenses for society to pollute, provided the damages created are less than the amount of compensation—in which case, compensation cannot be used to excuse the continuation of greenhouse gas emissions. Irreversible damages that will occur regardless of greenhouse gas emissions reductions would require compensation. In order to protect the future from potential infringements upon this right, actions with uncertain intertemporal consequences would have to be avoided and environmentally benign production and consumption processes encouraged.

Due to the cost of enforcing the rights of future generations to remain unharmed, the current generation has a vested interest in denying those rights. Continuing to emit greenhouse gases at current rates denies the future the right to remain undamaged and asserts the dominance of the current generation. The current generation is then being asked to change the present rights structure, as found within society, in a manner detrimental to its own interest. The dictatorship of the current generation allows the imposition of damages regardless of the gain now and the extent of future damages.

4.5. FUTURE PROSPECTS

Cost-benefit analysis runs into problems due to uncertainty in the estimation of benefits, attitudes toward future generations and, more fundamentally, the very size of the problem (there is a point at which marginal welfare analysis loses its theoretical basis). These problems prevent a clear answer as to what should be done, and economics cannot, of course, provide a complete answer. The costs of reducing carbon dioxide emissions may be quite high, but because the benefits of reducing emissions are beyond economists' ability to estimate, the extent to which control options should be adopted, on efficiency grounds alone, is unknown. Thus, a practical way forward is to adopt "no regret" or "double dividend" policies. These are actions that can be justified on their own account but that also reduce global warming. Such policies include solving Third World food insecurity, increasing energy efficiency, cutting CFC emissions, preventing deforestation, and encouraging reforestation. Similarly, if energy prices are below their marginal social cost (excluding global warming impacts), then raising energy prices will make utilization more efficient and reduce greenhouse gas emissions.

The economists' appeal to cost-benefit analysis attempts to take losses and gains of controlling harmful activities directly into account. In doing so, the rights of future generations are violated when the costs of controlling the greenhouse effect are deemed to exceed the benefits of that control. The use of cost-benefit analysis therefore denies the existence of inalienable rights because harm and good are seen as equivalent. However, harm is recognizably different from good, and the deliberate infliction of harm is morally objectionable, as recognized in modern democracies. If remaining unharmed is defined as a set of rights given to future individuals, actual compensation is required if these rights are violated. If at all possible, these rights should not be violated and people should be freed from actions that deliberately externalize the risk of damages by imposing it upon others. These issues begin to reflect upon the role of cost-benefit analysis and some of the problems apparent with WTA measures where a structure of rights enforces a compensation principle.

The task of defining harms will be difficult, but as suggested earlier, the United Nations charter of human rights provides guidance. A further difficulty arises in being uncertain as to when an action might result in the violation of such rights. In terms of the greenhouse effect, there is a strong case to believe that numerous contraventions of these basic rights will occur. The point here is to emphasise a fundamental basis for human action in morality.

5. A Third World Perspective

If there is one thing the Earth Summit brought home to the Third World, it was the Machiavellian primacy of politics over ethics. For in the final analysis, it seemed that the more powerful interests of the industrialized nations prevailed and that recommendations were based more on politics than on considerations of justice and ethics (see, e.g., Johnson 1993). Building upon previous sections of this chapter, we herein more particularly focus on the following implications of climate change from the perspective of developing nations: (1)the burden of risk and the price of change, (2)equity-based ecological development, (3)intergenerational responsibility, (4)environmental and financial debt, and (5)environmental rights and ecological duties.

One response to the possibility of global climate change would be to do nothing but to collect data and analyse it in the hope that we can further reduce our scientific uncertainty and only act when we have sufficient certainty with regard to the effects our interventions may have. Obviously such an option supports the present status quo and those privileged by it. A second response would be to implement a global effort to reduce greenhouse gas emissions, since we know that these have the capacity to affect global climate, and the probabilities are that the resulting changes will disproportionately affect the poorer and more vulnerable countries adversely. Clearly such a response favors those people who are least able to cope with the consequences of climate change.

Obviously, different responses to global climate change will affect groups of people differently. The essential question with regard to risk management in this situation is whether and to what extent risks and costs of climate change and mitigation policies should be borne by those most vulnerable and least able to afford them or whether they should be borne by people in affluent nations who are more able to afford the costs and who also are benefiting by the present status quo. Will it be the political rather than the ethical implications of the question that will decide our response?

We herein opt for an ethical perspective consequent on the earlier part of this paper. The issues we raise in this section are of course not exclusive to climate change. Rather, this is an area that helps to illustrate well the global dimensions of the world's ecological crisis. In other words, when we have a global crisis, only a global response can meet it, and for this we need to act as a global community. Ecological thinking forces us to this conclusion.

Moreover, our vantage point is that of the southern or developing nations. Nevertheless, we are aware that there are conditions of poverty in more developed nations as well as pockets of affluence in developing nations. The homeless shivering in the cold whom one sees in New York and the mansions gleaming in the sun in Delhi are surely telling images of this anomaly. Our discussion could be further refined to take cognizance of such situations. However, at the risk of over-generalization, we are confining ourselves here to the broader aspects of climate change issues between developed and developing nations.

5.1. THE BURDEN OF RISK AND THE PRICE OF CHANGE

Who should bear the burden of risk and who should pay the price of climate change? If we wait for more scientific data before adopting effective measures to mitigate climate change, then we are not reducing the risk of climate change and its consequences, but rather we are increasing it. The longer we delay implementation of effective mitigation policies, the more difficult it will be to reduce risk or ameliorate adverse consequences of climate change at some future date. If we want effectively to reduce the risk of climate change, then we must limit the emissions of greenhouse gases sooner as opposed to later.

The very complexities and uncertainties make a cost-benefit analysis of the risks involved inadequate and unfeasible. The use of cost-benefit analysis is tantamount to basing decisions (in large part) on economic calculations and political priorities (Ghosh and Jaitly 1993). The political resolution to risk, change, and the sharing of economic costs typically is dependent on the bargaining power of the parties involved and usually ends up with the weakest bearing the burden of risk and the poorest paying the price of change.

On the other hand, an ethical resolution of the question of who should bear the burden of risk and who should pay the price of climate change would be value-based and rather different. An ethical management of risk would require first that risk be minimized and then redistributed equitably, if indeed we are to face risk as a community and not as isolated individuals—for a community can hardly be considered ethical if it protects the powerful to the neglect of the powerless. In reality, the most effective indicator of equity in a community is not how the strongest fare but rather how the weakest are able to cope.

Furthermore, risk reduction and its equitable distribution in the context of the global climate system will obviously demand change, both in human consumption patterns and in production technology. With regard to the first, for the poor this will mean an increase in consumption to meet their basic needs and to improve their quality of life to acceptable levels. Allowing these basic needs to remain at the subsistence level not only is ethically unjustifiable but also is ecologically unsound. We shall return to this point later.

For the rich, changes in their consumption patterns will mean a reduction or at least a restriction of affluent wants. This can actually lead to, or at least it has the potential for, an enhancement of wealthy persons' quality of life, even at the cost of a reduction in their standard of living. As Birch (1976) notes: "The rich must live more simply so that the poor can simply live." Indeed, this is a crucial issue in the whole sustainability debate, but a thorough examination of it would take us beyond the scope of this paper, though it does need to be developed elsewhere to deepen this discussion.

With regard to the second factor, changes in production technologies for the poor, who are surviving at subsistence levels, this must mean an increase in productivity. One can hardly in good conscience urge the people of developing nations to forego development programs that represent their only chance to escape from the poverty to which they are subject. But if this is to be done in an environmentally friendly manner without externalizing the costs, as happened with the first industrial revolution that was the basis of the present development and affluence of the First World, then there must be a change toward more environmentally friendly technologies. Unfortunately, at present the developing nations do not seem to have the resources to buy such technologies from the more developed nations, or the research and development capabilities to implement them on their own. For the rich, changes in production technology are concerned more with decreasing waste while at the same time expanding employment and other benefits. While these forms of new technologies are being developed, their transfer to poorer nations still remains a much-disputed and problematic area.

Globally speaking, achieving sustainable development will very much depend upon how such problems are resolved. A power-based political approach will only postpone and accentuate the already significant risks of global climate change. In our opinion, what is required really is a structural adjustment on a global scale, not only of the economic structures of our societies, which might affect the developing nations more, but more particularly in our lifestyles as well, and this concerns the more developed nations most urgently.

In other words, we need to change the manner and the kind of the goods and services that are provided with regard both to the way they are produced and the way they are consumed. We must realize that ecological productivity differs from productivity in the economic sense, because the economic utilization of resources through extraction under certain conditions undermines and destroys vital ecological processes, leading to heavy but hidden diseconomies (Goodland et al. 1993). Further, the nature of these diseconomies can be understood only through the understanding of ecological processes operating under conditions relatively undisturbed by humans (Angermeier and Karr 1994).

We realize that we cannot cope with the problem of distributing the risks and cost of climate change except as a global community bound together by a common destiny. A failure of the world community to take decisive action now to mitigate the risks of climate change will require even more urgent and drastic action later, if indeed it is not too late by then. Is it not curious, though, that some would want scientific certainty to be established before taking actions to mitigate global climate change, while at the same time certainty is never demanded of economic policy interventions, even though these are based on statistical probabilities? But then too often such interventions are dictated by the market rather than ethically derived from commitments to members of the global community.

5.2. EQUITY-BASED ECOLOGICAL DEVELOPMENT

Agenda 21 recommendations reflect the widely held view that equity is integral to achieving sustainability. Indeed, if it is not the sufficient condition, it certainly is a necessary one, the sine qua non for sustainable development. Granted that certain kinds of development can be unecological, we still have to face the fact that in the struggle for survival within a resource-poor or limited environment, poverty and pollution are inexorably linked. If the poor have no sense of opportunity in the future, then one can hardly expect that they will sustain and renew their environment in the present. When involuntary poverty becomes the poor's *fait accompli,* then by necessity they struggle simply for day-to-day survival, a struggle in which they many times do not succeed. All too often they are caught in a downward spiral of marginalized people trapped into marginalized areas.

Given the present technological and other capabilities that exist on a worldwide basis, it should be possible to alleviate conditions of poverty significantly for many people, if only the necessary political will for the task could be mustered. Further, it is ethically unacceptable that concerns for humans be displaced by an inequitable distribution of the goods of this world. Indeed, inequality only sharpens the sense of relative deprivation that the poor feel when they find themselves in want in the midst of the plenty of the affluent. Thus, if sustainability were imposed on the developing nations at the cost of their development, then this means that those nations would remain impoverished in order to sustain the more affluent nations. This situation, were it to occur, would be based more on political and economic power than on an ethical response to the problem. Sustainable development must as a minimum meet the economic challenges of providing for basic needs for all people.

It is also ethically unacceptable that our concern for nature be allowed to negate fundamental human rights. Indeed, a true concern for nature cannot set humans and nature in opposition. Rather, humans must be perceived as a part of nature that preserves, protects, and restores ecological integrity. In fact, only when human and nonhuman nature are in harmony can both be protected. Ecologically based thinking necessarily leads to an awareness of interdependent communities, as Gandhi envisaged, in ever increasing and inclusive circles, to include the human, the biotic, and the cosmic as well, and even the transcendent (Ramamurthy 1986). Of course, there is a danger that humans will become too anthropocentric in their thinking (see, e.g., Devall and Sessions 1985). Yet, ways must be found to accommodate the needs of both human and nonhuman nature.

Global climate change is a problem that transcends national boundaries. Even if it were possible to achieve sustainable development in one nation at the cost of unsustainability in another, as happens all too often in exchange relations between developed and developing nations, this would do little to mitigate the problem of global climate change, because it is a transboundary problem. Unfortunately, national sovereignty is often used to thwart remedial action, infringe upon environmental rights, and negate ecological concerns (Johnson 1993). Using national sovereignty to obfuscate ecological concerns or human rights is not, of course, the prerogative of any single nation, whether developed or developing. But when the more powerful nations, who are the least in danger of having their sovereignty threatened, indulge in such obscurantism, it is all the more galling. Thus, when then President Bush of the United States said at the Earth Summit Conference that nothing would make him compromise his nation's way of life, when that lifestyle threatens the global environment, such a statement may be good domestic politics, but it is from an international perspective grossly unethical. Certainly, such positions cannot be the starting point for any international measures to mitigate the problem of global climatic change.

Equity demands a reduction of the gap between the rich and the poor both intranationally and internationally. If this reduction is to be done within the carrying capacity of the earth, then further problems arise. If the poor of the developing nations aspire to reach the same consumption levels of the developed nations, then this cannot be accomplished within the earth's potential carrying capacity as we know it, in spite of any technological advances or institutional changes we may realistically hope for. To be sure, it seems improbable to narrow this gap solely by reducing the consumption of the rich, though this would surely be fairer than restraining the development of the poor. Is it realistic to expect a person to be elected to political office in a developed nation on the promise of reducing consumption? And yet the imperative to live within ecological limits and the ethical mandate for equity seem to demand that leaders of developed nations promulgate programs to reduce consumption of natural resources and the generation of harmful chemicals such as greenhouse gases.

Accordingly, some kind of redistribution seems to be warranted. A more equitable distribution of consumption and production between developed and developing nations, in a manner that will allow both to become sustainable, seems to be necessary. But just as sustainable development must meet the ecological necessity of containing itself within the carrying capacity of the earth, it also must meet the ethical imperative of equity among nations. Some kind of planetary bargain between the rich and poor nations for a more stable and sustainable world would seem to be called for rather than waiting for poverty and environmental degradation in less developed nations to pose a threat to the more developed nations before appropriate action is taken.

A beginning for such a bargain with regard to greenhouse gas emissions would be to consider quotas based on a per capita basis and not on an aggregated national one. This would be an equitable way of fixing the responsibility for change on the polluters, who must pay the price for it. National emission quotas would then be fixed not in terms of present levels of pollution but in terms of population size (which needs to be limited) on a per capita calculation, not an aggregated nationwide one (Agarwal and Narain 1991). Those countries not using their quotas could then trade them in with those unable or unwilling to limit themselves to theirs. While greenhouse gas emission must be reduced in the long term, in the short term such trade-offs could be used for a transfer of technology and resources that would lead to a more equitable development now and a more sustainable one later. Moreover, such transactions would be a matter of trade and not aid. This would make for less unequal exchange between industrialized and nonindustrialized countries. Indeed, until such unequal and unfair exchange between rich and poor peoples, nations, and regions, both intranationally and internationally, is remedied, there seems little possibility of sustainable, let alone regenerative, development on the global scale we so urgently need.

On the other hand, the suggestion of tradeable carbon permits is highly controversial. The permits raise the question of how far responsibility for pollution is to be allowed if a nation (or corporation) can purchase the right to pollute versus the point of view that responsibility for limiting pollution should be based upon moral as opposed to economic or political grounds or bargaining ability.

5.3. INTERGENERATIONAL RESPONSIBILITY

Although philosophers are divided on the exact nature of responsibilities to future generations, the view that present generations have responsibility or obligation to future generations is gaining more widespread acceptance and is, of course, reflected in many recommendations contained in Agenda 21. In the final analysis, such a responsibility must be based on a sense of bonding across generations (Care 1982). If we feel this bonding with the future, should we not feel the same with past generations as well? If we are responsible to the future, are we not also responsible for the past? Responsible not in terms of feeling guilty for what our ancestors may have done, but rather in the sense of feeling responsible to address the adverse consequences of their actions that still affect us, especially if we have been advantaged by their actions. In other words, can we accept the benefits left to us and not make remuneration for the harm they might have done to others?

An ecological principle now gaining acceptance is that "the polluter pays." If the polluter pays for the pollution caused in the present, who should pay for the pollution caused in the past and that still affects us now? While present people may not be guilty of causing past pollution, should they accept the advantages obtained from such past actions without making remuneration for them? Would not this be like someone keeping stolen property even though that person actually may not have been guilty of the theft? And if, as we know, some people's ancestors because of their unecological development have in the past borrowed from our common future, should their descendants now refuse to remunerate in the present those who are being affected adversely by this? Consequently, if present members of industrialized nations enjoy and accept a certain amount of affluence because of past development that has led to high levels of greenhouse gases in the atmosphere, thereby threatening other people spatially and temporally, should they not also be responsible for mitigating the harm caused by such development and affluence that they have accepted?

In a sense, the past still exists in the present, for no present can escape its historical context. Indeed, there can be no intergenerational responsibility without such a context. The irony, of course, is that those nations and peoples whose prodigality in the past has degraded the global environment now are urging restraint on those who have been frugal, out of necessity perhaps, but who now aspire to a higher standard of living. In fact, the powerful governments and multinational corporations of developed nations responsible for significant global environmental degradation are using the economic levers of aid, trade, and debt to enforce environmental discipline in the developing nations that have little political clout to use against them (Agarwal and Narain 1992). Such a situation could easily degenerate into a new sort of imperialism or colonialism, as the Indian finance minister has cautioned (Singh 1993).

Consequently, a certain alarm has been expressed at the rapid industrialization of some developing countries in Asia. If every Chinese person or household has a refrigerator, what will happen to the ozone layer, especially if the Chinese continue to use older refrigeration technology? But when there were two cars in many Americans' garages— both adding carbon to the greenhouse effect—few if any governmental leaders acknowledged the role of American technology and consumption in contributing to global problems such as climate change, much less took steps to mitigate the problem. Obviously, concern for unecological development in Asia only can become legitimate by an equal concern for the unecological effects of development in other countries, not excluding their prodigal past.

It would seem that a community cannot be built unless and until people come to terms with their past. Unless past actions are redeemed, at least in the sense of remunerating those who have suffered or will suffer because of advantages accrued to present people due to unecological past actions, it is unlikely that a sustainable future will be created. Only when a global community transcends both space and time will a prospect exist that the global ecological crisis will be dealt with effectively.

5.4. ENVIRONMENTAL AND FINANCIAL DEBT

Chapters 33 and 34 of Agenda 21 deal respectively with financial resources and mechanisms to promote sustainable development and environmental protection and with issues relating to the transfer of environmentally sound technology. The transfer of more appropriate technology to mitigate global climate change is, of course, dependent upon the developing nations' acquisition of necessary financial resources. Chapter 33 identifies multilateral development banks and funds, specialized agencies and other United Nations agencies and international organizations, multilateral institutions for capacity-building and technical cooperation, bilateral assistance programs, private funding, investment, innovative financing, and debt relief as the primary sources and means of financial support for implementation of Agenda 21 recommendations. Interestingly, during the Agenda 21 deliberations, the head of the World Bank stated that the bulk of developing countries' investment needs for environmental purposes must come from savings that they achieve through improved economic policies, from private sector sources, and from improved trade, although some recognition was given to the need for increased aid from developed to developing nations (Johnson 1993).

Many developing countries have significant financial debts to other governments or world lending institutions, and many are selling off natural resources with little environmental regulation in order to raise income to finance their debt burden (Goodland et al. 1993). Financial borrowing mortgages the future of the next generation of a group by making them debtors to the creditors of this one. National financial debts are not written off if a government fails or a generation passes. The debtor pays, or the debtor's children, for such financial debts are inherited. The debt burden is forced onto the next generation by international financial agencies. The agencies often justify this by the need to support the international global economic order, which they claim would collapse without such accountability. International financial bodies may reschedule payments or make structural adjustments, but there is no reprieve from such debt—there is no free lunch.

Financial borrowing, then, is living beyond one's financial means, but there is an ecological parallel. There is an ecological borrowing, which involves living beyond the limits of one's ecological resources—that is, utilizing natural resources at a rate that exceeds their rate of regeneration, externalizing costs, polluting the global commons, and incurring a debt with nature that future generations will have to pay for. This situation is tantamount to a Faustian bargain between humanity and nature that leaves little possibility of appealing for debt relief, rescheduling, or default (Korten 1992).

If a financial debt is to be taken seriously, as it is by lending governments and international agencies (in other words, the debtor must pay), then why should environmental debts not be taken just as seriously (in other words, the polluter must pay)? If there is no such thing as an economically free lunch for anyone, why is it that there seems to be an ecologically free dinner for some? Why should not structural adjustments be made for past polluters to help them undo the damage done by the pollution they have caused and thus repay the environmental debt that they owe to the global community, especially the poor, who suffer most from such environmental degradation? Repayment of environmental debts by the rich are unlikely to the extent that decisions are based on political and economic power as opposed to ethical reasoning, because the poor of this world have little bargaining power in the international political arena and economic markets.

One way of paying an environmental debt would be the transfer of technology and resources to the less developed countries from the more developed ones responsible for past pollution. This could be a feasible way of reversing the transfer of assets from the less developed to the industrialized countries, as is happening at present and which perpetuates the debt crisis. This could also help the less developed countries to bypass the polluting first stage in the industrialization process, which the present industrialized countries went through, to environmentally cleaner and ecologically more friendly technologies. Such a transfer of technology then is not a matter of aid with all of its political implications but rather a matter of right, of ethical demands, and of ecological urgency. To this extent, the resource transfers could be interpreted as polluters' dues made toward repayment of environmental debts. International agencies could cost the environmental debts of the industrialized countries and suggest how they could be written off against the financial debts of the less developed countries. International agencies have been established to deal with the financial debts of less developed countries. If the global community takes the ecological crisis seriously, then international bodies should be established to deal with the environmental debts of developed nations as well. The creation of such international bodies also would seem to parallel the globalization of the world's economy. If there is to be a single global financial community with greater interdependence, this must in turn call for a single global ecological community with correspondingly greater reciprocity as well.

5.5. ENVIRONMENTAL RIGHTS AND ECOLOGICAL DUTIES

We suggest that the issues raised in this chapter call for the development of a new social contract, not just to enforce legal conventions between nations but also to foster a global community for the global environmental crisis and guarantee further environmental rights for individual persons and local communities. In other words, action is needed not only at the international and national levels but also at the local community level. For the only sound way of building an effective global community is with a bottom-up process, although this may need some top-down facilitation. Of course, this suggestion is hardly new (Uphoff 1982).

Indeed, Gandhi's decentralized logic of a "consociational" democracy of interdependent but self-reliant local communities makes more sound ecological sense than the highly centralized model so prevalent in modern nations. Accordingly, nations should derive some of their authority from local communities, while some of their sovereignty should be yielded to the global community, because the nation-state is too large for effective local community management and too small for effective global management (Agarwal and Narain 1992).

Environmental rights must include not just the right to a clean and productive environment, which is the concern of the rich, but more importantly the right of survival and subsistence with dignity for all persons and communities, which is the preoccupation of the poor (Guha 1988). Further, ecological duties and citizenship responsibilities also must include community obligations at the local, national, and global levels. Legal conventions among nations not founded on human rights and civic duties at more local levels only legalize injustice and institutionalize ecological degradation, which already is creating environmental refugees and may spawn ecological terrorists out of desperation. So also does administrative control that is insensitive to the needs of the underprivileged and the powerless in a country. Indeed, the question of legal liability and/or administrative regulation with regard to environmental issues remains very problematic, especially at the global level (Ghosh and Jaitly 1993).

5.6. PRESENT PERCEPTIONS AND FUTURE PROMISE

The ecological crisis, as exemplified in global climate change, forces us to quest for a community that is equitable, sustainable, and participative, even as it stretches across geographical space and different generations, and increasingly interdependent at the local, national, and global levels. This quest becomes more crucial in our anomic and alienating society that is so unequally divided between the affluent and the impoverished. In fact, it is community that is the answer to both the alienation of poverty and the anomie of affluence (Moltmann 1989). Further, this extensive community also must have its intensive dimensions, embracing the human, the biotic, and the cosmic and even opening to the transcendent. Toward this end, humans must be perceptive to the development of new and appropriate ethical norms and worldviews to serve as guides to mitigate problems such as global climate change. How the world responds to the problem of global climate change inevitably will define our future in irrevocable ways. Indeed, the present is but a parable of promise and anticipation for the future.

6. Conclusion

Agenda 21 recommendations to protect the earth's atmosphere focus, in part, on: (1)strengthening the scientific basis for sustainable management, (2)enhancing scientific understanding, (3)improving long-term scientific assessment, and (4)building up scientific capacity and capability. Importantly, Agenda 21 recommends adoption of a precautionary approach to protect the atmosphere by stating that in the face of threats of irreversible environmental damage, lack of full scientific understanding should not be an excuse for postponing actions that are justified in their own right. It states further that the precautionary approach could provide a basis for policies relating to complex systems that are not yet understood fully and whose consequences of disturbances cannot yet be predicted.

These types of recommendations reflect the scientific uncertainties surrounding the causes of climate change, the rate and magnitude of change, and the consequences of this change to ecosystems and human health. Of course, a recommendation to adopt a precautionary approach creates a public policy dilemma. Scientists are unable to make reasonably accurate predictions of future climate, yet without such predictions, the consequences of climate change and the societal responses and alternative choices of action cannot be assessed fully. Yet it is known that there is a substantial risk of climate change and that the consequences will affect countries and regions differently, as well as future generations. Thus, from the scientific understanding of the problem of climate change, including the problem of uncertainty, flows the ethical problems of: (1)whether to take action to mitigate the problem despite the uncertainties, or whether to delay action until more information is obtained; and (2)how to distribute risks and burdens both spatially and temporally. In addition, while Agenda 21 recommends the use of cost-benefit analysis and other forms of economic valuation and methods to assess the consequences of climate change, it must be recognized that such analyses and methods are value-laden and do not take into account sufficiently how costs and benefits are distributed, including across generations. Further, they do not take into account the protection of nonhuman nature adequately. Thus, their application should be viewed as an ethical problem requiring analysis and resolution lest decisions be based on economics alone and not on ethical reasoning.

The adoption of a precautionary approach as suggested by Agenda 21 would seem to be most consistent with reducing human health and environmental risks, would be based upon ethical reasoning as opposed to economic considerations and political power among countries, and would favor protecting developing countries that are least able to bear the costs of climate change. If the precautionary approach is adopted, especially by the developed countries, governmental, corporate, and personal behaviors regarding energy use and consumption would have to change in order to lessen the risk of climate change and its consequences. From the perspective of developing countries based upon ethical reasoning as opposed to decisionmaking based upon traditional economic analysis and political power, developed countries would be required to pay a so-called environmental debt caused by their historical emissions of greenhouse gases in the name of their economic development to developing countries in the form of technology transfer and debt relief in order that the latter countries would be able to provide for an appropriate quality of life for their people on a sustainable basis. This approach also would require a reduction in the per capita consumption of greenhouse gases emitted by developed countries as well as limits on population growth in the most heavily populated and affluent countries, respectively. Clearly, all of these problems cannot be understood or resolved unless the relationships among scientific, economic, and ethical knowledge and methods are understood.

Fundamentally, the problem of climate change is a problem of global ecology, but not solely in the strict scientific sense. The word *ecology* is derived from the Greek *oikos*, meaning "home" or "dwelling." In fact, ecology is all about being at home in our world, but today many humans seem to be homeless and neither at peace with themselves nor in harmony with their environment. As Ward and Dubos (1983) have pointed out, as a community of nations, we are not as yet a civilized world, even though we all have only one earth to share and care for. But to solve problems such as global climate change, the earth must truly become a common home in which an acceptable quality of life and dignity are provided for all, humans and nonhumans, a home in which we all share the promise of our common future together.

7. References

Adamowicz, W.L., V. Bhardwaj, and B. Macnab. 1993. Willingness to Pay and Willingness to Accept Compensation. *Land Economics* 69(4): 416-427.

Adams, R.M. 1989. Global Climate Change and Agriculture: An Economic Perspective. *American Journal of Agricultural Economics* 71(5): 1272-1279.

Agarwal, A., and S. Narain. 1991. *Global Warming in an Unequal World.* Centre for Science and Environment, New Delhi.

Agarwal, A., and S. Narain. 1992. *Towards a Green World: Should Global Environmental Management Be Built on Legal Conventions or Human Rights?* Centre for Science and Environment, New Delhi.

Angermeier, P.L., and J.R. Karr. 1994. Biological Integrity Versus Biological Diversity as Policy Directives. *BioScience* 44: 690-697.

Ayres, R.U., and J. Walter. 1991. The Greenhouse Effect: Damages, Costs and Abatement. *Environmental and Resource Economics* 1(3): 237-270.

Baes, C.F., Jr., H.E. Goeller, J.S. Olson, and R.M. Rotty. 1976. The Global Carbon Problem. ORNL-5194, Oak Ridge National Laboratories, Oak Ridge, TN.

Barry, B. 1983. Intergenerational Justice in Energy Policy. In *Energy and the Future*, D. Maclean and P.G. Brown, eds. Rowan and Allanheld, Totowa, NJ, pp. 15-30.

Batie, S.S., and H.H. Shugart. 1989. The Biological Consequences of Climate Changes: An Ecological and Economic Assessment. In *Greenhouse Warming: Abatement and Adaptation*, N.J. Rosenberg, W.E. Easterling, P.R. Crosson, and J. Darmstadter, eds. Resources for the Future, Washington, DC, pp. 121-132.

Bennett, J. 1981. Morality and Consequences. In *The Tanner Lectures on Human Values,* Vol. II, S. McMurrin, ed. University of Utah Press, Salt Lake City. pp. 45-116.

Bentham, J. (1843) 1954. The Psychology of Economic Man. *The Economic Writings of Jeremy Bentham,* Vol. 3, W. Stark, ed. Allen and Unwin, London, pp. 421-450.

Bentley, C.R. 1980. Response of the West Antarctic Ice Sheet to CO_2-Induced Climatic Warming. Report of DOE-sponsored meeting at Orono, ME, June.

Bergman, L. 1991. General Equilibrium Effects of Environmental Policy: A Cge-Modelling Approach. *Environmental and Resource Economics* 1(1): 43-62.

Birch, C. 1976. Creation, Technology and Human Survival. *Ecumenical Review* (January): 289.

Braden, J., and Kolstad C. 1991. *Measuring the Demand for Environmental Quality.* Elsevier North Holland, Amsterdam.

Brown, A.W. 1977. Yellow Fever, Dengue and Dengue Hemorrhagic Fever. In *A World Geography of Diseases,* G.M. Howe, ed. Academic Press, London.

Brown, L., H. Kane, and D. Roodman. 1994. *Vital Signs 1994.* W.W. Norton & Company, New York.

Cairns, J., Jr., and B.R. Niederlehner. 1993. Ecological Function and Resilience: Neglected Criteria for Environmental Impact Assessment and Ecological Risk Analysis. *The Environmental Professional* 15: 116-124.

Care, N.S. 1982. Future Generations, Public Policy, and the Motivation Problem. *Environmental Ethics* 4: 195-214.

CCTA/WHO. 1963. *African Conference on Ancyclostomiasis.* Technical Report Serial No. 255, World Health Organization, Geneva.

Chameides, W.L., P.S. Kasibhatla, J. Yienger, and H. Levy II. 1994. Growth of Continental-Scale Metro-Agro-Plexes, Regional Ozone Pollution, and World Food Production. *Science* 264: 74-77.

Clark, J.S. 1991. Ecosystem Sensitivity to Climate Change and Complex Responses. In *Global Climate Change and Life on Earth*, R.L. Wyman, ed. Routledge, Chapman and Hall, New York, pp. 65-98.

Clark, W.C., ed. 1982. *Carbon Dioxide Review: 1982.* Oxford University Press, New York.

Cline, W.R. 1992. *The Economics of Global Warming.* Longman, Harlow, Essex.

Cohn, J.P. 1989. Gauging the Biological Impacts of the Greenhouse Effect. *BioScience* 39: 142-146.

Cousteau Society. 1991. Protecting the Rights of Future Generations. *Calypso Log*, August.

Crosson, P.R. 1989. Climate Change: Problems of Limits and Policy Responses. In *Greenhouse Warming: Abatement and Adaptation*, N.J. Rosenberg, W.E. Easterling, P.R. Crosson, and J. Darmstadter, eds. Resources for the Future, Washington, DC, pp. 69-82.

Cumberland, J.H., J.R. Hibbs, and I. Hoch, eds. 1982. *The Economics of Managing Chlorofluorocarbons: Stratospheric Ozone and Climate Issues.* Johns Hopkins Press, Baltimore.

Daily, G.C., and P.R. Ehrlich. 1990. *An Exploratory Model of the Impact of Rapid Climate Change on the World Food Situation.* The Morrison Institute for Population and Resource Studies, Stanford University, Stanford, CA.

Daily, G.C., P.R. Ehrlich, H.A. Mooney, and A.H. Ehrlich. 1991. Greenhouse Economics: Learn Before You Leap. *Ecological Economics* 4: 1-10.

d'Arge, R.C., ed. 1975. Economic and Social Measures of Biological and Climatic Change, Vol. 6, Climate Impact Assessment Program. U.S. Department of Transportation, Washington, D.C.

d'Arge, R.C., and C.L. Spash. 1991. Economic Strategies for Mitigating the Impacts of Climate Change on Future Generations. In *Ecological Economics: The Science and Management of Sustainability*, R. Costanza, ed. Columbia University Press, New York, pp. 367-383.

Dave, J.M. 1988. *Policy Options for Development in Response to Global Atmospheric Changes: Case Study for India for Greenhouse Effect Causes.* Nehru University, New Delhi.

Devall, B., and G. Sessions. 1985. *Deep Ecology.* Peregrine Smith Books, Layton, UT.

Dudek, D.J. 1991. The Nexus of Agriculture, Environment, and the Economy Under Climate Change. In *Global Climate Change and Life on Earth*, R.L. Wyman, ed. Routledge, Chapman and Hall, New York, pp. 180-200.

Ehrlich, P.R. 1989. Attributes of Invaders and the Invading Processes: Vertebrates. In *Biological Invasions: A Global Perspective*, J.A. Drake, J.A. Mooney, F. di Castri, R.H. Groves, F.J. Kruger, M. Rejmanek, and M. Williamson, eds. Wiley, New York, pp. 315-328.

Ehrlich, P.R., and A.H. Ehrlich. 1990. *The Population Explosion.* Simon and Schuster, New York.

Ehrlich, P.R., and E.O. Wilson. 1991. Biodiversity Studies: Science and Policy. *Science* 253: 758-762.

El-Sayed, M. Kh. 1991. Implications of Climate Change for Coastal Areas Along the Nile Delta. *The Environmental Professional* 13: 59-65.

(EPA) Environmental Protection Agency. 1987. Assessing the Risks of Trace Gases That Can Modify the Stratosphere. EPA 400/1-87/001A-H. Government Printing Office, Washington, DC.

Faber, M., R. Manstetten, and J.L.R. Proops. 1992. Humankind and the Environment. *Environmental Ethics* 1(3): 217-241.

Fitzroy, F. 1992. Economic Aspects of Global Warming: A Comment. *Green Values: Scottish Environmental Economics Discussion Group Newsletter* 5, Stirling, Scotland.

Flavin, C., and N. Lenssen. 1990. *Beyond the Petroleum Age: Designing a Solar Economy.* Worldwatch Institute, Washington, DC.

Flohn, H. 1982. Climate Change and an Ice-Free Arctic Ocean. In *Carbon Dioxide Review: 1982,* W. C. Clark, ed. Oxford University Press, New York, pp. 143-179.

Freeman, A.M. 1993. *The Measurement of Environmental and Resource Values: Theory and Methods.* Resources for the Future, Washington, DC.

Fujii, Y. 1990. *An Assessment of the Responsibility for the Increase in the CO_2 Concentration and Inter-Generational Carbon Accounts.* International Institute for Applied Systems Analysis, Laxenburg, Austria.

George, S. 1992. *The Debt Boomerang: How Third World Debt Harms Us All.* Westview Press, Boulder, CO.

Ghosh, P., and A. Jaitly. 1993. Legal Liability Versus Administrative Regulation: The Problem of Institutional Design in Global Environmental Policy. In *The Road From Rio,* P. Ghosh and A. Jaitly, eds. Tata Energy Research Institute, New Delhi, pp. 193-213.

Goodland, R.J.A., H.E. Daly, and S. El Serafy. 1993. The Urgent Need for Rapid Transition to Global Environmental Sustainability. *Environmental Conservation* 20: 297-309.

Goodpaster, K. 1994. On Being Morally Considerable. In *Reflecting on Nature: Readings in Environmental Philosophy,* L. Gruen and D. Jamieson, eds. Oxford University Press, New York, pp. 55-64.

Gornitz, V., and S. Lebedeff. 1987. Global Sea Level Changes During the Past Century. In *Sea Level Fluctuation and Coastal Evolution,* Special Publication 41, D. Nummedal, O. Pilkey, and J. Howard, eds. The Society of Economic Paleontologists and Mineralogists, Tulsa, OK, pp. 3-16.

Gravelle, H., and R. Rees. 1992. *Microeconomics.* Longmans, Harlow, Essex.

Gruen, L., and D. Jamieson, eds. 1994. *Reflecting on Nature: Readings in Environmental Philosophy.* Oxford University Press, New York.

Guha, R. 1988. Ideological Trends in Indian Environmentalism. *Economics and Political Weekly* 22(49): 2578-2581.

Hanley, N., and C.L. Spash. 1993. *Cost-Benefit Analysis and the Environment.* Edward Elgar, Aldershot, UK.

Hansen, J., and S. Lebedeff. 1987. Global Trends of Measured Surface Air Temperature. *Journal of Geophysical Research* 92: 133345-13372.

Hansen, J.E., I. Fung, A. Lacis, D. Rind, S. Lebedeff, R. Ruedy. and G. Russell. 1988. Global Climate Changes as Forecast by Goddard Institute for Space Studies Three-Dimensional Model. *Journal of Geophysical Research* 93: 9341-9364.

Hansen, J., D. Rind, A. Del Genio, A. Lacis, S. Lebedeff, M. Prather, R. Ruedy, and T. Karl. 1989. Regional Greenhouse Climate Effects. In *Proceedings of the Second North American Conference on Preparing for Climate Change.* Climate Institute, Washington, DC, pp. 68-81.

Harrison, P. 1990. Too Much Life on Earth? *New Scientist* 126: 28-29.

Hoeller, P., A. Dean, and J. Nicolaisen. 1991. Macroeconomic Implications of Reducing Greenhouse Gas Emissions: A Survey of Empirical Studies. *OECD Economic Studies* 16: 45-78.

Hoffman, J., D. Keyes, and J.G. Titus. 1983. *Projecting Future Sea-Level Rise: Methodology, Estimates to the Year 2100, and Research Needs*. Environmental Protection Agency, Washington, DC.

Hogan, K.B., J.S. Hoffman, and A.M. Thompson. 1991. Methane on the Greenhouse Agenda. *Nature* 354: 181-182.

Houghton, R.A. 1991. The Role of Forests in Affecting the Greenhouse Gas Composition of the Atmosphere. In *Global Climate Change and Life on Earth*, R.L. Wyman, ed. Routledge, Chapman and Hall, New York, pp. 43-55.

Houghton, R.A., and D.L. Skole. 1990. Changes in the Global Carbon Cycle Between 1700 and 1985. In *The Earth Transformed by Human Action*, B.L. Turner, ed. Cambridge University Press, Cambridge, UK, pp. 43-55.

(IEA) International Energy Authority. 1989. Energy and the Environment. International Energy Authority, Paris.

(IGBP) International Geosphere-Biosphere Programme. 1990. *Global Change, Report 12*. International Council of Scientific Unions, Stockholm, Sweden.

Ingham, A., and A. Ulph. 1991. Carbon Taxes and the UK Manufacturing Sector. In *Environmental Policy and the Economy*, F. Dietz, F. Van Der Ploeg, and J. Van Der Straaten, eds. Elsevier, Amsterdam, pp. 197-240.

(IPCC) International Panel on Climate Change. 1990. *Climate Change: The IPCC Scientific Assessment*, J.T. Houghton, G.J. Jenkins, and J.J. Ephraums, eds. Cambridge University Press, Cambridge, UK.

(IPCC) International Panel on Climate Change. 1992. *Climate Change 1992: The Supplemental Report to the IPCC Scientific Assessment*, J.T. Houghton, G.J. Jenkins, and J.J. Ephraums, eds. Cambridge University Press, Cambridge, UK.

Jamieson, D. 1992. Ethics, Public Policy, and Global Warming. *Science, Technology and Human Values* 17:139-153.

Jamieson, D. 1994. Global Environmental Justice. In *Philosophy and the Natural Environment*, R. Attfield and A. Belsey, eds. Cambridge University Press, Cambridge, UK. pp. 199-210.

Johnson, S.P. 1993. *The Earth Summit, The United Nations Conference on Environment and Development (UNCED)*. Graham & Trotman/Martinus Nijhoff, London.

Kerr, R.A. 1993. No Way to Cool the Ultimate Greenhouse. *Science* 262: 648.

Kerr, R.A. 1994. Did Pinatubo Send Climate Warming Gases into a Dither? *Science* 263: 1562.

Keynes, J.M. (1921) 1973. *A Treatise on Probability*. Macmillan, London.

Klein, W.H. 1982. Detecting Carbon Dioxide Effects on Climate. In *Carbon Dioxide Review: 1982*, W.C. Clark, ed. Oxford University Press, New York, pp. 213-241.

Knetsch, J. 1990. Environmental Policy Implications of the Disparity Between Willingness to Pay and Compensation Demanded. *Journal of Environmental Economics and Management* 18: 227-237.

Korten, D.C. 1992. *Getting to the Twenty-First Century: Voluntary Action and the Global Agenda*. Oxford University Press, New Delhi.

Lashof, D.A., and D.R. Ahuja. 1990. Relative Contributions of Greenhouse Gas Emissions to Global Warming. *Nature* 344: 529.

Lashof, D.A., and D.A. Tirpak. 1989. Policy Options for Stabilizing Global Climate. U.S. Environmental Protection Agency, Washington, DC.

Leatherman, S.P. 1991. Climate-Induced Sea Level Rise. In *Global Climate Change and Life on Earth*, R.L. Wyman, ed. Routledge, Chapman and Hall, New York, pp. 170-179.

Lemons, J. 1985. Carbon Dioxide and the Environment: A Problem of Uncertainty. *The Journal of Environmental Sciences* 28: 60-66.

Lemons, J., ed. 1991. Preparing for Climate Change. *The Environmental Professional* 13: 1-92.

Lodge, D.M. 1993. Species Invasions and Deletions: Community Effects and Responses to Climate and Habitat Change. In *Biotic Interactions and Global Change*, P.M. Kareiva, J.G. Kingsolver, and R.B. Huey, eds. Sinauer Associates, Sunderland, MA, pp. 367-387.

Longstreth, J.A. 1990. Human Health. In *The Potential Effects of Global Climate Change on the United States*, J.B. Smith and D.A. Tirpak, eds. Hemisphere Publishing Corporation, New York, pp. 525-556.

MacLean, D. 1983. A Moral Requirement for Energy Policy. In *Energy and the Future*, D. MacLean and P. Brown, eds. Rowman and Littlefield, Totowa, NJ. 180-196.

Malinvaud, E. 1972. *Lectures on Microeconomic Theory*. North Holland, Amsterdam.

Manabe, S., and R.J. Stouffer. 1993. Century-Scale Effects of Increased Atmospheric CO_2 on the Ocean-Atmosphere System. *Nature* 364: 215-218.

Manne, A., and R. Richels. 1991. Global CO_2 Emission Reductions: The Impacts of Rising Energy Costs. *The Energy Journal* 12(1): 87-102.

Meier, M.F. 1990. Reduced Rise in Sea Level. *Nature* 343: 115-116.

Meier, C., and A. Randall. 1991. Use Value Under Uncertainty. *Land Economics* 67(4): 379-389.

Meyers, N. 1993. The Questions of Linkages in Environment and Development. *BioScience* 43: 302-310.

Mitchell, J.M. 1982. Commentary (a reply to H. Flohn). In *Carbon Dioxide Review: 1982,* W.C. Clark, ed. Oxford University Press, New York, p. 180.

Moltmann, J. 1989. *Creating a Just Future: The Politics of Peace and the Ethics of Creation in a Threatened World.* SCM Press, Philadelphia.

Nash, R.F. 1989. *The Rights of Nature: A History of Environmental Ethics*. University of Wisconsin Press, Madison.

Nordhaus, W. 1982. How Fast Should We Graze the Global Commons? *American Economic Review Association Papers and Proceedings* 72(2): 242-246.

Nordhaus, W. 1991a. To Slow or Not to Slow: The Economics of the Greenhouse Effect. *Economic Journal* 101: 920-938.

Nordhaus, W. 1991b. A Sketch of the Economics of the Greenhouse Effect. *American Economic Review* 81(2): 146-150.

Novelli, P.C., K.A. Masarie, P.P. Tans, and P.M. Lang. 1994. Recent Changes in Atmospheric Carbon Monoxide. *Science* 263: 1587-1589.

Nozick, R. 1974. *Anarchy, State, and Utopia*. Basic Books, New York.

Oppenheimer, M., and R.H. Boyle. 1990. *Dead Heat: The Race Against the Greenhouse Effect*. Basic Books, New York.

Pace, M.L. 1993. Forecasting Ecological Responses to Global Change: The Need for Large-Scale Comparative Studies. In *Biotic Interactions and Global Change*, P.M. Kareiva, J.G. Kingsolver, and R.B. Huey, eds. Sinauer Associates, Sunderland, MA, pp. 356-366.

Parry, M. 1990. *Climate Change and World Agriculture*. Earthscan Publications, London.

Peters, R.L., J.J. Armesto, B. Boeken, J.J. Cole, C.T. Driscoll, D.M. Duarte, T.M. Frost, J.P. Grime, J. Kolasa, E. Prepas, and W.G. Sprules. 1991. On the Relevance of Comparative Ecology to the Larger Field of Ecology. In *Comparative Analysis of Ecosystems*, J. Cole, G. Lovett, and S. Findlay, eds. Springer-Verlag, New York, pp. 46-64.

Pimentel, D. 1989. Ecological Systems, Natural Resources, and Food Supplies. In *Food and Natural Resources*, D. Pimentel and C.W. Hall, eds. Academic Press, San Diego, pp. 1-29.

Quine, W. 1960. *Word and Object*. MIT Press, Cambridge.

Ramamurthy, A. 1986. Gandhian Perspective on the Reconstruction of Indian Polity. In *Contemporary Crisis and Gandhi*, R. Ramashray, ed. Discovery Publications, Delhi, pp. 146-164.

Rawls, J. 1971. *A Theory of Justice*. Harvard University Press, Cambridge, MA.

Rogot, E., and S.J. Padgett. 1976. Associations of Coronary and Stroke Mortality with Temperature and Snowfall in Selected Areas of the United States 1962-1966. *American Journal of Epidemiology* 103: 565-575.

Rolston, H., III. 1988. *Environmental Ethics: Duties to and Values in the Natural World.* Temple University Press, Philadelphia.

Root, T.L. 1988. Environmental Factors Associated with Avian Distributional Boundaries. *Journal of Biogeography* 15: 489-505.

Root, T.L., and S.H. Schneider. 1993. Can Large-Scale Climatic Models Be Linked with Multiscale Ecological Studies? *Conservation Biology* 7: 256-270.

Ryther, J.H. 1969. Photosynthesis and Fish Production in the Sea. *Science* 166: 72-76.

Sagoff, M. 1988. *The Economy of the Earth*. Cambridge University Press, Cambridge, UK.

Schlesinger, M.E., and Z.-C. Zhao. 1989. Seasonal Climatic Change Induced by Doubled CO_2-Induced Climate Change: Preliminary Results from the OSU Coupled Atmosphere-Ocean General Circulation Model. In *Coupled Ocean-Atmosphere Models*, J.C.J. Nihoul, ed. Elsevier Oceanographic Series 40: 447-478.

Schneider, S.H. 1991. Three Reports of the Intergovernmental Panel on Climate Change. *Environment* 33: 25-30.

Schneider, S.H. 1992. Introduction to Climate Modeling. In *Climate System Modeling*, K.E. Trenberth, ed. Cambridge University Press, Cambridge, UK, pp. 3-26.

Schneider, S.H., P.H. Gleick, and L.O. Mearns. 1990. Prospects for Climate Change. In *Climate Change and U.S. Water Resources*, P.E. Waggoner, ed. John Wiley, New York, pp. 41-74.

Schwartz, T. 1978. Obligations to Posterity. In *Obligations to Future Generations*, R. Sikora and B. Barry, eds. Temple University Press, Philadelphia, pp. 3-13.

Sidgwick, H. 1907. *The Methods of Ethics.* Macmillan, London.

Singer, P. 1975/1990. *Animal Liberation*. Avon Books, New York.

Singh, M. 1993. Valedictory Address. In *The Road from Rio,* P. Ghosh and A. Jaitly, eds. Tata Energy Research Institute, New Delhi, pp. 15-22.

Smith, K.R. 1991. Allocating Responsibility for Global Warming: The Natural Debt. *Ambio* 20: 95-96.

Spash, C.L. 1993. Economics, Ethics and Long-Term Environmental Damages. *Environmental Ethics* 15(2): 117-132.

Spash, C.L. 1994. Double CO_2 and Beyond. *Ecological Economics* 10(1): 27-36.

Spash, C.L., and A. Clayton. In press. The Maintenance of Natural Capital: Motivations and Methods. *Environmental Values*.

Spash, C.L.,, and R.C. d'Arge. 1989. Compensation of Future Generations for Adverse Climatic Changes. Presented at the Western Economic Association International Meeting, Lake Tahoe, CA, June 18-22, 37pp.

Spash, C.L., and N. Hanley. 1994a. Cost-Benefit Analysis and the Greenhouse Effect. Ecological Economics Discussion Paper 94/6, Environmental Economics Research Group, Economics Department, University of Stirling, Scotland.

Spash, C.L., and N. Hanley. 1994b. Preferences, Information and Biodiversity Preservation. Ecological Economics Discussion Paper 94/1, Environmental Economics Research Group, Economics Department, University Of Stirling, Scotland.

Spash, C.L., and A. Young. In press. Environmental Imperatives and Renewable Sources of Energy. In *Global Oil Markets*, S. Shojai, ed. Praeger Publishers, New York.

Stevens, T., J. Echevarria, R. Glass, T. Hager, and T. More. 1991. Measuring the Existence Value of Wildlife: What Do Cvm Estimates Really Show? *Land Economics* 67(4): 390-400.

Streets, D. 1990. *Equity and Equality in the Greenhouse*. Argonne National Laboratory, Argonne, IL.

Sulzman, E.W., K.A. Poiani, and T.G.F. Kittel. In press. Modeling Human-Induced Climatic Change. *Environmental Management*.

Taylor, P. 1986. *Respect for Nature*. Princeton University Press, Princeton, NJ.

Tickell, C. 1992. The Quality of Life: What Quality? Whose Life? *Interdisciplinary Science Reviews* 17: 19-25.

Tolba, M.K. 1991. Preparing for Climate Change. *The Environmental Professional* 13: 2-7.

Trenberth, K.E., ed. 1992. *Climate System Modeling*. Cambridge University Press, Cambridge, UK.

(UNPF) United Nations Population Fund. 1991. *Population, Resources and the Environment*. UNFPA, New York.

Uphoff, N.T., ed. 1982. *Rural Development and Local Organization in Asia, Vol. I, Introduction and South Asia*. Macmillan, Delhi, 1982.

Waggoner, P.E. 1983. Agriculture and a Climate Changed by More Carbon Dioxide. In *Changing Climate*. National Academy Press, Washington, D.C.

Walker, J., and J. Kasting. 1992. Effects of Fuel and Forest Conservation on Future Levels of Atmospheric Carbon Dioxide. *Global Planet Change* 5: 151-189.

Wang, W.-C., M.P. Dudek, X.-Z. Liang, and J.T. Kiehl. 1991. Inadequacy of Effective CO_2 as a Proxy in Simulating the Greenhouse Effect of Other Radiatively Active Gases. *Nature* 350: 573-577.

Ward, B., and R. Dubos. 1983. *Only One Earth: The Care and Maintenance of a Small Planet*. Norton, New York.

Washington, W.M. 1992. Climate-Model Responses to Increased CO_2 and Other Greenhouse Gases. In *Climate System Modeling*, K.E. Trenberth, ed. Cambridge University Press, Cambridge, UK, pp. 643-668.

Washington, W.M., and G.A. Meehl. 1984. Seasonal Cycle Experiment on the Climate Sensitivity Due to a Doubling of CO_2 with an Atmosphere General Circulation Model Coupled to a Simple Mixed-Layer Ocean Model. *Journal of Geophysical Research* 89: 9475-9503.

Watt, W.B. 1992. Eggs, Enzymes and Evolution—Natural Genetic Variants Change Insect Fecundity. *Proceedings of the National Academy of Sciences*. National Academy Press 89: 10608-10612.

Webb, T., III. 1992. Past Changes in Vegetation and Climate: Lessons for the Future. In *Global Warming and Biological Diversity*, R.L. Peters and T.E. Lovejoy, eds. Yale University Press, New Haven, CT, pp. 59-75.

Weihe, W. 1979. Climate, Health, and Disease. In *Proceedings of the World Climate Conference*, WMO No. 537. World Meteorological Organization, Geneva, pp. 313-368.

Weiss, E. 1988. *In Fairness to Future Generations: International Law, Common Patrimony, and Intergenerational Equity*. Transnational Publishers, Dobbs Ferry, NY.

Wetherald, R.T., and S. Manabe. 1986. An Investigation of Cloud Cover Change in Response to Thermal Forcing. *Climate Change* 10: 11-42.

White, J.C., ed. 1989. *Global Climate Change Linkages*. Elsevier, New York.

Whiteman, A. 1991. A Comparison of the Financial and Non-Market Costs and Benefits of Replanting Lowland Forests. Presented to the Scottish Environmental Economics Discussion Group, Stirling, Scotland.

WHO. 1977. *Infectious Diseases: Cases and Deaths*. World Health Organization, Geneva.

Willig, R. 1976. Consumers' Surplus Without Apology. *American Economic Review* 66: 589-597.

Wilson, C.A., and J.F.B. Mitchell. 1987. A Doubled CO_2 Climate Sensitivity Experiment with a Global Climate Model Including a Simple Ocean. *Journal Geophysical Research* 92: 13315-13343.

WMO. 1993. *Scientific Assessment of Stratospheric Ozone, 1991*. World Meteorological Organization, Geneva.

Woodwell, G.M., R.H. Whittaker, W.A. Reilners, G.E. Likens, C.C. Delwiche, and D.B. Botkin. 1978. The Biota and the World Carbon Budget. *Science* 199: 141-146.

World Bank. 1992. *World Development Report 1992: Development and Environment*. Oxford University Press, New York.

(WRI) World Resources Institute. 1992. *World Resources 1992-1993: A Guide to the Global Environment*. Oxford University Press, New York.

Wynne, B. 1992. Uncertainty and Environmental Learning. *Global Environmental Change*: 111-127.

Chapter 8
PROTECTION OF MARINE AND FRESHWATER RESOURCES

Larry Canter[1]
Konrad Ott[2]
Donald A. Brown[3]

1. Scientific Issues in Sustainable Water Resource Programs

1.1. INTRODUCTION

The United Nations Conference on Environment and Development (UNCED) held in Rio de Janeiro, Brazil, in June 1992 led to several international agreements and protocols. One resulting document, Agenda 21, included program actions for various natural resources to guide the international community in environmental management and sustainable development. Agenda 21 recognizes that human activities can affect physical, biological, and socioeconomic dimensions of the environment; such effects need to be incorporated in strategies for resource management.

Both marine and freshwater resources are essential components of the earth's hydrosphere, and they are interconnected with both aquatic and terrestrial ecosystems. Society depends on these hydrological resources for many things, including water supply, navigation, food supplies, and recreation. Agenda 21 included marine and freshwater resources as environmental components needing rational planning and use, including resource protection. The general objective is to make certain that adequate supplies of water of good quality are maintained for the entire population of this planet, while preserving the hydrological, biological, and chemical functions of ecosystems, adapting human activities within the capacity limits of nature, and combating vectors of water-related diseases (United Nations 1992: 334). Of increasing importance are transboundary water resources, including freshwater, coastal, and marine resources.

Protection and management of marine and freshwater resources require the unprecedented understanding of numerous complex scientific and technical issues in addition to legal, policy, and ethical issues. This chapter addresses scientific and technical issues related to the Agenda 21 program for marine and freshwater resources. Emphasis is given to freshwater resources, although the enunciated principles and concepts can also be applied to marine resources. While much is known about the scientific basis of water resources management, numerous uncertainties are related to many specific topical issues.

---

[1]Environmental and Groundwater Institute, 200 Felgar St., Norman, OK 73019, U.S.A.; [2]Center for Ethics in Science and Humanities, University of Tubingen, D-7400 Tubingen, Germany; [3]Bureau of Hazardous Sites and Superfund Enforcement, Pennsylvania Department of Environmental Resources, Commonwealth of Pennsylvania, 400 Market St., Harrisburg, PA 17101-2301, U.S.A.

J. Lemons and D. A. Brown (eds.), Sustainable Development: Science, Ethics, and Public Policy, 158–214.

This chapter begins with a brief review of background information related to the hydrological cycle and pollution; this is followed by a review of seven program areas for freshwater resources as included in Agenda 21. Scientific and technical uncertainties related to water resources planning and assessment are then described, along with specific examples related to human health and climate change. Next, a 12-element water resources management strategy is presented as a means to integrate the Agenda 21 program areas. A concluding section deals with ethical issues that must be faced in water resource management programs.

1.2. BACKGROUND INFORMATION ON FRESHWATER RESOURCES

When considering surface water quantity and/or quality, it is important to understand the processes that create surface water bodies (rivers, streams, lakes, etc.). Surface water is replenished by rainfall that ends up in runoff and groundwater that discharges into a surface water. Rainfall can infiltrate into the subsurface, be intercepted by foliage (initial abstraction), or result in runoff. The rainfall may subsequently evapotranspirate (evaporate naturally or through vegetative growth), enter the groundwater, and/or result in surface water flow. The runoff flows downgradient (usually from higher to lower elevation) into creeks, streams, lakes, and rivers and eventually into the oceans (unless it evaporates, infiltrates, or is withdrawn along the way). The rainfall that infiltrates into the subsurface and becomes groundwater may discharge into a surface water at some other location; the surface water is referred to as a receiving stream, and the discharging groundwater is referred to as baseflow. Baseflow accounts for the flow in streams, rivers, and so on between rainfall/runoff events. Once the rainfall has reached the oceans (or anytime between), evaporation can return the runoff to the atmosphere for a future rainfall event (likely at some other location), and the cycle starts over again; thus, the system is referred to as the hydrologic cycle.

Surface water pollution can be defined in a number of ways; however, the basic features of most definitions address excessive concentrations of particular substances for sufficient periods of time to cause identifiable effects. Water quality can be defined via the physical, chemical, and biological characterization of the water. Physical parameters include color, odor, temperature, solids (residues), turbidity, oil, and grease. The characterization of solids can be further subdivided into suspended and dissolved solids (size and settleability) as well as organic (volatile) and inorganic (fixed) fractions. Chemical parameters associated with the organic content of water include biochemical oxygen demand (BOD), chemical oxygen demand (COD), total organic carbon (TOC), and total oxygen demand (TOD). It should be noted that BOD is a measure of the organics present in the water; it is determined by measuring the oxygen necessary to biostabilize the organics (the oxygen equivalent of the biodegradable organics present). Inorganic chemical parameters include salinity, hardness, pH, acidity, alkalinity, iron, manganese, chlorides, sulfates, sulfides, heavy metals (mercury, lead, chromium, copper, and zinc), nitrogen (organic, ammonia, nitrite, and nitrate), and phosphorus. Bacteriological parameters include coliforms, fecal coliforms, specific pathogens, and viruses.

In evaluating surface water pollution impacts associated with the construction and operation of a potential development project, two main sources of water pollutants should be considered: nonpoint and point sources. Nonpoint sources are also referred to as area or diffuse sources. Nonpoint pollutants refer to those substances that can be introduced into receiving waters as a result of urban area, industrial area, or rural runoff—e.g., sediment, pesticides, or nitrates entering a surface water due to surface runoff from agricultural farms. Point sources are related to specific discharges from municipalities or industrial complexes—e.g., organics or metals entering a surface water due to wastewater discharge from a manufacturing plant. In a given surface water body, nonpoint pollution can be a significant contributor to the total pollutant loading, particularly with regard to nutrients and pesticides.

The effects of pollutants on receiving water quality are manifold and dependent upon the type and concentration (Nemerow and Dasgupta 1991). Soluble organics, as represented by high BOD wastes, cause depletion of oxygen in the surface water. This can result in fish kills, the growth of undesirable aquatic life, and undesirable odors. Trace quantities of certain organics cause undesirable taste and odors, and certain organics may be biomagnified in the aquatic food chain. Suspended solids decrease water clarity and hinder photosynthetic processes; if solids settle and form sludge deposits, changes in benthic ecosystems result. Color, turbidity, oils, and floating materials are of concern due to their aesthetic undesirability and possible influence on water clarity and photosynthetic processes. Excessive nitrogen and phosphorus can lead to algal overgrowth with concomitant water treatment problems resulting from algae decay and interferences with treatment processes. Chlorides cause a salty taste to be imparted to water, and in sufficient concentration limitations on water usage can occur. Acids, alkalies, and toxic substances have the potential for causing fish kills and creating other imbalances in stream ecosystems. Thermal discharges can also cause imbalances as well as reductions in stream waste assimilative capacity. Stratified flows from thermal discharges minimize normal mixing patterns in receiving streams and reservoirs. Table 1 summarizes the impacts of certain pollutants in relation to impairments to the uses of the water (Chapman 1992).

Table 1. Limits of Water Uses Due to Water Quality Degradation (Chapman 1992).

	Use						
Pollutant	**Drinking Water**	**Aquatic Wildlife, Fisheries**	**Recreation**	**Irrigation**	**Industrial Uses**	**Power and Cooling**	**Transport**
Pathogens	xx	0	xx	x	xx[1]	na	na
Suspended solids	xx	xx	xx	x	x	x[2]	xx[3]
Organic matter	xx	x	xx	+	xx[4]	x[5]	na
Algae	x[5,6]	x[7]	xx	+	xx[4]	x[5]	x[8]
Nitrate	xx	x	na	+	xx[1]	na	na
Salts[9]	xx	xx	na	xx	xx[10]	na	na
Trace elements	xx	xx	x	x	x	na	na
Organic micropollutants	xx	xx	x	x	?	na	na
Acidification	x	xx	x	?	x	x	na

xx Marked impairment causing major treatment or excluding the desired use
x Minor impairment
0 No impairment
na Not applicable
\+ Degraded water quality may be beneficial for this specific use
? Effects not yet fully realized
[1] Food industries
[2] Abrasion
[3] Sediment settling in channels
[4] Electronic industries
[5] Filter clogging
[6] Odor, taste
[7] In fish ponds, higher algal biomass can be accepted
[8] Development of water hyacinth (*Eichhornia crassipes*)
[9] Also includes boron, fluoride, etc.
[10] Ca, Fe, Mn in textile industries, etc.

Examples of development projects that create impact concerns for the surface water environment (rivers, lakes, estuaries, or oceans) include: (1)industrial plants or power plants withdrawing surface water for use as cooling water (this may be a particular concern during low flow conditions); (2)power plants discharging heated wastewater from their cooling cycle; (3)industries discharging process wastewaters either from routine operations or as a result of accidents and spills; (4)municipal wastewater treatment plants discharging primary, secondary, or tertiary treated wastewaters; (5)dredging projects in rivers, harbors, estuaries, and/or coastal areas (increased turbidity and releases of sediment contaminants may occur); (6)surface mining projects with resultant changes in surface water hydrology and nonpoint pollution; and (7)construction of dams for purposes of water supply, flood control, and/or hydropower production.

1.3. SUMMARY OF AGENDA 21 PROGRAM AREAS

Agenda 21 included seven program areas related to freshwater resources: (1)integrated water resources development and management; (2)water resources assessment; (3)protection of water resources, water quality, and aquatic ecosystems; (4)drinking-water supply and sanitation; (5)water and sustainable urban development; (6)water for sustainable food production and rural development; and (7)impacts of climate change on water resources. Key scientific/technical information related to the objectives, activities, and/or implementation of these areas are highlighted herein.

Integrated water resources development and management has as its overall objective the satisfaction, at the national level, of freshwater needs for sustainable development. Four related specific objectives include (United Nations 1992: 335):

1. to promote a dynamic, interactive, iterative, and multisectoral approach to water resources management, including the identification and protection of potential sources of freshwater supply, that integrates technological, socioeconomic, environmental, and human health considerations;
2. to plan for the sustainable and rational utilization, protection, conservation, and management of water resources based on community needs and priorities within the framework of national economic development policy;
3. to design, implement, and evaluate projects and programs that are both economically efficient and socially appropriate within clearly defined strategies, based on an approach of full public participation, including that of women, youth, indigenous people, and local communities, in water management policymaking and decisionmaking; and
4. to identify and strengthen or develop, as required, in particular in developing countries, the appropriate institutional, legal, and financial mechanisms to ensure that water policy and its implementation are catalysts for sustainable social progress and economic growth.

Table 2 includes a summary of scientific and technological means for implementing the integrated water resources development and management program area (United Nations 1992: 337).

Water resources assessment, including the identification of potential sources of freshwater supply, comprises the continuing determination of sources, extent, dependability, and quality of water resources and of the human activities that affect those resources (United Nations 1992: 338). Table 3 summarizes some scientific information and research needs related to this program area (United Nations 1992: 340-341). Five specific objectives related to water resources assessment from a global perspective are (United Nations 1992: 339):

1. to make available to all countries water resources assessment technology that is

appropriate to their needs, irrespective of their level of development, including methods for the impact assessment of climate change on fresh waters;
2. to have all countries, according to their financial means, allocate to water resources assessment financial resources in line with the economic and social needs for water resources data;
3. to ensure that the assessment information is fully utilized in the development of water management policies;
4. to have all countries establish the institutional arrangements needed to ensure the efficient collection, processing, storage, retrieval, and dissemination to users of information about the quality and quantity of available water resources at the level of catchments and groundwater aquifers in an integrated manner; and
5. to have sufficient numbers of appropriately qualified and capable staff recruited and retained by water resources assessment agencies and provided with the training and retraining they will need to carry out their responsibilities successfully.

Table 2. Scientific and Technological Means of Implementation of Program Area on Integrated Water Resources Development and Management (United Nations 1992: 337).

1. The development of interactive data bases, forecasting methods, and economic planning models appropriate to the task of managing water resources in an efficient and sustainable manner will require the application of new techniques such as Geographical Information Systems (GIS) and expert systems to gather, assimilate, analyze, and display multisectoral information and to optimize decisionmaking. In addition, the development of new and alternative sources of water-supply and low-cost water technologies will require innovative applied research. This will involve the transfer, adaptation, and diffusion of new techniques and technology among developing countries, as well as the development of endogenous capacity, for the purpose of being able to deal with the added dimension of integrating engineering, economic, environmental, and social aspects of water resources management and predicting the effects in terms of human impact.

2. Pursuant to the recognition of water as a social and economic good, the various available options for charging water users (including domestic, urban, industrial, and agricultural water-user groups) have to be further evaluated and field-tested. Further development is required for economic instruments that take into account opportunity costs and environmental externalities. Field studies on the willingness to pay should be conducted in rural and urban situations.

3. Water resources development and management should be planned in an integrated manner, taking into account long-term planning needs as well as those with narrower horizons; that is to say, they should incorporate environmental, economic, and social considerations based on the principle of sustainability; include the requirements of all users as well as those relating to the prevention and mitigation of water-related hazards; and constitute an integral part of the socioeconomic development planning process. A prerequisite for the sustainable management of water as a scarce vulnerable resource is the obligation to acknowledge in all planning and development its full costs. Planning considerations should reflect benefits investment, environmental protection, and operation costs, as well as the opportunity costs reflecting the most valuable alternative use of water. Actual charging need not necessarily burden all beneficiaries with the consequences of those considerations. Charging mechanisms should, however, reflect as far as possible both the true cost of water when used as an economic good and the ability of the communities to pay.

4. The role of water as a social, economic, and life-sustaining good should be reflected in demand management mechanisms and implemented through water conservation and reuse, resource assessment, and financial instruments.

5. The setting afresh of priorities for private and public investment strategies should take into account (a)maximum utilization of existing projects, through maintenance, rehabilitation, and optimal operation; (b)new or alternative clean technologies; and (c)environmentally and socially benign hydropower.

Protection of water resources, water quality, and aquatic ecosystems is related to minimizing stresses on the aquatic environment that might preclude its usage for water supply or other purposes. All countries must recognize and address such concerns. Major problems affecting the water quality of rivers and lakes arise, in variable order of importance according to different situations, from inadequately treated domestic sewage, inadequate controls on the discharges of industrial wastewaters, loss and destruction of catchment areas, ill-considered siting of industrial plants, deforestation, uncontrolled shifting cultivation, and poor agricultural practices. This gives rise to the leaching of nutrients and pesticides. Aquatic ecosystems are disturbed and living freshwater resources are threatened. Aquatic ecosystems can also be affected by development projects such as dams, river diversions, water installations, and irrigation schemes. Erosion, sedimentation, deforestation, and desertification have led to increased land degradation and adverse effects on ecosystems (United Nations 1992: 342). Three objectives that need to be pursued concurrently to integrate water quality elements into water resource management include (United Nations 1992: 342):

1. maintenance of ecosystem integrity, according to a management principle of preserving aquatic ecosystems, including living resources, and of effectively protecting them from any form of degradation on a drainage basin basis;
2. public health protection, a task requiring not only the provision of safe drinking water but also the control of disease vectors in the aquatic environment; and
3. human resources development, a key to capacity-building and a prerequisite for implementing water quality management.

Detailed information on selected technical activities related to this program area is in Table 4 (United Nations 1992: 343-345). One of the necessary tools for effectively conducting such activities is monitoring, with such monitoring encompassing water quality, point and nonpoint pollution sources, and aquatic ecology. Table 5 includes a summary of technical means for implementing the protection programs, including needs for research and monitoring (United Nations 1992: 345-346). Additional information related to monitoring is included later in this chapter.

Safe water supplies are a necessary requisite for local, national, and global sustainable development. The overall objective for this program area is the provision, on a sustainable basis, of access to safe water in sufficient quantities and the provision of proper sanitation. Examples of associated technical activities related to these health concerns include (United

Table 3. Scientific and Technological Means of Implementation of Program Area on Water Resources Assessment (United Nations 1992: 340-341).

1. Important research needs include: (a)development of global hydrologic models in support of analysis of climate change impact and of macroscale water resources assessment; (b)closing of the gap between terrestrial hydrology and ecology at different scales, including the critical water-related processes behind loss of vegetation and land degradation and its restoration; and (c)study of the key processes in water-quality genesis, closing the gap between hydrologic flows and biogeochemical processes. The research models should build upon hydrologic balance studies and also include the consumptive use of water. This approach should also, when appropriate, be applied at the catchment level.

2. Water resources assessment necessitates the strengthening of existing systems for technology transfer, adaptation, and diffusion and the development of new technology for use under field conditions, as well as the development of endogenous capacity. Prior to inaugurating the above activities, it is necessary to prepare catalogues of the water resources information held by government services, the private sector, educational institutes, consultants, local water-use organizations, and others.

Table 4. Examples of Technical Activities Related to Protection of Water Resources, Water Quality, and Aquatic Ecosystems (United Nations 1992: 343-345)

Water pollution prevention and control

1. Application of the "polluter pays" principle, where appropriate, to all kinds of sources, including on-site and off-site sanitation.
2. Promotion of the construction of treatment facilities for domestic sewage and industrial effluents and the development of appropriate technologies, taking into account sound traditional and indigenous practices.
3. Establishment of standards for the discharge of effluents and for the receiving waters.
4. Introduction of the precautionary approach in water-quality management, where appropriate, with a focus on pollution minimization and prevention through use of new technologies, product and process change, pollution reduction at source and effluent reuse, recycling and recovery, treatment, and environmentally safe disposal.
5. Mandatory environmental impact assessment of all major water resource development projects potentially impairing water quality and aquatic ecosystems, combined with the delineation of appropriate remedial measures and a strengthened control of new industrial installations, solid waste landfills, and infrastructure development projects.
6. Use of risk assessment and risk management in reaching decisions in this area and ensuring compliance with those decisions.
7. Identification and application of best environmental practices at reasonable cost to avoid diffuse pollution, namely, through a limited, rational, and planned use of nitrogenous fertilizers and other agrochemicals (pesticides, herbicides) in agricultural practices.
8. Encouragement and promotion of the use of adequately treated and purified waste waters in agriculture, aquaculture, industry, and other sectors.

Development and application of clean technology

1. Control of industrial waste discharges, including low-waste production technologies and water recirculation, in an integrated manner and through application of precautionary measures derived from a broad-based life-cycle analysis.
2. Treatment of municipal wastewater for safe reuse in agriculture and aquaculture.
3. Development of biotechnology, *inter alia*, for waste treatment, production of biofertilizers and other activities.
4. Development of appropriate methods for water pollution control, taking into account sound traditional and indigenous practices.

Groundwater protection

1. Development of agricultural practices that do not degrade groundwaters.
2. Application of the necessary measures to mitigate saline intrusion into aquifers of small islands and coastal plains as a consequence of sea level rise or overexploitation of coastal aquifers.
3. Prevention of aquifer pollution through the regulation of toxic substances that permeate the ground and the establishment of protection zones in groundwater recharge and abstraction areas.
4. Design and management of landfills based upon sound hydrogeologic information and impact assessment, using the best practicable and best available technology.
5. Promotion of measures to improve the safety and integrity of wells and wellhead areas to reduce intrusion of biological pathogens and hazardous chemicals into aquifers at well sites.
6. Water-quality monitoring, as needed, of surface waters and groundwaters potentially affected by sites storing toxic and hazardous materials.

Protection of aquatic ecosystems

1. Rehabilitation of polluted and degraded water bodies to restore aquatic habitats and ecosystems.

Table 4. Continued on next page.

Table 4. Continued.

Protection of aquatic ecosystems

2. Rehabilitation programs for agricultural lands and for other users, taking into account equivalent action for the protection and use of groundwater resources important for agricultural productivity and for the biodiversity of the tropics.
3. Conservation and protection of wetlands (owing to their ecological and habitat importance for many species), taking into account social and economic factors.
4. Control of noxious aquatic species that may destroy some other water species.

Protection of freshwater living resources

1. Control and monitoring of water quality to allow for the sustainable development of inland fisheries.
2. Protection of ecosystems from pollution and degradation for the development of freshwater aquaculture projects.

Nations 1992: 347): (1)establishment of protected areas for sources of drinking water supply; (2)sanitary disposal of excreta and sewage, using appropriate systems to treat wastewaters in urban and rural areas; (3)expansion of urban and rural water supply and development and expansion of rainwater catchment systems, particularly on small islands, in addition to the reticulated water supply system; (4)building and expansion, where appropriate, of sewage treatment facilities and drainage systems; (5)treatment and safe reuse of domestic and industrial wastewaters in urban and rural areas; and (6)control of water-associated diseases. Table 6 highlights some technologies and technology transfer needs related to safe drinking water supplies and sanitation (United Nations 1992: 349).

The objective of the water and sustainable urban development program area is to support local and national governmental efforts and capacities to sustain national development and productivity through environmentally sound management of water resources for urban use. This objective will require the identification and implementation of strategies and actions to ensure the continued supply of affordable water for present and future needs and

Table 5. Scientific and Technological Means of Implementation of Program Area on Protection of Water Resources, Water Quality, and Aquatic Ecosystems (United Nations 1992: 345-346).

1. States should undertake cooperative research projects to develop solutions to technical problems that are appropriate for the conditions in each watershed or country. States should consider strengthening and developing national research centers linked through networks and supported by regional water research institutes. The North-South twinning of research centers and field studies by international water research institutions should be actively promoted. It is important that a minimum percentage of funds for water resource development projects is allocated to research and development, particularly in externally funded projects.

2. Monitoring and assessment of complex aquatic systems often require multidisclipinary studies involving several institutions and scientists in a joint program. International water quality programs, such as GEMS/WATER, should be oriented toward the water quality of developing countries. User-friendly software and Geographical Information Systems (GIS) and Global Resource Information Database (GRID) methods should be developed for the handling, analysis, and interpretation of monitoring data and for the preparation of management strategies.

to reverse current trends of resource degradation and depletion (United Nations 1992: 350). Table 7 emphasizes implementation needs as related to developing countries (United Nations 1992: 352).

Adequate water resources are a basic requisite for both food production and development programs and projects in rural areas. Water can be used for crop irrigation, as aquatic habitat for fisheries and other harvested food products, and as a supply for humans and livestock living in rural areas. Selected technical activities included in this program area are summarized in Table 8 (United Nations 1992: 355-357). Table 9 highlights implementation needs in terms of monitoring, research, and technology transfer.

Finally, the objectives of the program area on impacts of climate change on water resources include (United Nations 1992: 359): (1)to understand and quantify the threat of the impact of climate change on freshwater resources; (2)to facilitate the implementation of effective national countermeasures, as and when the threatening impact is seen as sufficiently confirmed to justify such action; and (3)to study the potential impacts of climate change on areas prone to droughts and floods. Table 10 summarizes science and technology needs basic to this program area (United Nations 1992: 359-360).

1.4. UNCERTAINTIES RELATED TO PROTECTION OF FRESHWATER RESOURCES

Numerous scientific and technical uncertainties can be identified for the seven program areas described in Agenda 21. The purpose herein is to highlight examples of such uncertainties; the discussion should not be perceived as all-inclusive. The examples are grouped into those related to water resources planning and technical analyses. Specific illustrations are then included for the human health impacts of water resources projects and the effects of global climate change on water resources.

1.4.1. *Uncertainties in the Planning Process*

Effective planning for maintaining or enhancing the quality of freshwater resources while at the same time considering project developments or modifications does not occur by accident, nor should it occur as an afterthought. One way to provide a positive opportunity for such planning is to include relevant considerations in an overall planning (or conceptual) model. Numerous models have been described for water resources planning; examples

Table 6. Scientific and Technological Means of Implementation of Program Area on Drinking Water Supply and Sanitation (United Nations 1992: 349).

To ensure the feasibility, acceptability, and sustainability of planned water-supply services, adopted technologies should be responsive to the needs and constraints imposed by the conditions of the community concerned. Thus, design criteria will involve technical, health, social, economic, provincial, institutional, and environmental factors that determine the characteristics, magnitude, and cost of the planned system. Relevant international support programs should address the developing countries concerning, *inter alia*:

1. Pursuit of low-cost scientific and technological means, as far as practicable;
2. Utilization of traditional and indigenous practices, as far as practicable, to maximize and sustain local involvement; and
3. Assistance to country-level technical/scientific institutes to facilitate curricula development to support fields critical to the water and sanitation sector.

include EOP (1983), Lyon et al. (1984), Petersen (1984), Grigg (1985), Stakhiv (1986), and Dzurik (1990). These planning models are similar in concept.

The planning process as applied to water resources is comparable to other types of resource planning. It involves a logical series of steps, beginning with identification of needs, proceeding to recommendations for action, and culminating in implementation and monitoring. The planning process shown in Figure 1 has the following components: (1)problem identification; (2)data collection and analysis; (3)development of goals and objectives; (4)clarification and diagnosis of the problem or issues; (5)identification of alternative solutions; (6)analysis of alternatives; (7)evaluation and recommendation of actions; (8)development of an implementation program; and (9)surveillance and monitoring. Components 3, 5, 6, 7, and 8 are at the heart of the process and are commonly known as the rational planning model (Dzurik 1990). Each of the components listed should be self-explanatory.

Risks and uncertainties are inherent throughout the water resources planning process, as shown in Figure 1. For example, assume that an analysis of flooding problems in a geographical area is being conducted and that the results will be utilized in the engineering design of a flood control dam. The stochastic nature of flood events and associated measurements may include many sources of hydrologic uncertainty; examples of such sources include (IWR 1992: 38): (1)data availability; (2)data error; (3)accuracy and imprecision of measurements and observations; (4)sampling uncertainty, including the choice of samples and appropriate sample size; (5)selection of an appropriate probability distribution to describe the stochastic events; (6)estimation of the hydrological and statistical parameters in flow and/or water-quality models; (7)low-probability flood extrapolation, e.g., tail problems of frequency curves; (8)modeling assumptions; and (9)the characterization of river basin parameters.

The U.S. Army Corps of Engineers has developed guidelines for conducting and incorporating risk and uncertainty analyses in the water resources planning process (IWR 1992: 1-2). The guidelines are intended to illustrate the background and principles involved in risk and uncertainty analysis in order to provide direction in conducting such an analysis. Emphasis is given to identifying potential sources of risk and uncertainty in each step of the planning process, evaluating alternatives for minimizing or analyzing the risk and uncertainty, and implementing one of the alternatives for dealing with risk and uncertainty (IWR 1992: 17-21). Risk assessment principles can also be applied in impact studies for specific development projects (Canter 1993b).

In water resources management and planning, uncertainties may be associated with alternatives for meeting identified planning needs. For example, it may be necessary to: (1)develop new water supply and/or flood control options via the building of dams; (2)analyze the potential impacts of water usage and/or wastewater discharges on existing surface water and/or groundwater resources; and/or (3)identify and evaluate options for

Table 7. Scientific and Technological Means of Implementation of Program Area on Water and Sustainable Urban Development (United Nations 1992: 352).

The 1980s saw considerable progress in the development and application of low-cost water supply and sanitation technologies. The program envisages continuation of this work, with particular emphasis on development of appropriate sanitation and waste disposal technologies for low-income high-density urban settlements. There should also be international information exchange, to ensure a widespread recognition among sector professionals of the availability and benefits of appropriate low-cost technologies. The public-awareness campaigns will also include components to overcome user resistance to second-class services by emphasizing the benefits of reliability and sustainability.

Table 8. Examples of Technical Activities Related to Water for Sustainable Food Production and Rural Development (United Nations 1992: 355-357).

Water-use efficiency

1. Increase efficiency and productivity in agricultural water use for better utilization of limited water resources.
2. Strengthen water and soil management research under irrigation and rain-fed conditions.
3. Monitor and evaluate irrigation project performance to ensure, *inter alia*, the optimal utilization and proper maintenance of the project.
4. Support water-users groups with a view to improving management performance at the local level.
5. Support the appropriate use of relatively brackish water for irrigation.

Waterlogging, salinity control, and drainage

1. Introduce surface drainage in rain-fed agriculture to prevent temporary waterlogging and flooding of lowlands.
2. Introduce artificial drainage in irrigated and rain-fed agriculture.
3. Encourage conjunctive use of surface waters and groundwaters, including monitoring and water-balance studies.
4. Practice drainage in irrigated areas of arid and semi-arid regions.

Water supply for livestock

1. Improve quality of water available to livestock, taking into account their tolerance limits.
2. Increase the quantity of water sources available to livestock, in particular those in extensive grazing systems, in order both to reduce the distance needed to travel for water and to prevent overgrazing around water sources.
3. Prevent contamination of water sources with animal excrement in order to prevent the spread of diseases, in particular zoonosis.
4. Encourage multiple use of water supplies through promotion of integrated agro-livestock-fishery systems.
5. Encourage water spreading schemes for increasing water retention of extensive grasslands to stimulate forage production and prevent runoff.

Inland fisheries

1. Develop the sustainable management of fisheries as part of national water resources planning.
2. Study specific aspects of the hydrobiology and environmental requirements of key inland fish species in relation to varying water regimes.
3. Prevent or mitigate modification of aquatic environments by other users or rehabilitate environments subjected to such modification on behalf of the sustainable use and conservation of biological diversity of living aquatic resources.
4. Develop and disseminate environmentally sound water resources development and management methodologies for the intensification of fish yield from inland waters.
5. Establish and maintain adequate systems for the collection and interpretation of data on water quality and quantity and channel morphology related to the state and management of living aquatic resources, including fisheries.

Aquaculture development

1. Develop environmentally sound aquaculture technologies that are compatible with local, regional, and national water resources management plans and take into consideration social factors.
2. Introduce appropriate aquaculture techniques and related water development and management practices in countries not yet experienced in aquaculture.
3. Assess environmental impacts of aquaculture with specific reference to commercialized culture units and potential water pollution from processing centers.
4. Evaluate economic feasibility of aquaculture in relation to alternative use of water, taking into consideration the use of marginal-quality water and investment and operational requirements.

Table 9. Scientific and Technological Means of Implementation of Program Area on Water for Sustainable Food Production and Rural Development (United Nations 1992: 357).

1. There is an urgent need for countries to monitor water resources and water quality, water and land use, and crop production; compile inventories of type and extent of agricultural water development and of present and future contributions to sustainable agricultural development; evaluate the potential for fisheries and aquaculture development; and improve the availability and dissemination of data to planners, technicians, farmers, and fishermen. Priority requirements for research are as follows:
   a. Identification of critical areas for water-related adaptive research;
   b. Strengthening of the adaptive research capacities of institutions in developing countries; and
   c. Enhancement of translation of water-related farming and fishing systems research results into practical and accessible technologies and provision of the support needed for their rapid adoption at the field level.

2. Transfer of technology, both horizontal and vertical, needs to be strengthened. Mechanisms to provide credit, input supplies, markets, appropriate pricing, and transportation must be developed jointly by countries and external support agencies. Integrated rural water-supply infrastructure, including facilities for water-related education and training and support services for agriculture, should be expanded for multiple uses and should assist in developing the rural economy.

enhancing surface water and/or groundwater resources relative to existing pollution problems. A number of additional water resources-related problems (or planning objectives) are listed in Table 11 (Petersen 1984), and various management measures that could be used to address these problems are delineated. A common feature in water resources management and planning, irrespective of the specific issue being addressed, is the need to identify the problem and delineate and systematically evaluate alternatives (options) for meeting such needs.

The need to address transboundary concerns in water resources planning can also increase uncertainties, particularly as related to describing baseline conditions, identifying influencing projects and hydrological and quality impacts, and applying pertinent low flow

Table 10. Scientific and Technological Means of Implementation of Program Area on Impacts of Climate Change on Water Resources (United Nations 1992: 359-360).

1. Monitoring of climate change and its impact on freshwater bodies must be closely integrated with national and international programs for monitoring the environment, in particular those concerned with the atmosphere as discussed under other sections of Agenda 21, and the hydrosphere, as discussed earlier herein. The analysis of data for indication of climate change as a basis for developing remedial measures is a complex task. Extensive research is necessary in this area, and due account has to be taken of the work of the Intergovernmental Panel on Climate Change (IPCC), the World Climate Program, the International Geosphere-Biosphere Program (IGBP), and other relevant international programs.

2. The development and implementation of response strategies requires innovative use of technological means and engineering solutions, including the installation of flood and drought warning systems and the construction of new water resource development projects such as dams, aqueducts, well fields, wastewater treatment plants, desalination works, levees, banks, and drainage channels. There is also a need for coordinated research networks such as the International Geosphere-Biosphere Program/Global Change System for Analysis, Research, and Training (IGBP/START) network.

and/or water quality standards. Transboundary impacts of development projects on the surface water environment are becoming increasingly important due to (Canter and Vieux 1993):

1. population growth and industrial expansion in countries sharing surface water resources within and between their borders;
2. the introduction of greater quantities of heavy metals, organic micropollutants, and other potentially toxic contaminants into receiving bodies of water;
3. the creation of intergovernmental, including international, organizations whose purpose is to "manage" water quality, and the development of water quality-related goals or objectives by these organizations;
4. increased water quality, flow, and biological monitoring programs which are enabling the development of more quantitative information on river systems, water usage, and point and nonpoint sources of water pollution; and
5. the adoption of more stringent in-stream water quality standards and effluent discharge standards by countries sharing surface water resources.

Figure 1. General Planning Model (Dzurik 1990).

Problem Identification

↓

Data Collection and Analysis

↓

Development of Goals and Objectives*

↓

Clarification and Diagnosis of the Problem or Issues

↓

Identification of Alternative Solutions*

↓

Analysis of Alternatives*

↓

Evaluation and Recommendation of Actions*

↓

Development of an Implementation Program*

↓

Surveillance and Monitoring

*Denotes components of what is frequently called the rational planning model.

Table 11. Examples of Conceptual Management Measures for Water Resources Planning Objectives (Peterson 1984).

Planning Objective	Measure
Flood damage reduction	Reservoirs Levees Channel modification Flood warning and floodplain evacuation Floodproofing Floodplain zoning Watershed management
Water supply	Surface water reservoirs Groundwater reservoirs Interbasin transfers Precipitation Wastewater reclamation Water conservation Reallocation Watershed management Water harvesting Desalinization New technologies
Water-oriented recreation	Reservoirs Regulated streamflows Parks Trails
Water quality improvement	Stream flow augmentation Land-use zoning Improved agricultural practices Pollutant diversion Selective withdrawal from reservoirs
Fish and wildlife preservation	Stream flow augmentation Gravel management Fish hatchery Wildlife areas Wildlife habitat improvement
Hydroelectric power	Reservoirs Steam diversion Run-of-river projects
Groundwater overdraft correction	Recharge Restriction of pumping Modified cropping pattern

A seven-step methodology has been proposed for addressing transboundary impacts on surface water resources (Canter and Vieux 1993). The methodology is flexible and can be adapted to various project types by modification, as required, to enable the addressing of specific concerns related to projects in unique locations. It could be adapted to address transboundary groundwater impacts; also, it could also be applied to plans, programs, and regulatory actions. Each step in the methodology has associated uncertainties. The seven generic steps include:

1. identification of the types and quantities of water pollutants to be introduced, or water quantities to be withdrawn, or other impact-causing factors;
2. preparation of a description of the existing environmental setting, in terms of river flow patterns, water quality characteristics, existing/historical pollution problems, influential meteorological factors (including precipitation, evaporation, and temperature), relationships to groundwater resources, existing point and nonpoint sources of pollution, pollution loadings, and existing water abstractions;
3. procurement of relevant laws, regulations, or criteria related to water quality and/or water usage, and any relevant agreements among countries or other entities related to transnational rivers, lakes, or coastal waters;
4. impact prediction, including the use of mass balances in terms of water quantity and/ or pollutant loading changes, mathematical models for relevant pollutant types (conservative, nonconservative, bacterial, nutrient, and thermal), aquatic ecosystem models to account for floral and faunal changes and nutrients/pollutants cycling, and/or qualitative predictions based on case studies and professional judgment;
5. the use of pertinent information from Step 3, along with professional judgment and public input, to assess the significance of anticipated beneficial and detrimental impacts;
6. identification and development of appropriate mitigation measures for the adverse impacts; and
7. preparation of a technical report summarizing the findings.

1.4.2. *Uncertainties in Technical Analyses*

There are also many uncertainties related to various technical analyses required in water resources planning and management. Examples to be mentioned are related to accounting for nonpoint sources of pollution, characterizing wastewater discharges, considering pollutant transport and fate, and determining water users in geographical areas. In addition, the identification of typical impacts associated with proposed development projects can be fraught with uncertainty and lack of specificity.

Nonpoint sources of water pollution from urban or rural runoff have been recognized as potential major contributors to the total waste load within the aquatic environment, and it is vitally important to consider nonpoint sources along with point sources of water pollution in water resources planning efforts. Agricultural pollutants have their origin in fertilizer use and pesticide applications, and generally, the primary causes are agricultural methods of disturbing soils by tillage (agricultural lands) or logging (silvicultural lands). Land uses that produce the most pollution per unit area are animal feedlot operations and farming on steep slopes. Forested lands and pastures, on the other hand, produce the least amount of pollution. Urbanization and related hydrologic modifications may cause increased pollution loadings above background levels. The sources of urban nonpoint pollutants vary widely; they include urban bird and pet populations, street litter accumulations, tire wear of vehicles, abrasion of road surfaces by traffic, street salting practices, and construction activities. Urban nonpoint pollution may contain contaminants such as lead, zinc, asbestos, PCBs (polychlorinated biphenyls), oil, and grease (Novotny and Chesters 1981). Quantitative information on unit waste generation factors for nonpoint sources as a function of land usage is increasing. More detailed and/or site-specific information could be developed by using the Universal Soil Loss Equation.

Information on the characteristics of wastewater discharges from municipalities and various industries may be needed in water resource assessments. For example, the typical composition of untreated domestic wastewater in the United States has been described (Metcalf and Eddy 1991). Several books are available that summarize the characteristics of wastewaters from various industries; examples include Nemerow (1978), Corbitt (1990), and Nemerow and Dasgupta (1991). Information on the types and quantities of water pollutants from major private-sector categories such as the apparel, food, materials, chemical, and energy industries is available (Nemerow and Dasgupta 1991). While considerable information may exist, there can be variations from plant to plant within an industrial category. In addition, the available information is typically more complete for conventional water pollutants than for potentially toxic pollutants such as metals and synthetic organics.

In addition to information on pollutant types and quantities, it may also be necessary to consider information on the transport and fate of specific pollutant materials. For example, information may be needed on the fate of petroleum products, other organics, nutrients, metals, and similar substances in the water environment. It is important to know whether the pollutant will partition between the water and sediment phases or become associated with aquatic flora and fauna. Metals can occur in surface water systems as both dissolved and particulate constituents. Biogeochemical partitioning of metals can yield a diversity of forms, including hydrated or "free" ions, colloids, precipitates, adsorbed phases, and coordination complexes with dissolved organic and inorganic ligands. An excellent review related to aluminum, arsenic, cadmium, chromium, copper, iron, lead, manganese, mercury, nickel, and zinc is in Elder (1988). Controlling factors for biogeochemical partitioning include pH, oxidation reduction potential, hydrologic features, sediment grain size, clay minerals, organic matter, and biological processes. While extensive information exists on transport and fate concerns, it is certainly not complete relative to all pertinent concerns and combination effects in different water environments.

It may also be necessary to consider joint toxicity effects when several chemicals are present in the surface water environment, and toxicity testing using juvenile fathead minnows or other appropriate organisms may be necessary (Broderius 1990). Toxicity testing has typically been associated with single chemicals; thus there are many unknowns related to potential combination effects of toxic compounds. Uncertainties also exist relative to toxicity testing protocols and statistical interpretation of the results.

Current and historical stream flow information is typically needed in water resources planning and project design. An issue that may need to be addressed is the relationship between surface water and groundwater resources, particularly if groundwater provides the base flow for the stream segment. One of the key concerns with regard to stream flow is the flow frequency that is utilized for determination of compliance with water quality standards. In some instances, the seven-day, two-year low flow is utilized; in other cases, the seven-day, ten-year low flow is required. The phrase "seven-day, two-year low flow" indicates that this is the minimum flow that occurs over a seven-day period on a frequency of once every two years. Historical records of stream flow are necessary in the development of flow frequency information; such records may not exist in many geographical areas, thus increasing uncertainties related to planning, project design, and analysis.

It is appropriate in technical studies of freshwater resources to consider what other potential and actual sources of surface water pollution already exist in the study area and also to consider current and potential future usage of the surface water resource(s) for various purposes, including water supply. The emphasis should be to assemble sufficient information to enable the determination of the level of other sources of pollution and water uses in the study area. This does not necessarily require the conduction of a detailed surface water monitoring program. Pertinent information could be assembled on items such as the types of land usage (and size of each land use), storage areas (stockpiles, tailing piles), and so on within the study area. For example, if nitrates or pesticides are of concern, the land area within the study area associated with agricultural usage, including the types, quantities, and timing of fertilizer and/or pesticide applications, should be established.

A detailed review of the current numbers of surface water users and the quantities associated with such uses may also be needed. The types of information that may be needed include general estimates of the number of users of the surface water (private, public, industrial); types of water uses (drinking water, recreation, cooling water, etc.); the location and rates of existing surface water withdrawals; and the location, quantity, and quality of existing discharges into the surface water. Water quantity concerns are of major importance in water-deficient areas. The types of water uses are important, since quality requirements may vary for different uses.

To serve as a final example, a large number of environmental impact issues can be identified for dam/reservoir projects; Table 12 includes a summary listing. A pragmatic approach for addressing impacts on physical resources, ecological resources, human-use values, and quality-of-life values from irrigation projects is included in a recent United Nations report (ESCAP 1990). Irrigation systems can cause adverse environmental effects which may be classified in terms of: (a)soil modifications, including water logging, salinization, and alkalization; (b)water quality and quantity modifications; (c)effects on ecology; (d)public health impacts; and (e)other socioeconomic impacts. The approach for addressing these issues is summarized in Table 13; the table also represents a good summary of impact issues for irrigation projects. While qualitatively listing impacts and issues to be addressed is useful, all such lists may be incomplete for a given location. Also, the quantification of baseline conditions and forecasted impacts can be subject to inaccuracies and uncertainty.

1.4.3. *Uncertainties in Forecasts*

One of the key technical elements in the water resources planning process is the forecasting (predicting) of impacts (effects) for both the without-project and the with-project conditions. Such forecasting is basic to decisionmaking relating to the environmental viability of a proposed project and also to the selection of a proposed plan to meet identified needs, along with the inclusion of relevant mitigation measures. The Water Resources Council in the United States delineated several forecasting approaches that could be utilized; they include (WRC 1983): (1)adoption of forecasts made by other agencies or groups; (2)use of scenarios based on differing assumptions regarding resources and plans; (3)use of expert group judgment via the conduction of formalized Delphi studies or the usage of the nominal group process; (4)extrapolation approaches based upon the use of trends analysis and/or simple models of environmental components; (5)analogy and comparative analyses that involve the use of look-alike resources and/or projects and the application of information from such look-alike conditions to the planning effort.

Additional techniques for forecasting (or making predictions of future conditions) include checklists, matrices, and networks. Several types of checklists have been developed, ranging from simple listings of anticipated impacts by project type to questionnaire checklists incorporating detailed questions which provide a structure to the impact prediction activity. Some checklists have been extended to include the use of scaling/rating/ranking of the anticipated impacts of alternatives and the incorporation of relative-importance weights to the individual environmental or other decision factors. These types of decision-focused checklists can be utilized to aggregate the impacts of a project into a final index or score which can then be used for purposes of comparisons of alternatives.

Interaction matrices include simple x-y matrices to identify impacts and to provide a basis for further evaluation of such impacts in terms of their magnitude and importance. Stepped matrices have also been developed wherein secondary and tertiary consequences of project actions can be delineated. The most sophisticated types of matrices are those referred to as networks or impact trees wherein systematic approaches are utilized to trace the consequences of a given project or activity. The key point to note relative to both checklist and matrix methods is that they tend to be qualitative in terms of the actual forecasted impacts; however, they do represent useful tools for purposes of impact prediction.

Environmental indices represent another category of impact prediction approaches (Ott 1978). An environmental index refers to a mathematical and/or descriptive presentation of information on a series of factors which can be used for purposes of classification of environmental quality and sensitivity and for predicting the impacts of a proposed project or activity. The basic concept for impact prediction would be to anticipate and quantify (if possible) the change in the environmental index as a result of the project or activity and to then consider the difference in the index from the with- and without-project conditions as one

Table 12. Typical Impacts of Dams and Reservoirs (not in priority order).

1. Change in quality of impounded water (seasonal).
2. Water loss due to evaporation (seasonal).
3. Downstream effects in terms of decreased (and more uniform) flow into estuaries, thus causing changes in saltwater intrusion patterns and changes in estuarine fisheries.
4. Changes in local groundwater levels and quality.
5. Due to water pressure, possible in-reservoir landslides and/or increase seismic activity in the area.
6. Changes in microclimate of area—more wind, humidity, and/or precipitation.
7. Inundation of mineral resources.
8. Changes to number and types of fish—from cold-water to warm-water fishery.
9. Prevention of movement of migratory fish (salmon on Columbia River are an example).
10. Fish destruction in turbines and pumps (use protective screens).
11. Possible creation of new reservoir fishery for positive impact.
12. Increase in areas for breeding of mosquitoes and related insects—and their public health implications (e.g., malaria and schistosomiasis).
13. Promotion of growth of aquatic weeds such as water hyacinths.
14. Change in habitat in inundated area and wildlife associated with habitat.
15. Change in waterfowl habitat from shallow, flowing habitat to deeper lakes; might impact migratory birds.
16. Impacts on rare, threatened, endangered, unique floral and faunal species.
17. Decreased waste assimilative capacity of river segment.
18. Inundation of historical/cultural/archaeological/religious resources.
19. People relocation/resettlement (and possible change in style of life).
20. Influx of construction workers and associated social, infrastructure, and health impacts.
21. Increased tourism around reservoir.
22. Downstream effects on traditional floodplain cultivation; reduced flood delivery of nutrients to downstream fields.
23. Developments in catchment area due to roads and other associated increases in sediment/nutrients into reservoir.

Table 13. Approach for Addressing the Environmental Impacts of Irrigation Projects (ESCAP 1990).

Soil modifications

1. Waterlogging: Waterlogging, or saturation of the plant root zone owing to the rise of the water table, can be a severe problem in low-lying areas and may be caused not only by over-irrigation but also by seepage losses from unlined canals and reservoirs. Over-irrigation may also leach nutrients, especially nitrates, from the root zone.

2. Soil salinization: This problem may occur in any irrigation system, but it is common in arid and semiarid zones with limited rainfall or fresh water available for flushing the soil. Even with irrigation water of good quality (low salinity, say 200 ppm), the amount of salt added to the soil per year with normal water application can be 0.2 tons per hectare. The problem of salinization is more serious if coupled with waterlogging.

3. Soil alkalization: Soil alkalization (or "secondary soil salinization") is caused by alkaline groundwater or irrigation water, i.e., waters with excess sodium/calcium ratios, resulting in progressive decreases in soil permeability.

4. Soil texture: New tillage practices introduced by irrigation may result in excessive reworking of the soil, creating fine particles that can be lost by wind erosion. Also, turbidity contained in irrigation water may change the soil texture (and surface evaluation) over a period of time.

5. Soil fertility: Continuing irrigation together with other local agricultural practices may result in progressive loss of soil fertility. An important consideration is the progressive loss in soil organic matter (hence in soil fertility) caused by the progressive substitution of commercial chemical fertilizer for organic matter over the past two decades, and while this has not yet had serious effects in reducing crop yields, such an outcome seems likely within the next few decades.

6. Coastal swamp areas: Reclaimed coastal swamp areas contain sulphur compounds which, when exposed to air, are oxidized, resulting in acid sulphur soils, especially in areas covering former mangroves.

Water quality

Water quality is an important parameter in irrigation planning, both with respect to effects of water quality on the crops and soils and with respect to effects of irrigation return flows on downstream water quality.

1. Effects of mineral constituents: Waters used for irrigation must not contain soluble constituents above the limits that can be tolerated by the particular crop, including total salinity and specific toxic substances like boron, arsenic, and biocides. Also, as noted above, the sodium/calcium ratio must not be excessive. Water quality criteria for irrigation have been established by various agencies, including the United States Department of Agriculture and the Food and Agriculture Organization of the United Nations. In addition to considering the present status of the quality of the irrigation water supply, consideration must be given to possible alterations from future stream water uses.

2. Return flows: Of the total irrigation water delivered to the farm fields, some 20 to 60 percent leaves the fields as surface runoff, percolates through the soil to the groundwater, or is removed by engineered drainage systems. Because of evapotranspiration, its mineral content will be sizably increased (in the range of 2 to 10 times). In addition, the return flow will include agricultural chemical residuals including fertilizers and toxic biocides. These added constituents will tend to impair downstream beneficial water uses because of increased mineralization, increased nutrients causing eutrophication, and increased toxicity.

3. Hydrological changes: Use of water for irrigation decreases the flow available downstream and thus may impair downstream uses. In addition, if groundwater is being used, such use will, of course, subtract from the groundwater resource available for other purposes.

Table 13. Continued on next page.

Table 13. Continued.

Ecology (flora and fauna)

1. Effects on flora and fauna in irrigation vicinity: The change in soil moisture conditions in the vicinity of the irrigation area will result in changes in the existing ecosystem in the vicinity of the irrigated plots. For example, this tends to favor the more aggressive plant and animal species, including plants with highly mobile seeds (including weeds) and insect, rodent, and seed-eating birds. Irrigation scheme planning rarely takes the role of the natural predators into account.

2. Effects on paddy fishery: Use of toxic biocides may impair or even eliminate the natural paddy fishery which has been an important traditional source of protein food for farm families in many regions.

3. Effects on downstream ecology: The reduced downstream flow volumes and increased concentrations of salts, toxic substances, and nutrients will, of course, considerably alter the downstream aquatic ecology, depending upon the extent of the changes. This will include downstream fisheries and wildlife dependent on the stream and the riverine fishery, and possibly the estuarine and even the marine fishery associated with the river system. Many of these relationships are very complex; hence, they can hardly be completely evaluated or provided for in the project plan and can only be evaluated by periodic monitoring following commencement of project operations.

4. Use of degradable pesticides: In earlier years, it was common practice to utilize persistent pesticides (such as DDT and other chlorinated hydrocarbons), which resulted not only in immediately observable adverse effects such as downstream fish kills but also in accumulation of these materials in the environment. This included accumulation in soils and river bottom muds and accumulations in flora and fauna, sometimes with marked impairment of adult birds and fish. Only appropriate "soft" degradable biocides should be used, with this use kept to a minimum.

Public health

1. Water-oriented communicable diseases: Because of the close association between water and many communicable diseases, any project that modifies water balances and use practices may also result in serious new disease hazards (as well as eliminating or minimizing existing disease hazards). Water is not only a favorable medium per se for spread of disease but also important in the transmission of diseases by disease-carrying vectors.

2. Other health impacts: Other public health hazards involved in irrigation practices include hazards to personnel who handle toxic substances, which should be regulated with instructions for safe use furnished to the farm users, and pollution of downstream community water supply systems.

Socio-economics

In addition to the public health impacts noted above, irrigation projects may involve numerous socioeconomic disturbances as well as benefits. Socioeconomic surveys are especially important for irrigation projects, first for establishing the baseline conditions in the project vicinity (service area and adjacent nearby areas) of family income amounts/sources prior to project implementation, and for permitting evaluations of project impacts by future periodic surveys. Also, these data serve to help plan the project to give optimal distribution of project benefits. In addition to income, the surveys should include other basic quality-of-life values such as public health and nutrition.

1. Resettlement: This problem is similar to that for dam/reservoir projects. Farm families displaced by the new facilities should be fairly provided for.

2. Land tenure: Where land titles are "obscure" (as they often are in the developing countries), the result of the project may be unfair displacement of families who have been farming their plots for centuries; hence, the project plan should give careful consideration to management of the land title problem so

Table 13. Continued on next page.

Table 13. Continued.

that the problem is resolved before the project increases the land values. Where irrigation projects are established in regions with a high percentage of tenant farmers, careful and adequate planning should be conducted in order to prevent or mitigate the unfair treatment of the tenant farmers.

3. Farm pattern changes and institutional requirements: Irrigation projects may require extensive changes in farmer behavioral patterns, including abandoning traditional farmer methods in favor of new technologies with which the farmer may not be familiar. Also, the project may reduce the farmer's freedom to function on his own, i.e., he may become closely bound by decisions made by the larger irrigation social unit. The project may also involve new work/housing patterns. Hence, the project plan may require use of demonstration plots on a pilot scale before full-scale implementation, together with extension services for guiding changeover to the new technologies. New institutions may be needed, such as farmer organizations for maintaining the irrigation canals and other facilities and for ensuring fair delivery allocations and farmer cooperatives for handling such activities as provision and sale of inputs and of marketing.

   With respect to agricultural cooperatives, care must be exercised in the basic assumptions made in planning such institutions, to be sure they are appropriate to local conditions, including traditional farmer behavioral and work ethic patterns. The record in the developing countries over the past two decades is replete with failures of cooperative organizations which initially work well with outside assistance but then gradually wither away without it.

4. Credit problems: Implementation of irrigation will mean higher operating expenses for the farm family, and some form of financial credit system will probably be necessary. Careful management of this is needed if the farmer is not to become victimized by the system—for example, through unavailability of credit when he needs it and unfair interest rates.

5. Community participation: Because of the many complexities involved in irrigated agriculture as compared with rainfed agriculture, including technical, socioeconomic, and political considerations, the success of irrigation projects depends critically on the willingness of the farmers to work with the level of cooperation needed for success of the enterprise. This requires consultation with the community itself by the project planners at all stages of project planning, beginning with preparation of a community socioeconomic profile at the prefeasibility stage which describes the social, cultural, economic, and political aspects of the affected farm families. The profile must reflect the needs, priorities, and interests of these people so that a project can be planned that will merit community support.

Agro-industries

Potentials for agro-industrial development in the irrigation areas and plans for assisting in implementing such projects should be explored, to take advantage of the ready availability of local agricultural products.

Wildlife

Discuss possibilities of significant impairment by canals to wildlife movement in the vicinity, and if this is a problem, review remedial measures.

Security of water supply

In some developing countries, insufficient attention has been paid to the limits of the total available water supply in a basin, so that individual projects are implemented with cumulative uses that may exceed the basin's actual yield, especially over a series of dry years. As a result, investments in downstream irrigation (and other water-consuming) projects have been jeopardized.

measure of impact. Numerous environmental indices have been developed, including those for water quality, aquatic or terrestrial biological diversity, aquifer vulnerability assessment, and hazardous waste disposal (Canter and Sabatini 1990; CTAGWV 1993; Canter 1995). One type of index that has received wide usage is based on habitat considerations and the utilization of Habitat Evaluation Procedures or a Habitat Evaluation System which is primarily based on the development of a numerical index to describe habitat quality and geographical extent. A key advantage of index approaches for impact prediction is that they can be related to available information, and they provide a systematic basis upon which to consider the potential consequences of a project or activity.

Another category of impact prediction approaches includes experimental methods which could encompass the conduction of experiments from specific laboratory ones to develop factors or coefficients for mathematical models to large-scale field experiments to measure changes in environmental features as a result of system perturbations. In addition, physical models have been utilized to examine impacts related to hydrodynamics and ecological changes within microcosms of environmental settings. Experimental methods are primarily useful in dealing with physical/chemical components and/or biological features of the environmental setting.

The most sophisticated approach for impact prediction involves the use of mathematical models. Numerous types of mathematical models have been developed to account for hydrologic regimes and pollutant transport and fate within the environmental setting. In addition, models have been developed for describing environmental features and the functioning of ecosystems. Anderson and Burt (1985) noted the following about hydrologic modeling:

> All models seek to simplify the complexity of the real world by selectively exaggerating the fundamental aspects of a system at the expense of incidental detail. In presenting an approximate view of reality, a model must remain simple enough to understand and use, yet complex enough to be representative of the system being studied.

Anderson and Burt classified hydrological models into three types:

1. Black-box models—these models contain no physically based transfer function to relate input to output, depending instead upon establishing a statistical correspondence between input and output.
2. Conceptual models—these models occupy an intermediate position between the deterministic approach and empirical black-box analysis; they are formulated on the basis of a simple arrangement of a relatively small number of components, each of which is a simplified representation of one process element in the system being modeled. Each element of the model will consist of a nonlinear reservoir in which the relationship between outflow (Q) and storage (S) is given by

$$S = K \cdot Q^n$$

   where K and n represent constants.

3. Deterministic models—these models are based on complex physical theory. However, despite the simplifying assumptions necessary to solve the flow equations, such models still have huge demands in terms of computational time and data requirements and are very costly to develop and operate.

Surface water and groundwater quality and quantity models are also plentiful, with major research developments within the last decade occurring in the realm of solute transport in subsurface systems. Surface water quality and quantity models range from one-dimensional steady-state models to three-dimensional dynamic models which can be utilized for rivers, lakes, and estuarine systems (Anderson and Burt 1985; COE 1987; Henderson-Sellers 1991; James 1993). Groundwater flow models have been recently modified to include subsurface processes such as adsorption and biological decomposition (Domenico and Schwartz 1990; WSTB 1990).

Simple mixing models include those models based on the assumption of uniform mixing of a contaminant in a defined area of the receiving environment. The resulting concentration of the contaminant is then equal to the quantity released divided by the size of the receiving unit, or some equivalent formulation. Examples of simple mixing models include (ERL 1982): (1)river mixing and dilution models which assume uniform mixing of conservative contaminants across the river cross section and, therefore, straightforward dilution of the effluent flow in the river flow; and (2)lake mixing models which assume uniform mixing of a conservative contaminant over the total depth of unstratified lakes, or over the depth of the upper or lower layer (depending on the discharge location) in stratified lakes. The principal advantage of simple mixing models is that they can be used quickly and easily with a calculator; however, they do require either some expertise or field studies to define the appropriate size of the receiving unit. Simple mixing models are often used as a basis for worst-case predictions. Instead of using average or typical values for the input variables, values are taken which together will generate the worst possible conditions.

Steady-state dispersion models provide steady-state predictions of environmental effects taking into account processes occurring in the environment that influence the behavior of contaminants. Examples of processes that may be addressed include advection (i.e., transport) in one, two, or three dimensions; diffusion in one, two, or three dimensions; transfer or partition among components of the environment (including living subjects); and physical, chemical, and biological transformation or removal ("sinks"). Examples of steady-state dispersion models that require manual or simple computer calculations and have low input data needs include: (1)for surface water quality—dissolved oxygen models for rivers based on Streeter-Phelps, simple segment models for rivers and estuaries, and lake circulation models; and (2)for groundwater hydraulics and quality—flow and time of travel models based on Darcy's Law. Usage of steady-state dispersion models may involve reading graphs and tables, manually solving mathematical equations, or applying packaged computer programs. The main advantage of these models is that they can be applied fairly simply and without major requirements for input data. Because these models only provide steady-state predictions, they are usually applied to predict either typical or, more often, worst-case conditions.

Complex mathematical models have fairly extensive input data requirements and are dependent upon computer solution. Complex models are dynamic in that they allow for variations in input variables with time. Some allow for variations over space and time, known usually as Lagrangian models, and a few allow for random fluctuations in model relationships, known as probabilistic or stochastic models (as opposed to deterministic models). Examples of complex models include river basin flow and quality models, marine and estuarine dispersion models, watershed run-off models, biological population or productivity models, nutrient cycling and eutrophication models, and groundwater hydraulic and solute transport models. The advantage of complex mathematical models is that computer application means that many more factors can be taken into account in modeling environmental behavior, by use of mathematical techniques that would be impractical in manual application. The main disadvantages of complex mathemathical models are their extensive requirements for input data and calibration, their need for computer facilities, and the necessity of careful interpretation of the results.

One surface water quality model being used by governmental regulatory agencies and consulting engineers in the United States is QUAL-IIE. QUAL-IIE can be used to develop or evaluate waste load allocations to rivers (Ray 1990). The major steps in the waste load allocation (WLA) process are the designation of desirable water uses and the corresponding water quality standards, a cause-effect analysis of projected waste inputs and the water quality response, and a projection analysis for achieving water quality standards under various levels of waste load input. Figure 2 illustrates the principal steps in waste load allocation modeling (Ray 1990).

The basic equation solved by QUAL-IIE is the one-dimensional advection-dispersion mass transport equation, which is numerically integrated over space and time for each water quality constituent. The constituents modeled by QUAL-IIE include (Ray 1990): (1)dissolved oxygen, (2)carbonaceous biochemical oxygen demand; (3)temperature; (4)algae as chlorophyll-a; (5)nitrogen species (organic, ammonia, nitrate, and nitrite); (6)phosphorus

Figure 2. Steps in Waste Load Allocation Modeling.

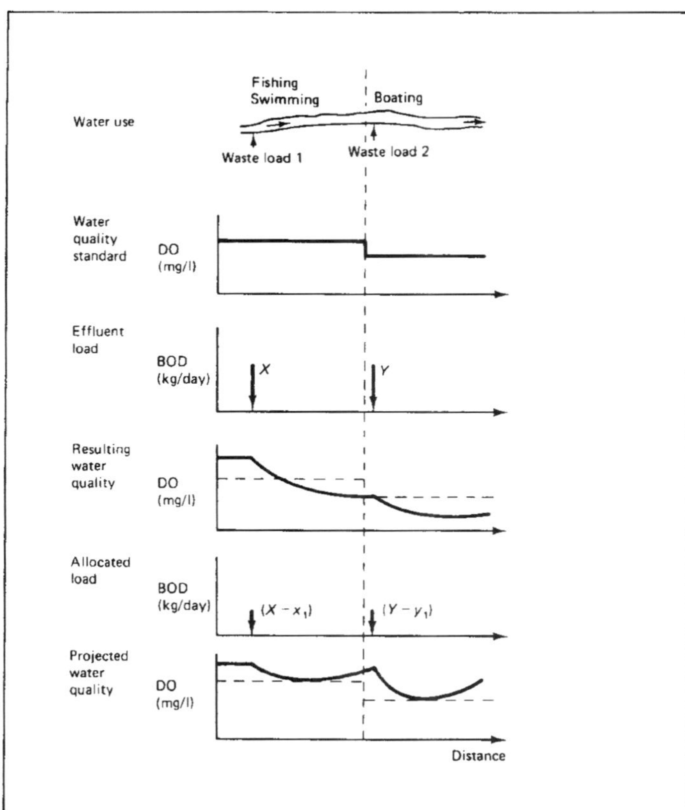

species (organic and dissolved); (7)coliforms; (8)one nonconservative constituent; and (9)three conservative constituents.

Several other considerations may be relevant to the prediction of surface water quantity/quality impacts. Examples include: (1)frequency distribution of decreased quality and quantity; (2)effects of sedimentation on stream bottom ecosystems; (3)fate of nutrients by incorporation into biomass; (4)reconcentration of metals, pesticides, or radionuclides into the food web; (5)chemical precipitation or oxidation/reduction of inorganic chemicals; and (6)anticipated distance downstream of decreased water quality and the implications for water users and related raw water quality requirements.

1.4.3.1. Reservoir Water Quality Modeling Complexities—An Example

Water quality modeling for human-made reservoirs created as a result of dam construction is an important issue in water resources planning. Reasons for this attention include: (1)the increasing use of multipurpose reservoirs with potentially conflicting objectives—i.e., reservoirs for hydroelectric power, flood control, and/or water supply; (2)the increasing adoption and enforcement of water quality standards in relation to in-reservoir and downstream water uses; and (3)the documentation of water quality changes based on information collected over the last 25 years (Canter 1995). An excellent book describing the water quality and aquatic ecology impacts of reservoirs in the Mekong River Basin in southeast Asia has been prepared (IMC 1982); information in this book has generic applicability. Since water quality changes are often temperature- and rate-dependent, they are of increased concern for reservoirs in more tropical developing countries due to higher temperatures and larger rate constants.

A number of complex and interrelated processes within a reservoir can influence water quality. These processes include thermal/density stratification, sedimentation, evaporation, chemical and/or biological cycling, bacterial die-away, and gas or nitrogen supersaturation. Summary facts regarding these processes are as follows:

1. Thermal/density stratification can occur, with the resultant promotion of epilimnetic and hypolimnetic processes (aerobic processes in epilimnion and anaerobic processes in hypolimnion).
   a. The density of water varies with the temperature of water; it is maximum at about 4°C.
   b. As a result of density differences, density-induced flow anomalies can occur in a reservoir (for example, warmer water flowing over colder water, colder water flowing under warmer water, or cooler water flowing between warmer and colder water).
   c. Because of the layering effect, reservoirs can thermally stratify and thus an epilimnion, thermocline, and hypolimnion can form — an illustration is in Figure 3 (Smalley and Novak 1978).
   d. The epilimnion is characterized by good oxygen transfer, good mixing, and diurnal temperature fluctuations; the thermocline is the transition zone; and the hypolimnion is characterized by low oxygen to anaerobic conditions, limited mixing, and nearly isothermal conditions.
   e. Stratification can lead to both lateral and vertical water quality changes within a reservoir.
2. Sedimentation of suspended materials (sediments) which flow into the reservoir can occur; suspended materials inputs are a function of drainage area size, soil characteristics, rainfall, and other factors.
   a. The sedimentation rate varies as a function of sediment size and other factors

Figure 3. Thermal Stratification of a Reservoir.

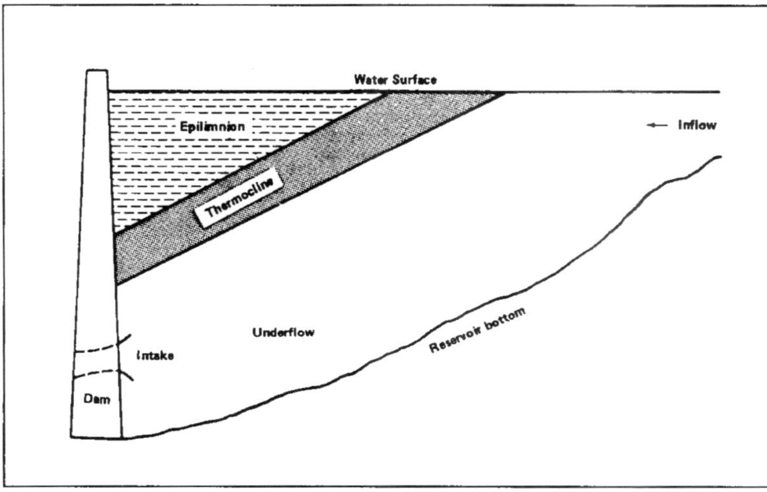

such as temperature and turbulence; sedimentation will increase with temperature and decrease with turbulence.

b. Excessive sedimentation can cause reservoir filling and thus reduce the usable life of the reservoir.

c. Sediments can adsorb dissolved constituents and transport them to the reservoir benthic zone.

d. Sediments can disrupt benthic organisms and cover previously deposited pollutants.

3. Evaporation of water from the reservoir will lead to higher concentrations of dissolved constituents; evaporation rates increase with increases in temperature.
4. Chemical and/or biological cycling of water quality constituents can occur; examples and processes include the following:
   a. nitrogen — inorganic and organic chemical forms, nitrification/denitrification, and biological uptake;
   b. phosphorus — inorganic and organic chemical forms, adsorption, ion exchange, precipitation, and biological uptake;
   c. iron and manganese — oxidation and precipitation, and reduction and dissolution;
   d. chromium — chemical forms, adsorption, oxidation and reduction;
   e. other metals — adsorption and/or precipitation;
   f. hydrogen sulfide — several chemical forms, adsorption, biological uptake, and stripping (due to anaerobic conditions and decaying vegetation);
   g. oxygen — used in decomposition, produced by algal photosynthesis, and promoted via reaeration;
   h. carbon — inorganic and organic forms, and biological uptake; and
   i. methane—may be formed from decaying vegetation under anaerobic conditions; can be stripped to the atmosphere over a spillway or by passage through a dam and/or turbines.

5. Bacterial die-away within the reservoir can occur as a function of temperature and residence time.
6. Gas or nitrogen supersaturation due to air entrainment in stilling basins can occur; high nitrogen gas concentration can cause fish impacts (Legg 1978).

Another basic issue related to water quality impacts of hydropower projects involves the consideration of reservoir input loadings of different potential pollutants. Key water pollutant sources and consequent loadings include the following: (1)point source discharges upstream of the reservoir, which can include treated and/or untreated discharges from urban areas and industrial activities; (2)nonpoint source discharges upstream of the reservoir, which can include runoff from urban, industrial, agricultural, and undeveloped areas, with some key pollutants including sediment, nutrients, organics, and pesticides (note—river basin deforestation may exacerbate these discharges); (3)in-reservoir decomposition of terrestrial flora left in place during reservoir filling, with anaerobic conditions near the bottom or in the hypolimnion causing problems with methane and hydrogen sulfide formation and then their subsequent release; (4)uses of water in the reservoir and resultant quality changes, which could cause concerns regarding organics, nutrients, and bacteria; (5)direct precipitation and/or dry deposition into the reservoir, with nutrients and metals as well as acid rain of possible concern; and (6)groundwater flow into the reservoir.

Finally, it is pertinent to consider which reservoir water quality parameters should be modeled. Typical parameters or characteristics subjected to modeling include: (1)dissolved oxygen and biochemical oxygen demand; (2)nutrients; (3)temperature; (4)others (pH, iron, solids, pesticides, fecal coliforms, methane, hydrogen sulfide, etc.); and (5)the "waste assimilative capacity" of the reservoir. It should be obvious from this discussion of reservoir biophysical and chemical complexities that modeling is subject to many unknowns and uncertainties.

1.4.3.2. Aquatic Ecosystem Modeling

Biological impact prediction models in the United States practice typically involve the use of habitat approaches. Specifically, a widely used impact prediction approach involves the Habitat Evaluation Procedures developed by the U.S. Fish and Wildlife Service (FWS 1980). In addition, the Habitat Evaluation System developed by the U.S. Army Corps of Engineers (COE 1980) has also been utilized in a number of resource planning studies. As noted earlier, these models involve the calculation of an index that incorporates both quality and quantity information. Prediction of changes involves determination of the index under baseline as well as future with and without conditions. Other types of biological impact models include species population models which may have been developed based on empirical approaches involving statistical correlations (Starfield and Bleloch 1986). The most sophisticated biological impact models are those involving energy system diagrams; these have been utilized in some resource management studies.

Several methods have been developed to quantify and assess biological impacts on aquatic resources resulting from hydrologic changes (Brookes 1988). One example is the Instream Flow Incremental Methodology (IFIM) (Brookes 1988). The IFIM was originally developed in the United States as a means of assessing how much water could be extracted from a river at various times of the year, but without adversely affecting the fishery resource. The approach is based on the concept that a particular species can be correlated with specific habitat requirements such as water quality, velocity, depth, substrate, temperature, and cover. If these requirements are known, then an assessment of the habitat suitability for a particular species can be made by measuring the quality of the available habitat. This approach has the advantage that it is easier to measure the existing habitat conditions than the fish population itself. It is also easier to predict the habitat conditions that will be produced by a proposed water extraction or flow alteration project (Brookes 1988). Numerous aquatic productivity models are also available for use in impact studies.

Risk assessment can be used to address human health and ecological risks. However, there are many uncertainties in risk assessment. For example, Figure 4 delineates the four major components of a chemical contaminant-based risk assessment, including various process limitations and uncertainties associated therewith (Paustenbach and Keenan 1989).

Figure 4. Examples of Possible Pitfalls in Conducting or Presenting Health Risk Assessments (Paustenbach and Keenan 1989).

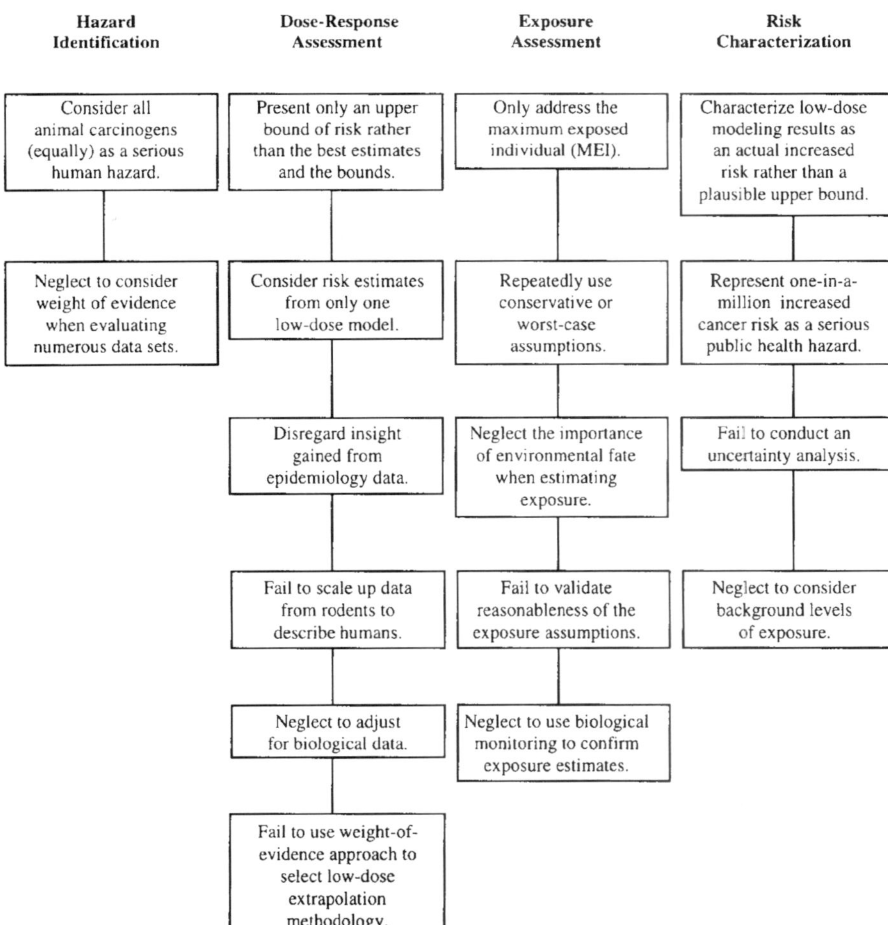

1.4.3.3. Uncertainties in Forecasting — A Summary

Forecasting is improving as a result of increased knowledge relative to the previous usage or availability of scientifically based impact prediction techniques. However, the identification and quantification of the sources of uncertainty should be an important step in the application of methods. The results of environmental predictions should indicate the margin of uncertainty involved. Further, despite the increasing number of relevant tools, scientific prediction of changes in the biophysical environment is often limited due to both technical and policy reasons. Examples of these reasons include (ERL 1982; Anderson and Burt 1985; and ECE 1992) the following:

1. Prediction often involves an attempt to predict change in a hydrological or ecological system that is complex, where many of the variables are imperfectly understood and where the system itself is constantly changing.
2. There are often time and money constraints in the planning study. Even when there is no shortage of money and other resources, it is likely that there will be insufficient time to study the environmental system in sufficient depth to provide more than a superficial understanding of the complexities.
3. Limitations arise due to inadequacies of current theory or to failure of a selected model to incorporate certain elements of current theory; such restrictions include the computational difficulties associated with solution of deterministic flow equations and the inability to incorporate variable source area concepts into drainage area runoff models.
4. Limitations are caused by the scarcity of appropriate field data for model calibration and use.
5. For many complex environmental problems, the prediction methods should take into account the interaction of various trophic levels and biotic/abiotic factors and be reliable and comprehensive. When possible, simple, quick, and inexpensive methods should be preferred.
6. Methods related to the transfer of pollutants from one environmental compartment to another should be improved. Many environmental concerns, including those in a transboundary context, demonstrate the need for transcompartmental prediction methods.

1.4.4. *Uncertainties Related to Monitoring*

Baseline hydrologic and water quality information is needed in water resources planning. Information gathering can be achieved through the usage of existing data collected by governmental agencies and/or the planning and conduction of specific baseline studies. Within the United States, the typical agencies that would have pertinent data include city, county, and state water resources agencies as well as federal agencies such as the Environmental Protection Agency, Geological Survey, Bureau of Reclamation, and Army Corps of Engineers. If data is not available, it may be necessary to plan and conduct specific baseline flow and quality studies.

Planning of baseline studies involves the development of a sampling network design, including selection of water quality and biological parameters for monitoring. Table 14 summarizes 12 steps that can be used for the design of a sampling network (Sanders 1980). The 12 steps listed should be viewed as general guidance as opposed to specific rules for design in every monitoring program. Steps 1 through 3 are related to defining the objectives of the sampling program. Steps 4 through 7 are self-explanatory, while step 8 involves revisions of sampling station locations and sampling frequencies based on information aggregated to that point. Step 9 on the development of operational plans includes consideration of sampling equipment, field or *in situ* monitoring, and laboratory analyses. Step 10 focuses on data formats and the associated data reporting. Step 11 relates to adjustment of the first ten steps for study compatibility. Finally, step 12 is associated with the preparation of a network design report.

Various water quality parameters that could be included in surface water monitoring programs are listed in Table 15 (IHD-WHO 1978). Biological surveys should be fully integrated with toxicity and chemical-specific testing methods in assessing attainment/nonattainment of water quality standards. Biological surveys can be used to detect impacts caused by: (1)pollutants that are difficult to identify chemically or characterize toxicologically; (2)complex or unanticipated exposures from spills; and (3)habitat degradation due to channelization, sedimentation, or historical contamination. Aquatic life uses, biological integrity, and biological criteria also need to be considered (EPA 1991).

Detailed information on the planning and conduction of water resources monitoring studies is contained elsewhere (USGS 1982; Canter 1985; Canter and Fairchild 1986; Canter 1993a). Detailed information related to selecting variables and biota and sediment sampling is in Chapman (1992). Selection of biological parameters is complicated due to system relationships in terms of materials and energy flows. Aquatic organisms include primary producers, plant eaters, meat eaters, and decomposers. Primary producers include algae, while plant eaters encompass zooplankton, fish, and benthic organisms. A monitoring program related to water pollution impacts should include consideration of sampling of various planktonic and benthic forms, as well as fish. The type of background data to be assembled will depend on the objectives of the planning effort and the sensitivity/usage of the surface water. Thus, impact factors will provide insight into the type of background data to be accumulated. Also, surface water characteristics that are especially sensitive and that may be influenced by proposed development projects should be evaluated.

A review of the technical requirements for planning and implementing baseline monitoring programs reveals many choices and associated uncertainties. Three central issues

Table 14. Steps Related to Sampling Network Design (Sanders 1980).

1. Determine monitoring objectives and relative importance of each.
2. Express objectives in statistical terms.
3. Determine budget available for monitoring and amount to be allocated for each objective.
4. Define the characteristics of the area in which the monitoring is to take place.
5. Determine water quality variables to be monitored.
6. Determine sampling station locations.
7. Determine sampling frequency.
8. Compromise previous objective design results with subjective considerations.
9. Develop operating plans and procedures to implement the network design.
10. Develop data and information-reporting formats and procedures.
11. Develop feedback mechanisms to fine-tune the network design.
12. Prepare a network design report.

Table 15. Selection of Parameters for River Water Quality Surveys (IHD-WHO 1978).

Type of Survey	Physical Parameters	Chemical Parameters			Biological Parameters	
		Inorganic	Organic	Nutrients	Microbiological	Hydrobiological
Proposed for inclusion in all surveys	Color pH Specific conductance Suspended solids Total solids		Chemical Oxygen Demand (COD) Total Organic Carbon (TOC)		Coliforms, total and fecal	
Recommended for collection of baseline data	Odor	Acidity Alkalinity Calcium, Ca Chlorides, Cl Dissolved oxygen Hardness Iron, Fe Magnesium, Mg Manganese, Mn Potassium, K Selenium, Se Silver, Ag Sodium, Na	Biochemical Oxygen Demand (BOD); immediate, 5-day, ultimate	Nitrate nitrogen, NO_3	Total plate count	
Recommended additional parameters where municipal and/or industrial pollution is expected	Floating solids	Arsenic, As Barium, Ba Beryllium, Be Boron, B Cadmium, Cd Chromium, Cr Copper, Cu Dissolved carbon dioxide, CO_2 Fluorides, F	Cyanide, CN Dissolved organic carbon Methylene Blue Active Substances (MBAS) Oil and grease Pesticides Phenolics	Ammonia nitrogen, NH_3 Nitrite nitrogen, NO_2 Organic nitrogen Soluble phosphorus Total phosphorus	Fecal streptococci Salmonella	Benthos Plankton counts

Table 15. Continued next page.

Table 15. Continued.

Type of Survey	Physical Parameters	Chemical Parameters			Biological Parameters	
		Inorganic	Organic	Nutrients	Microbiological	Hydrobiological
		Hydrogen sulfide, H_2S Lead, Pb Mercury, Hg Nickel, Ni Vanadium, V Zinc, Zn				
Optional parameters for surveys of special purpose	Bed load Light penetration Particle size Sediment concentrations Settleable solids	Aluminum, Al Sulfates	Carbon Alcohol Extract (CAE) Carbon Chloroform Extract (CCE) Chlorine demand	Organic phosphorus Orthophosphates Polyphosphates Reactive silica	Shigella Viruses: —Coxsackie A&B —Polio —Adenoviruses —Echoviruses	Chlorophylls Fish Periphyton Taxonomic composition

that should be considered include: (1)the representativeness of the samples/measurements at the geographical location and time of collection, (2)means of expressing natural and human-induced variations of flows and quality, and (3)necessary quality assurance/quality control (QA/QC) requirements for program operation.

1.4.5. *Uncertainties in Health Impact Issues*

Health impact concerns (disease transmission) for water resources projects are a result of construction worker influx, people relocation, and the construction area or resultant project area serving as a breeding place for mosquitoes and other disease factors. Table 16 lists some important disease vectors associated with water (WHO 1983a). These potential water-related health impacts are of particular concern for hydropower and/or irrigation projects in developing countries.

While this brief discussion does not thoroughly address how to deal with all health impact issues of water resources projects, example information on two issues is highlighted: (1)identifying the health impacts of concern for a proposed project and (2)evaluation of mitigation measures for reducing potential negative health impacts. One tool for determining the health impacts of concern is a questionnaire checklist. Table 17 includes such a checklist organized around both direct and indirect impacts for a broad range of water resources projects (WHO 1983b). Careful consideration of the relevance of each of these questions can

Table 16. Important Disease Vectors Associated with Water (WHO 1983a).

Culex quinquefasciatus, which breeds in organically polluted water and is associated with bancroftian filariasis and arbovirus transmission.

Culex tritaeniorhyncus, the rice-field-breeding mosquito known to be a vector of Japanese B encephalitis.

Aedes aegypti, the container-breeding mosquito vector of dengue and urban yellow fever.

Aedes simpsoni, breeding in banana, Colocasia, and pineapple leaf axils and incriminated in the sylvatic yellow fever cycle.

Aedes africanus, a tree-hole-breeding mosquito also incriminated in the sylvatic yellow fever cycle.

Mansonia ssp., breeding in ponds with Pistia, Salvinia, Eichornia, etc., whose roots provide its substrate; associated with brugian filariasis and Spondweni arbovirus transmissions.

Anopheles, breeding mostly in less polluted still- or slowly moving water with or without vegetation and shade; responsible for malaria and bancroftian filariasis transmission as well as Chikungunya arbovirus.

Cyclops, the tiny fresh water crustacean that thrives in ponds and is the intermediate host of dracontiasis (Guinea worm infections).

Glossina, the tsetse fly, associated with light forest type vegetation zones; certain species (*G. palpatis*) are even riverine; responsible for the transmission of African trypanosomiasis or sleeping sickness.

Simulium, the blackfly, breeding in fast-flowing clear waters attached to any suitable substratum; it is the vector of onchocerciasis and capable of long-distance migration.

Aquatic snails: bulinids - *Schistosoma haematobium* transmission; planorbids - *S. mansoni* transmission; and *Oncomelania*, the amphibious species - *S. japonicum* transmission. All these are aquatic snails found in still or very slow-moving waters and often associated with certain types of aquatic vegetation and muddy soil and capable of aestivation.

facilitate the determination of those potential health impacts of specific concern for a proposed project.

A number of design and operational measures can be used to mitigate potential undesirable vector-related impacts of water resources projects. Three major groups of measures can be considered (WHO 1983b): (1)environmental modification, that is, large-scale alterations to the form of the environment such as clearance of land before a project commences, drainage, and dewatering of areas around a project; (2)environmental manipulation, or smaller-scale control of the environment during the operational phase using physical, chemical, and biological methods; and (3)modification or manipulation of human behavior or habitats to reduce human-vector-pathogen contact. Table 18 illustrates some types of mitigation measures available for the control of vector-borne diseases associated with water resources projects.

Perhaps the greatest uncertainty related to human health impacts of water resources projects is associated with quantification of baseline conditions and forecasted effects. Considerable research is also needed on establishing cause-effect linkages between projects and resultant health consequences.

Table 17. Questionnaire Checklist of Potential Health Impacts of Concern for Water Resources Projects (WHO 1983b).

Direct Impacts on Project Area

- Will new diseases or new strains of the disease be introduced by immigrations of construction workers or new settlers? Will these affect new settlers or residents or both?
- Will relocated communities be exposed to diseases to which they have little or no immunity?
- Will new settlers be exposed to locally endemic diseases to which they have little or no immunity?
- Will food, waste, or water cycles aggravate sanitation and disease problems?
- Will housing and sanitary facilities become overburdened, misused, or not used at all, leading to conditions conducive to increases in water-washed diseases and spread of communicable diseases by the fecal-oral route?
- Will soil and water be contaminated by excreta, facilitating spread of communicable disease?
- Will introduction of migrant workers cause increases in venereal disease among workers and subsequently residents?
- Will new settlers and relocated communities be exposed to physical, social, and cultural changes leading to psychological strains and traumas? (These may include changes in lifestyles and employment.)
- Will changes in food supplies lead to possibilities of malnutrition, nutritional deficiencies, or toxic effects? These effects may occur because of:
    - introduction of Western-style convenience foods;
    - changes in staple foods—possibly using unfamiliar toxic plants as substitutes for usual foods;
    - contamination of soil or agricultural water supplies with toxic substances;
    - reduced productivity of soils caused by hydrological changes (waterlogging, etc.), mineralization, or pollution of groundwater and surface waters;

Table 17. Continued on next page.

Table 17. Continued.

- reduced productivity of fisheries caused by hydrological changes or water pollution;
- change in availability of trace metals in soils caused by hydrological changes (lowering or raising of water table, etc.).

- Will effluents and emissions or substances released intentionally into the environment (e.g., pesticides) pollute air or water or soil, presenting a threat to human health?

- Will irrigation of fields increase opportunities for human contact with water-borne, water-based and water-related disease?

- Will traffic in the area, and therefore road accidents, increase as a result of the development?

- Will new industries and similar activities attracted to the area by growth result in pollution of air, soil, or water or noise pollution, with subsequent impacts on human health?

Indirect Impacts Through Effects on Disease Vectors

- Will new vectors be introduced into the area from upstream as a result of hydrological changes?

- Will new vectors be introduced into the area on vehicles, animals, transplanted plants, soil, etc.?

- Will existing vectors be infected or reinfected by contact with infected humans coming into the area?

- Will the prevalence and distribution of existing infected vectors be changed by changes in the availability of suitable habitats for breeding and survival? These changes may result from hydrological changes (water velocities, temperature, depth, standing water, etc.), morphological changes (bank slopes, cover, etc.), climate changes (rainfall, humidity) and biological changes (vegetation, predators, etc.). They may affect presently infected or uninfected areas.

Direct Impact on Workers

- Will migrant workers be exposed to locally endemic diseases to which they have little or no immunity?

- Will migrant workers be exposed to psychological strains and traumas from changes in living and working conditions?

- Will workers be exposed to physical threats to their safety (injuries, deaths) or chemical and physical hazards to health (toxic substances, noise, vibration, radiation, high pressures, etc.)?

- Will workers be particularly exposed to contact with water and thus to water-associated disease during their work?

- Will workers be exposed to dangerous animals during their work (snakes, scorpions, etc.)?

- Will adequate supplies of food be provided to prevent malnutrition and minimize spread of disease (e.g., by use of itinerant food vendors)?

Impact on Health Services

- Will health and other social services be overburdened, with consequent effects on health of residents and workers?

Table 18. Mitigation Measures for the Control of Vector-Borne Diseases.

Vector or intermediate host	Diseases transmitted	Environmental modification						Environmental manipulation						Modification or manipulation of human habitation or behavior				
		Drainage (all types)	Total earth filling	Deepening and filling	Land grading	Velocity alteration	Impoundment	Clearing and burning of terrestrial vegetation	Shading or exposure to sunlight	Water level fluctuation	Sluicing for flushing	Aquatic vegetation control	Salinity regulation	Water supply and sewerage	Screening and bednets	Refuse collection and disposal	Land-use restriction	Improved housing
Anopheles mosquitoes	Malaria	++	++	++	++	+	-	+	+	+	+	+	++	+	+	+	+	+
Aquatic snails	Schistosomiasis	+	+	++	+	+	-	-	-	+	+	+	+	++	-	-	+	-
Culex and *Aedes* mosquitoes	Filariasis: viral and other diseases	++	+	+	+	+	-	+	+	+	+	+	+	+	+	+	+	+
Blackflies	Onchocerciasis	-	-	-	-	+	++[a]	-	-	-	+	-	-	+	-	-	+	-
Houseflies	Infantile diarrhea	-	-	-	-	-	-	-	-	-	-	-	-	++	+	++	-	+
Tsetse flies	African trypanosomiasis	-	-	-	-	-	-	++	-	-	-	-	-	-	-	-	+	-
Triatomid bugs	Chagas' disease	-	-	-	-	-	-	+	-	-	-	-	-	-	+	-	-	++
Rat fleas	Plague	-	-	-	-	-	-	-	-	-	-	-	-	-	-	++	-	++
Cyclops	Dracontiasis	-	-	-	-	-	-	-	-	-	-	-	-	-	++	-	-	-

Key: - = Little or no directly demonstrated value, or not applicable. ++ = Primarily effective (most species).
+ = Partially effective (some species). [a]Small dams = adverse effect; large dams = good effect.

1.4.6. *Uncertainties Related to Climate Changes*

The main consequences of climatic changes related to inland waters include (da Cunha 1988): (1)changes in the global amount of water resources available and in the spatial and temporal distribution of these resources; (2)changes in soil moisture; (3)changes in extreme phenomena related to water resources, i.e., floods and droughts; (4)changes in water quality; (5)changes in sedimentation processes; and (6)changes in water demand. Another consequence of the greenhouse effect and climate change could be the rise of the sea level due to the melting of the ice cover. The impact of a sea level rise would be quite different from one region to another, depending on such factors as land elevation, local morphologic and hydrologic conditions, degree of urban and economic occupation, and degree of development of flood protection schemes and water resources management. The main consequences of climatic change related to coastal waters are (da Cunha 1988): (1)flooding of urban, industrial, and service areas; (2)rendering inoperable civil engineering hydraulic works and other facilities; (3)intrusion of sea water in aquifers and estuaries; (4)changes in coastal sedimentation processes; and (5)general ecological changes in the coastal environment.

Since climate change is the initiating event for the above-summarized impacts on water resources, it is necessary to consider critically the prospects for such climate change. An excellent summary of the prospects is presented by Schneider et al. (1990). Overview information on climate change as related to water resources in the United States is in Waggoner (1990).

One difficulty in assessing the prospects for climate change is that such changes have occurred naturally over widely diverse scales of time and space. For example, although the hemispheric average of interannual temperature variability may be no more than 0.5°C, local variability can be larger. For example, warm or cold winters can differ locally by many degrees from averages, and interannual precipitation variability can be many times different from average variability. Another difficulty in assessing climate change retrospectively is the wide variability in temperature and precipitation measurement records. Variability exists temporally, spatially, and in the measurement techniques utilized. However, reconstructed data do suggest some longer-term rises in temperature (Schneider et al. 1990).

Climate models vary in their spatial resolution, that is, how many dimensions they simulate and the spatial detail they include (see Chapter 7). A simple model calculates only the average temperature of the earth, independent of time, as an energy balance between the earth's average reflectivity and the average greenhouse properties of the atmosphere. In contrast, three-dimensional climate models reproduce the variation of temperature with latitude, longitude, and altitude. The most complex models are known as General Circulation Models (GCMs); they can be used to predict the time evolution of temperature plus humidity, wind, soil moisture, sea ice, and other variables through three dimensions in space (Schneider et al. 1990). Even though numerous models have been developed, problems exist relative to their validation. Such problems have led to considerable debate in the scientific community over predicted changes and, more particularly, the time scales of the predicted changes. In order to mitigate climate change effects on water resources, reliable information is needed now; however, considerable uncertainty exists. This uncertainty tends to lead to postponements of necessary decisions regarding engineering works to minimize detrimental impacts on water resources.

Based on this brief review of water resources issues related to global climate change, the following summary comments can be made:

1. Global climate change caused by the greenhouse effect appears to be a reality, although major differences of scientific opinion exist as to the rates of change (NAS 1987; Waggoner 1990).

2. Numerous inland and coastal water resources management issues will be impacted when temperature and precipitation patterns change.
3. The global perspective presented in most studies is not specific enough for delineating water resources implications in regional and local areas.
4. Global climate change might be "controlled" by effective programs to reduce the atmospheric emissions of greenhouse gases.
5. Global climate change must rise higher on national and international political agendas before effective societal measures can be implemented for management of the water resources implications.

1.5. WATER RESOURCES MANAGEMENT STRATEGY

An effective and integrated water resources management program at the local, regional, or national scale must be based on multiple elements encompassing both technical and policy issues. Such a program must address both surface water and groundwater resources, water demand and usage limitations, pollution sources and control, and predecision analysis associated with land-use planning and development activities. Table 19 includes 12 suggested elements of a water resources management strategy (Canter 1991). It should be noted that while the nonprioritized listed elements may not be all-inclusive, they do represent the majority of pertinent issues/concerns.

Surface water and groundwater resources are typically interrelated in given geographical areas. During periods of high surface water flows, alluvial groundwater systems can be recharged, while during low surface flow periods, groundwater may be the dominant contributor to the stream base flow. Water usage from either surface water or groundwater resources may have implications in terms of the quantitative aspects of the abstracted resource as well as the related resource. Water withdrawals for usage and associated pollutant discharges can also influence the quality characteristics of both surface water and groundwater systems. Accordingly, an ideal water resources management program would encompass responsibilities for both surface water and groundwater systems; however, in many cases,

Table 19. Elements in an Integrated Water Resources Management Program.

1. Recognition of relationships between surface water and groundwater resources.
2. Statutory authority and water usage and quality standards.
3. Resource characterization studies (quantity and quality).
4. Resource usage studies.
5. Water usage conservation program.
6. Resource pollution studies.
7. Pollution source control and resource protection program (point and nonpoint sources).
8. Technically based decisionmaking regarding usage allocations, wastewater (point and nonpoint) discharges, and permitting program.
9. Monitoring and enforcement program.
10. Emergency response program.
11. Institutional coordination program (related to element 2 above).
12. Public awareness, public participation, and education program.

these are divided in an institutional structure. If surface water and groundwater management is separated, institutional coordination will obviously be necessary.

Statutory authority basically refers to the legal authority vested in the entity charged with the responsibility of water resources planning and management. To serve as an illustration of the potential complexity of this element, Table 20 lists federal statutes related directly or indirectly to water resources management in the United States. It should be noted that states and even local governmental agencies may have laws, regulations, and/or policies that are also directly or indirectly pertinent. It is highly unlikely that one agency would have the responsibility for administering and funding allocations for all pertinent laws, regulations, and/or policies; thus it becomes vitally important that institutional coordination occur on a planned basis.

Water usage standards refer to specific limitations on the quantities of withdrawal from surface water and/or groundwater systems. Usage limitations or restrictions may be based on the surface water flow conditions or on the amount of land ownership in relation to groundwater abstraction. For example, in Oklahoma in the United States, the maximum groundwater abstraction is two acre-feet/year/acre of land surface owned. To continue the illustration of water usage limitations, the rights to groundwater usage in the United States can be summarized in four categories (Bouwer 1978):

1. English Rule (Common Law): This approach is based on the premise that the groundwater belongs to the owner of the land above it.
2. American Rule (Reasonable Use): This approach is similar to the English Rule except that groundwater use is limited to "reasonable use." Reasonable use is a fairly ambiguous concept and is often defined in court through lawsuits.
3. Correlative Rights: This approach is a modification of the American Rule; it attempts to give equal distribution of groundwater among land owners based on land area owned.
4. Prior Appropriation: This approach is based on the concept of first-in-time, first-in-right. Thus, the first water user is guaranteed the groundwater that user has historically utilized, and users who come along later must rely on the remainder.

The more arid the region (and when alternate surface water or groundwater supplies are not readily available), the more important it becomes to have carefully developed regulatory approaches for water usage.

Surface water and groundwater quality standards are vital to an effective water resources management program. Standards are necessary for development and use of surface water or aquifer classification systems, pollution source evaluations, interpretation of water quality monitoring data, and the issuance of permits (Canter 1986). Surface water quality standards can be related to beneficial use designations of stream segments in a river system. For example, in the state of Oklahoma, the following beneficial use designations are pertinent:

- A - public and private water supplies;
- B - emergency public and private water supplies;
- C_1 - fish and wildlife propagation;
- C_2 - fish and wildlife propagation to the extent allowed by specifically stated water quality parameters;
- D - agriculture (includes livestock water and irrigation);
- E - hydroelectric power;
- F_1 - municipal and industrial cooling water;
- F_2 - receiving, transporting, and/or assimilating adequately treated waste;
- G_1 - recreation, primary body contact (includes recreational uses where the human body may come in direct contact with the water to the point of complete body submergence);

G_2 - recreation, secondary body contact (includes recreational uses, such as fishing, wading, and boating, where ingestion of water is not probable);
H - navigation;
I - aesthetics;
J - small-mouth bass fishery excluding lake water; and
K - trout fishery.

The important point to note is that specific numerical standards for water quality constituents will differ as a function of the pertinent beneficial use designation. In recent years, the number of constituents addressed in surface water quality standards has been increasing, and the numerical standards (in terms of maximum allowable concentrations) have been decreasing. Potentially toxic constituents are now being addressed in addition to traditional parameters such as residues and inorganic salts.

Groundwater quality standards have tended to be fewer in numbers in terms of the constituents addressed. For example, historical groundwater quality standards have primarily addressed inorganic constituents; however, with the current emphasis on synthetic organic and metal constituents, numerical standards for these components are being promulgated.

Discharge standards are also pertinent for point and nonpoint sources releasing effluents into surface bodies of water. In the United States, publicly owned treatment works (POTWs or municipalities) are subjected to secondary treatment standards (or even tertiary treatment standards depending upon the location in relation to other discharges and receiving water quality). Industries are subject to pretreatment standards if discharging into POTWs or to industry-specific effluent standards if discharging directly into a receiving body of water. Stormwater runoff (nonpoint source) discharge standards are also being developed.

Resource characterization studies must be planned and conducted in order to have a technical (and quantitative) basis for subsequent decisionmaking. These studies should focus on determining flow variations and quality characteristics of surface bodies of water and on determining developable groundwater supplies and their associated quality characteristics. Resource characterization studies should be carefully planned and implemented; in fact, they can include baseline studies as well as continuing water resources monitoring. In areas with minimal data, it should be noted that characterization studies can be both time-consuming and expensive, particularly if information is needed on the hydrogeological characteristics of aquifer systems. Finally, there is an expanding need for the development of compatible data bases and information management systems to facilitate information dissemination and use.

Resource usage studies are also vital in a water resources management program. Historical patterns of the usage of both surface water and groundwater resources should be established. Comparisons of current usage patterns with applicable usage limitations (based on resource availability and/or applicable governmental laws, regulations, or policies) may be necessary in developing an optimum and equitable water resources management plan. Depending on water usage patterns in relation to surface water and/or groundwater resource availability, it may be necessary to develop a water conservation and/or reuse program that can be implemented on an as-needed basis. The program may involve a public awareness effort, and it may need to include an enforcement component.

Resource pollution studies will also be needed in a comprehensive water resources management program. These studies should be directed toward identifying both point and nonpoint sources of pollution in relevant surface water and groundwater systems. With expanding documentation of surface water and/or groundwater pollution resulting from human-made sources, it is important to prioritize water pollution source categories and individual sources. Prioritization schemes may include consideration of the type of pollutant,

Table 20. Examples of Federal Statutes Related to Water Resources Management.

Agriculture and Food Act (Farmland Protection Policy Act) of 1981	7 U.S.C. 4201 *et seq.*
Anadromous Fish Conservation Act of 1965, as amended	16 U.S.C. 757a *et seq.*
Clean Air Act of 1972, as amended	42 U.S.C. 7401 *et seq.*
Clean Water Act of 1972, as amended	33 U.S.C. 1251 *et seq.*
Coastal Barrier Resources Act of 1982	16 U.S.C. 3501-3510
Coastal Zone Management Act of 1972, as amended	16 U.S.C. 1451 *et seq.*
Comprehensive Environmental Response, Compensation, and Liability Act of 1980	42 U.S.C. 9601
Deepwater Port Act of 1974, as amended	33 U.S.C. 1501
Emergency Flood Control Funds Act of 1955, as amended	33 U.S.C. 701m
Endangered Species Act of 1973, as amended	16 U.S.C. 1531
Estuary Protection Act of 1968	16 U.S.C. 1221 *et seq.*
Federal Environmental Pesticide Control Act of 1972	7 U.S.C. 138 *et seq.*
Federal Water Project Recreation Act of 1965, as amended	16 U.S.C. 4601
Fish and Wildlife Coordination Act of 1958, as amended	16 U.S.C. 661
Flood Control Act of 1944, as amended, Section 4	16 U.S.C. 460b
Land and Water Conservation Fund Act of 1965	16 U.S.C. 4601
Marine Mammal Protection Act of 1972, as amended	16 U.S.C. 1361
Marine Protection, Research, and Sanctuaries Act of 1972	33 U.S.C. 1401
Migratory Bird Conservation Act of 1928, as amended	16 U.S.C. 715
Migratory Bird Treaty Act of 1918, as amended	16 U.S.C. 703
National Environmental Policy Act of 1969, as amended	42 U.S.C. 4321 *et seq.*
Resource Conservation and Recovery Act of 1976	42 U.S.C. 6901-6987
River and Harbor Act of 1888, Section 11	33 U.S.C. 608
River and Harbor Act of 1899, Sections 9, 10, 13	33 U.S.C. 401-413
River and Harbor and Flood Control Act of 1962, Section 207	16 U.S.C. 460d
River and Harbor and Flood Control Act of 1970, Sections 122, 209, 216	33 U.S.C. 426 *et seq.*
Safe Drinking Water Act of 1974, as amended	42 U.S.C. 300f
Submerged Lands Act of 1953	43 U.S.C. 1301 *et seq.*
Superfund Amendments and Reauthorization Act of 1986	42 U.S.C. 9601
Surface Mining Control and Reclamation Act of 1977	30 U.S.C. 1201-1328
Toxic Substances Control Act of 1976	15 U.S.C. 2601
Water Resources Development Act of 1974, as amended	88 Stat. 12
Water Resources Development Act of 1976, Section 150	90 Stat. 2917
Water Resources Development Act of 1986	33 U.S.C. 2201 *et seq.*
Water Resources Planning Act of 1965, as amended	42 U.S.C. 1962a
Watershed Protection and Flood Control Act of 1954, as amended	16 U.S.C. 1001 *et seq.*
Wild and Scenic Rivers Act of 1968, as amended	16 U.S.C. 1271 *et seq.*

its transport and fate characteristics, and its potential for limiting water usage. Several methodologies have been developed to provide a systematic basis for source prioritization (Canter et al. 1987). In addition, it should be noted that specific pollutant source categories may be the administrative responsibility of differing agencies at local, state, or national governmental levels; accordingly, there can be overlap in information needs and prioritization methodologies used to address multiple source types in a given geographical area.

Development of source control measures, or preventive measures, directed toward minimizing future surface water and/or groundwater pollution from existing and/or new sources is important in water quality protection. Ongoing programs of source control and permitting continue to require considerable time and effort. Three examples of groundwater pollutant source categories that have been or are being considered for subjection to prevention programs in the United States include septic tank systems, underground storage tanks, and the use of agricultural chemicals (Canter 1986). Permitting may also involve hazardous waste sites, liquid impoundments, sanitary landfills, and deep injection wells for chemical wastes.

An example of a resource protection program is the wellhead protection program initiated by the 1986 Safe Drinking Water Act in the United States. The "wellhead protection area" means the surface and subsurface area surrounding a water well or wellfield, supplying a public water system, through which contaminants are reasonably likely to move toward and reach such water well or wellfield. The extent of a wellhead protection area necessary to provide protection from contaminants that may have any adverse effect on the health of persons is a function of factors such as the radius of influence around a well or wellfield, the depth of drawdown of the water table by such well or wellfield at any given point, the time or rate of travel of various contaminants in various hydrologic conditions, distance from the well or wellfield, or other factors affecting the likelihood of contaminants reaching the well or wellfield, taking into account available engineering pump tests or comparable data, field reconnaissance, topographic information, and the geology of the formation in which the well or wellfield is located (EPA 1987).

Examples of surface water pollution sources in the United States subjected to controls/ treatment measures via permitting programs include municipal and industrial wastewater point source discharges and nonpoint discharges (storm water) from urban and industrial areas. Wastewater treatment is typically required for point sources, and the utilized technologies can be considered in terms of primary, secondary, and tertiary (advanced) schemes. A wide range of such technologies is available for industrial wastewater, while a more narrow set of choices has been utilized for domestic wastewater. Examples of unit processes or operations include sedimentation, activated sludge, trickling filtration, air stripping, and chemical precipitation. Sludge handling and disposal may include dewatering, conditioning, digestion, and/or incineration. In addition to end-of-pipe or emission source control technologies, increasing attention is being given to waste minimization and pollution prevention measures. In industry, there is increasing attention to the use of "clean technologies" (less polluting). Also, wastewater recycling and reuse is critical in water-deficient areas.

Nonpoint source pollution can be managed through the application of Best Management Practices (BMP), which means a practice or combination of practices that is determined by a state (or designated areawide planning agency) after problem assessment, examination of alternative practices, and appropriate public participation to be the most effective practicable (including technological, economic, and institutional considerations) means of preventing or reducing the amount of pollution generated by nonpoint sources to a level compatible with water quality goals (Novotny and Chesters 1981). To serve as an example, several management measures are available for preventing, or at least minimizing, nitrate pollution of groundwater resulting from agricultural practices. Prevention measures can be considered in two groups: (1)fertilizer management measures; and (2)other management measures.

Table 21 contains a categorical listing of ten types of fertilizer management measures the most frequently utilized measures are associated with matching the timing and rate of fertilizer application to the needs of the crops being fertilized. Table 21 also contains a listing of seven other management measures representing a compendium of approaches that could be used to control nitrate pollution of groundwater from agricultural practices. The approaches can be used either singly or in various combinations.

An important element in an integrated water resources management program is technically based decisionmaking. Such decisionmaking may be required relative to surface water and/or groundwater usage allocations, waste load allocations and permits relative to point/nonpoint pollutant discharges, and/or technologies for pollution source control and/or environmental remediation projects. Fundamental tools to analyze problems or needs and the technical consequences of various alternatives being evaluated include surface water flow and quality models and/or groundwater flow and solute transport models. Risk assessment and environmental impact assessment studies may also be used to provide basic information for decisionmaking. Tools that can provide a structure for technically based decisionmaking include multiattribute or multicriteria decision models, multiattribute utility measurement, and/or decision analysis.

Continuing environmental monitoring will be a requirement in an integrated water resources management program. The monitoring should be targeted on critical water resources (in terms of quantity and/or quality) and on key pollutant sources. Data organization and presentation can include the use of numerical indices representing a composite of measured quality/quantity parameters. One such index is the Water Quality Index developed in the United States in 1970 (Ott 1978). Monitoring results should be integrated with an enforcement program, or at least they should be coordinated with enforcement effects. Such

Table 21. Examples of Fertilizer Management Measures.

Measure
Fertilizer Management
Match timing of fertilizer application to meet crop needs.
Match fertilizer application rate to crop needs.
Choose fertilizer application technique to minimize nitrogen losses.
Limit fertilizer application in well recharge areas.
Regulate crop types to those with minimal fertilizer requirements.
Implement management policies intended to minimize nitrogen losses.
Adopt legislation related to minimizing fertilizer usage.
Implement taxation plan focused on minimizing fertilizer usage.
Implement Best Management Practices (BMPs).
Control the usage of irrigation.
Other Management
Implement drinking water protection zones.
Consider soil organic nitrogen in determining fertilizer requirements.
Implement appropriate water management practices.
Choose irrigation method to minimize nitrogen losses.
Choose tillage practice and soil conditioning to minimize nitrogen losses.
Use nitrification inhibitors.
Use well packer for selective zoning of groundwater withdrawal.

efforts can include periodic inspections of pollutant discharges, annual or less frequent environmental regulatory auditing, renewable permits after three to five years, and the preparation of annual monitoring and enforcement reports by the responsible agency.

Surface water and groundwater protection programs should also address responses to both acute and chronic pollution problems. Acute problems are reflected by accidental chemical spills, and immediate and appropriate remedial action measures are necessary to minimize water pollution. Chronic groundwater pollution has occurred over time as a result of leachates from hazardous waste sites and landfills, as well as leakage from underground storage tanks and liquid impoundments. Remedial action efforts for chronic groundwater pollution are expanding, with the best-known efforts associated with the cleanup of hazardous waste sites through the federal Superfund program in the United States. A comprehensive water pollution protection strategy should include a contamination response program. Implementation of a pollution incident tracking program is also important. It is anticipated that there will be more incidents of water pollution as a result of both inadvertent waste disposal practices and the gathering of heretofore unknown information relative to both the surface water and groundwater quantity and quality. The contamination response program must be based in part upon a pollution incident tracking program.

Institutional coordination is needed among the various governmental entities with partial or full responsibilities for various functions and activities. Examples of functions and activities with overlapping agency responsibilities include water quality monitoring, aquifer mapping, pollution source prioritization, source control programs, permits and enforcement, and use of data bases (Canter 1986). These overlapping responsibilities typically result from single-purpose laws and administrative responsibilities. A common need in a water quality protection program is for technical as well as policy-oriented information. Since multiple agencies may be involved in collecting relevant information for their use, as well as use by others, there is a major need for the provision of information access. In addition, it is desirable to optimize data and information exchange among agencies at a given governmental level and among different governmental levels. Due to multiple agency involvement and continuing responsibilities related to various elements in water quality protection programs, it may be desirable to consider joint funding of some elements and sharing of technical and policy personnel for achieving selected functions and activities.

A final element for a water resources management program is associated with information transfer and education (Canter 1986). Education and training needs can be viewed from both the perspective of staffs involved in program planning and implementation, and the perspective of the general public. One of the frequently identified issues related to water resources, and particularly to groundwater, is the need for technically trained individuals. Technical staffs, as well as policy and legal staffs, are necessary in the planning and implementation of water quality protection programs. In addition to technical and policy-related education needs, there is a major need for increasing public awareness and participation relative to the importance of surface water and groundwater as a resource and appropriate measures for protection and remediation. Individuals employed by local, state, and national government agencies, as well as elected officials and their staffs, have needs for training relative to resource awareness and general policy implications.

2. Ethical Issues in Sustainable Water Resources

2.1. ETHICAL PRINCIPLES IN AGENDA 21 PROVISIONS DEALING WITH WATER RESOURCES

Part 1 of this chapter identified many scientific issues that must be dealt with in management of water resources. This section discusses several ethical matters that must be faced in implementing water resources policy. By "ethics" is meant that domain of inquiry that attempts to define what is right or wrong, obligatory or nonobligatory. Ethical statements about water resources are prescriptive, while scientific statements are usually understood to be descriptive of water resource "facts," including facts about causes of water resource degradation. It is generally accepted that science cannot deduce prescriptive statements from facts. That is, one cannot deduce "ought" from "is" without supplying a new minor premise. One cannot introduce an evaluative term, such as "acceptable risk," into the conclusion of an argument if the prior premises of that argument are entirely nonevaluative (e.g., probability of water being contaminated by a project). Although a description of certain facts may suggest an ethical position, one cannot through a description of the facts alone deduce an ethical conclusion. An ethical system such as utilitarianism may provide the minor premise needed for ethical reasoning. For instance, if one concludes that water resources program option A will create the greatest happiness, by applying the utilitarian maxim that one should choose the option that creates the greatest happiness, one can conclude that option A is the optimal solution. From a proposition that such-and-such a problem creates a particular risk, one cannot, however, deduce whether that risk is acceptable without first deciding on certain criteria for acceptability. Therefore, on this largely traditional view of the logic of ethics, science cannot answer ethical questions all by itself.

The general objective of Agenda 21 relating to water resources is "to make certain that adequate supplies of water of good quality are maintained for the entire population of this planet, while preserving the hydrogeological, biological, and chemical functions of the ecosystems, adapting human activities within the capacity limits of nature..." (United Nations 1992: Sec. 18.2). Thus, Agenda 21 identifies two main goals for sustainable water resources programs: (1)to meet human needs and (2)to preseve the functions of ecosystems. These dual objectives of Agenda 21 for water resources decisionmaking is compatable with all commonly accepted ethical or moral positions. That is, no common moral or religious conviction or ethical theory opposes the objective of making water available to meet human needs while managing water so that ecosystem functions are protected. Meeting human needs and preserving the functions of ecosystems is, for example, supportable under Kantian, Rawlsian, utilitarian, Aristotelian, and discourse-based ethical theories.

The goal of preserving the functions of ecosystems should, however, be distinguished from the duty to protect ecosystems as such. Preserving the function of ecosystems is supportable by all Western anthropocentric ethical systems because humans need ecosystems to live. Protecting ecosystems without regard to their use to humans, however, is a goal not necessarily supported by Western ethical systems. A person operating within a Western anthropocentric ethical framework might value an ecosystem highly but feel no obligation to protect it.

The Agenda 21 objectives for water resources have also been formally endorsed by international institutions other than the United Nations, such as the World Bank. Thus, Agenda 21's general objectives for water resources are widely accepted as a matter of ethical theory or institutional policy throughout the world. Therefore, the ethical controversies that are entailed by implementation of sustainable water resource programs under Agenda 21 are not created by lack of coherent general policy objectives but will arise in interpretation and application of these general objectives in concrete situations.

The ethical issues that must be faced in implementation of Agenda 21 provisions of water resources include, among others:

1. resolving conflicts between human demand for water resources and needs of water for ecosystem protection in situations of scarcity;
2. developing just allocation schemes within and among nations for limited water resources;
3. determining rights of future generations in relation to water resource availability;

4. dealing with human actions that may impact water resources in the face of scientific uncertainty about those impacts; and
5. applying economic analyses to evaluate alternatives to water resource projects.

These and other ethical issues will need to be faced in the years ahead if governments throughout the world are to make water resource projects sustainable. These issues are particularly important and urgent in areas of the world where water is scarce and demand is growing, such as the Middle East or in parts of Africa.

2.2. HUMAN VERSUS ECOSYSTEM NEEDS

Water resource projects often must face conflicts between competing potential uses of water where water is desired to meet expanding human demands but where water is also necessary to maintain ecosystems. Thus, water projects or proposed projects that have effects on water resources may create ethical questions relating to human responsibilities to nonhuman species. Although water in itself may have little intrinsic value, it serves as a necessary means to animals, certain plant species, and ecosystems beings, which have intrinsic value or inherent worth according to many philosophers. Therefore, water resource projects must resolve conflicts between rights of humans to fulfill basic needs and arguable rights of nonhuman species.

Even those ethicists, such as utilitarians, who deny the existence of rights for humans or nonhuman species agree that important ethical issues arise when projects must face competing water uses. For instance, water resource projects are often justified on cost-benefit analysis, which justifications are often argued to be a species of preference utilitarian ethical legitimization by some philosophers, yet such justifications raise a host of ethical controversies.

Many philosophers assert that humans have a right to fulfill their basic needs. The ethical argument that supports such a right takes its starting point from the assumptions that humans possess certain basic rights, including rights to liberty and political rights. Because the possibility to exercise these rights rests on the fulfillment of basic needs, then people also have rights to all those commodities that are related to basic needs (such as fresh water). According to those philosophers who assume that humans possess certain basic rights (see, e.g., Dworkin 1978), access to fresh water presumably belongs to a class of personal human rights to share in the just distribution of common goods.

However, even if one starts with the assumption that persons have some basic rights to meet basic needs, it does not necessarily follow that humans have a right to water resources according to all interests in water. Not all uses of water serve the fulfillment of primary human needs. Although water for drinking and eating is necessary to support human life, water used to irrigate domestic lawns is not. Therefore, if one assumes that nonhuman species and ecosystems have inherent worth, the use of water for other than fulfillment of basic needs that damages environmental entities may be criticized on ethical grounds. Along this line, water resource allocation schemes that assume that water will be provided to meet human demand without regard to needs to provide water to protect ecosystems can be criticized on ethical grounds. It does not, however, follow that all "luxury" water usages in the rich, "wet" countries or where water is not scarce are ethically wrong. Many usages of water for "luxury" needs in the wet and rich countries may be sustainable, while in arid zones, practices that serve more urgent needs might not be sustainable. Therefore, the ethical appropriateness of any water resources project may depend upon the particular location of the project.

Yet some types of activities are always objectionable because they are incompatible with the concept of sustainability. For example, some human projects significantly reduce the amount or the quality of fresh water, including some types of "poor" land use such as deforestation, excessive groundwater mining, pollution of aquatic systems, using rivers and lakes as sewers for discharge, and certain types of wasteful irrigation. To the extent that these practices damage ecosystems, they are incompatible with the notion of sustainability and are usually ethically objectionable. Some practices may be ipso facto incompatible, while others may be incompatible only if done in an "inadequate" way. The notion of sustainability entails, for example, that extraction rates of groundwater should not excessively exceed recharge rates or that both rates should be balanced in a state of equilibrium.

Where conflicts exist among different desired uses of water, the "best" uses of water must be understood to create prescriptive values questions as distinguished from descriptive scientific or technical questions. To answer such values questions, no amount of value-neutral descriptive analysis can logically certify a prescriptive course of action. Therefore, as we shall discuss in more detail below, no amount of economic analysis by itself can logically determine best use. Managers may decide to use market mechanisms to allow willingness to pay to determine the best uses of water, but such price-based determinations should not be understood to be ethically neutral determinations of value. The most valuable use of water raises questions about relative distribution, reallocation, sustainability of existing supplies, social unrest, and governments' search for the "good life" (Priscoli 1994). In a world of limits, therefore, not only do sustainable development decisions about water resources raise ethical questions, but also these decisions must be understood to be positions taken about ethics and values.

2.3. INTRANATIONAL AND INTERNATIONAL DISTRIBUTIVE JUSTICE AND WATER RESOURCES

The previous section of this chapter examined ethical questions that arise in decisions about water resources where there are conflicts between human demand for water and the need of water to protect ecosystems. This section discusses ethical issues that arise when different people desire limited water. When water is scarce and not all uses of water can be met, decisionmakers should look to principles of distributive justice when attempting to determine how benefits and burdens should be distributed among different people. Principles of distributive justice assert that benefits and burdens should be distributed according to concepts of equality or merit or some combination of these two. Where basic human needs are at stake, principles of distributive justice demand that humans be treated equally but allow merit to be considered as a distribution consideration after basic needs have been met.

The geographical distribution of water throughout the world is extremely unequal. No one can be morally blamed for this inequality itself, since it results from the prehuman history of the earth. Therefore, in water resource allocation, injustice has to be distinguished from inequality. This unavoidable distinction is a challenge for any ethical analysis because the principles of distributive justice have to be applied under circumstances that do not allow identification of inequality with injustice. For instance, it is not obviously unjust that some people who live in an arid part of a nation have less water than those in wetter areas. Yet it may be unjust to allocate water unequally among persons living in the arid area or to allow water used by some to be polluted by an industrial project while others receive the benefits of that project.

In a world of limits, if nations can no longer assume they are able to solve problems of poverty by simply expanding the economic pie, limited water resources raise urgent new questions of distributive justice between rich and poor within nations. That is, limited water resources will force nations to decide who shall receive the benefits of water resource projects. Should limited water resources be used for irrigation so that farmers benefit or for industrial development, which benefits city dwellers and investors in industrial development projects? These kinds of questions may force some nations to consider anew questions of social justice within their borders as demand for limited water increases.

In addition to these questions of intranational distributive justice, scarce water resource problems often raise issues of international fairness. Because water basins or groundwater aquifers do not conform to national boundaries, a nation's use of water within its jurisdiction often raises questions about fairness in relation to other nations that share the water resources. Throughout the world there are areas with limited water and growing demand. This situation often causes serious conflicts among nations about water usage. In other parts of the world, international conflict over water has been avoided through the creation of international bodies that allocate water cooperatively.

Where international water bodies have been degraded by one nation's activities, principles of distributive justice require that that the polluting nation bear the costs of remedial action. Moreover, if water resource problems are to be solved by new technologies, the poor nations cannot rely on expensive technology to solve environmental problems, because they are already struggling to survive (Heredia 1994). Therefore, increases in technology costs mean that the poor lose unless the richer nations accept responsibility for the pollution they have created. Many economists assert that the solution to water resource problems lies in making sure that all human activities are forced to internalize full environmental costs. However, full costs are high when a nation is poor and low when a nation's ability to pay is high. Thus, an important question that must be paid attention to in resolving international water disputes is: "Who pays for water resources protection or development?"

2.4. FUTURE GENERATIONS AND WATER RESOURCES

Agenda 21's general objective for sustainable water resources discussed above includes providing water resources for the "entire" human population. An important interpretation question for Agenda 21 implementation is whether the word "entire" in this objective means just currently living human beings or also future generations and, if it does encompass persons who will live in the future, does it mean all possible future generations (including those people who will live 5,000 years in the future)? If so, both water resource projects with dramatic immediate effects and those with small but cumulatively adverse impacts—for instance, projects that slowly increase the amount of salt in surface water or exceed the recharge rate in an aquifer—are objectional as a violation of the rights of future generations. In addition to this question of interpretation, the duty of current people to future generations is also an important ethical question recognized by environmental ethicists.

It has been argued that if the principle of sustainability is to be taken seriously, the protection of resources for future generations must be a goal of sustainable development programs, because the notion of sustainability itself entails protection of resources for the future (Ott 1994). Sustainability requires adopting patterns of human behavior that can be continued over a long range of time without causing well-known ecological problems such as erosion, salinization, pollution, and desertification.

A commonly shared ethical position among environmental ethicists holds that future generations should find living conditions similar to those we take for granted in our daily practices. However, two arguments are often made in opposition to the notion that future generations have rights that should not be ignored by existing persons. The first argument states that nonexisting persons have no rights here and now because a right requires an existing right-bearer. This argument is usually dismissed on several grounds, including arguments asserting that rights of future generations are not derived from the temporal existence of the right-holder but rather from the ability of a person to reason and suffer. The more serious argument against obligations of existing populations to future generations asserts that a right of future generations might exist, but it is unclear which obligations follow because of scientific uncertainties about whether harm will exist in the future. That is, harm might not exist in the future, because humans may invent technologies that will replace damaged water resources or remediate damaged water resources. However, most philosophers doubt whether potential technological fixes allow persons to damage ecologic life-support systems, because existing generations have no right to gamble about matters that harm future generations. Partridge has argued very convincingly that all the well-known arguments by which rights of future generations can be denied or rejected are conceptually wrong or will fail for several reasons (Partridge 1990).

2.5. SCIENTIFIC UNCERTAINTY AND WATER RESOURCES PROJECTS

Part 1 of this chapter identified numerous scientific and technical uncertainties that must be faced in the management of water resources. These included scientific uncertainties in planning, technical analysis, forecasting water resource impacts, aquatic ecosystem and other modeling, health impact prediction, and climate change impacts. This section examines some of the ethical questions that are created by these uncertainties.

In the face of scientific uncertainty about water resources impacts, decisionmakers must decide whether they will err on the side of protecting the resource or on the side of not imposing overly protective measures that limit the current use of water resources. Such decisions must be understood to create ethical questions because the decisionmakers must decide what should be done in the face of uncertainty. Science attempts to describe the probability of any event happening but cannot prescribe what should be done about the risk. Therefore, the question of what should be done is an ethical matter. Although science may not derive obligations from facts about risks, scientific facts may be used in ethical reasoning about what to do.

As stated more fully in Chapter 2, scientific norms followed in many scientific endeavors may not be appropriate in some public policy matters. That is, where the purpose of scientific research is to enlarge the store of human knowledge, scientists are taught to be silent until scientific norms of proof are achieved. For instance, scientists are taught, in normal scientific endeavors, to use a 95 percent confidence level of correlation between hypothetical cause and effect as the standard to define whether cause and effect have been proven. But where facts are uncertain, values in dispute, stakes high, and decisions urgent, scientists may need to follow methods that are inappropriate in other scientific endeavors (Funtowicz and Ravetz 1991). For instance, if there is a small probability that locating a hazardous waste disposal facility may degrade an aquifer potentially usable for drinking water, it may be appropriate to decide not to allow the location of the facility even though the probability of the facility damaging the aquifer does not reach a 95 percent confidence level, a level of correlation necessary to demonstrate proof of harm in other scientific matters. What level of probability should be used as the basis for denying permission to locate the facility is a question of values, not of science.

The Rio Declaration adopted at the Earth Summit at the same time as Agenda 21 included the precautionary principle, a normative rule that is meant to guide nations when faced with scientific uncertainty in sustainable development matters. The precautionary principle states that where there is the reasonable basis to assume that serious or irreversible damage to environmental resources may occur, the absence of scientific certainty shall not be used as an excuse for failing to take cost-effective protective action. The precautionary principle is based on the notion that when one is deciding what to do about actions that may have adverse impacts upon environmental resources, the decisionmaker should err on the side of environmental protection, especially where the action has potentially serious or irreversible adverse environmental impacts.

If decisionmakers follow the precautionary principle in water resource management, they will err on the side of protecting water resources when facing scientific uncertainties in planning, predicting the consequences of, and making decisions about water resources projects. This duty to err on the side of water resources protection is in proportion to what is at stake. If a proposed action could possibly seriously damage ecosystems, then the duty to err on the side of protection is great.

Scientists are taught to deal with uncertainty by assigning probability statements to various outcomes. To assign probability statements, one must have some reliable way of determining the probability of future events. In determining the probability of the meltdown of a nuclear power plant, for instance, engineers can rely on empirical records of valve failures to determine the probability of some events that may lead to a meltdown. However, because many ecological mechanisms are not understood, assigning probabilities to environmental impacts of proposed human actions often requires high degrees of speculation. Where objective empirical evidence of future events is not available, scientists are taught to assign subjective probabilities. However, the less the underlying ecological mechanisms are understood in predicting impacts, the more the assignment of subjective probabilities is similar to untutored speculation. In particular, in these circumstances, the identification of environmental impacts is often influenced by value judgments, although these value judgments are rarely disclosed in the quantification of the probability. If decisionmakers are to follow the precautionary principle in predicting the environmental consequences of water resources projects, they should not only err on the side of protection but also identify all uncertainties and disclose what values influenced the assignment of probabilities. The need to err on the side of protection in identifying impacts is particularly important where the consequences of any decision may create serious or irreversible impacts on water resources or on the beings that depend on water resources.

Jonas (1984) has argued that in predicting long-range consequences of human actions where there are serious possible consequences, humans should be guided by the "heuristic of fear" in predicting consequences. That is, humans should give preference to the bad over the good predictions. Particularly where there are possible serious irreversible consequences from the use of technology, where the stakes are high, decisionmakers should give more weight to prognosis of doom than of bliss. The philosophical reason for the duty to give more weight to the prediction of harm is, according to Jonas, premised on the notion that present generations don't have a right to gamble with the interests of others or to act so that life on earth is jeopardized.

2.6. ECONOMIC ANALYSIS OF WATER RESOURCE PROJECTS

When policymakers recognize that particular value questions have to be considered in decisionmaking about water resources, the values are often discussed in terms of economic considerations, in terms of costs and benefits, efficient markets, or changes to gross national product. Although these analyses often appear to be value-neutral quantifications, as more fully set out in Chapter 4, they are value-laden throughout and often controversial from an ethical perspective.

Agenda 21 calls for the use of certain economic tools to manage water resources projects. They include: (1)economic instruments that take into account opportunity costs and environmental externalities; (2)tests for the value of water resources based on willingness to pay; (3)pricing mechanisms that reflect both the true cost of water and the ability of communities to pay; and (4)planning mechanisms that reflect benefits investment, environmental protection, and operation costs, as well as the opportunity costs reflecting the most valuable alternative use of water (United Nations 1992: Secs. 18.15, 18.16). Several ethical controversies are associated with the use of these economic tools, including: (1)using willingness to pay as a way of valuing natural resources or measuring market externalities, (2)conflicts between ability to pay and the need to protect ecosystems, and, (3)certain problems with cost-benefit analysis.

2.6.1. *Market Externalities and Willingness to Pay*

Economists understand market failures as costs to society that do not show up in economic exchanges and are not borne by the parties to the exchange. Programs directed at fixing market failures are often identified as programs that "internalize" the "externalities," that is, as programs that force certain costs to be implicitly considered within market decisions that would otherwise be imposed on others than the participants in the market exchange. Externalities are sometimes defined as unintentional side-effects of production and consumption that affect a third party positively or negatively.

The classic example of market failures is pollution because, although the prices of industrial commodities cover costs of labor, materials, and a return on investment, they usually fail to include costs of environmental damage. Although market failures are understood to occur whenever prices fail to cover costs not considered in an exchange, market failure is a particularly serious problem in environmental matters. As one commentator has noted, "Market exchanges forget that they are tied to the biosphere" (Brown 1994). For example, the price charged for steel does not cover costs of the damage to the water resources if any pollution is allowed. In this situation, the price of steel does not internalize all the costs of making steel.

Agenda 21 calls for pricing mechanisms that consider the externalities and for that reason can be understood to deal with one of the ethical criticisms often leveled at how market mechanisms deal with environmental resources. Yet Agenda 21 seems to endorse calculating the damage to the external resource on the basis of individuals' willingness to pay for the resource. A serious limitation to neoclassical economic theory applied to environmental problems is how value is assigned to environmental entities. Because individual preferences measured in economic exchanges are what determine value in neoclassical economic theory, economists usually assign value to environmental entities by measuring the willingness of humans to pay for those entities. Thus, the value of scenery or the noise of an industrial plant is determined by comparing real estate prices of affected properties with nonaffected properties. The value of fish killed by an oil spill is the price per pound of fish at a local fish market. In this way, economic tests tend to treat animals, plants, and other environmental entities as commodities whose value is determined solely by their use to humans. Therefore, a strict market approach offers no limits as to what should be sold in the market because value is determined solely by the prices individuals are willing to pay (Brown 1994).

Some philosophers assert that nonhuman sentient beings and other environmental entities have a right to exist that transcends their use value to humans. From this perspective, the fate of environmental entities should not be determined by human subjective preferences. In addition, some economists and philosophers argue that values of the environment should be understood to transcend market preferences (Sagoff 1981). If animals or plants have intrinsic or other nonmonetary value, market valuations may understate or ignore these values. Because market-based prices only measure the strength of human desires, they do not reflect values that are not dependent on subjective human preferences. For this reason, using willingness to pay as a measure of external costs of damage to important environmental entities is very controversial from an ethical perspective.

2.6.2. *Conflicts Between Ability to Pay and the Need to Protect Ecosystems*

Agenda 21 provisions on water resources planning calls for planning pricing mechanisms for water that reflect both the true cost of water and the ability of communities to pay. Because the call for true cost pricing can be understood as a desire to assure that all transactions relating to water reflect all external costs, this provision is ethically preferable to pricing mechanisms that don't include the costs of externalities. However, Agenda 21 seems to authorize willingness to pay as the measure of value, a measurement technique that is ethically controversial, as explained above. Moreover, Agenda 21 appears to modify this call for true cost pricing through the call that prices should also be based on ability of communities to pay. The inclusion of an "ability of communities to pay" consideration in pricing creates the potential that the ability of a community to pay may become the basis for adjusting prices so that they don't cover damages to water resources or the beings that depend on them. If this is done, all environmental costs will not be internalized by the price mechanisms.

. This conflict created by the notion of true or full costs and ability to pay embedded in Agenda 21's pricing mechanism was probably overlooked by the drafters of Agenda 21 because of the desire to deal with another ethical consideration, namely, the desire to assure that the poor would not be burdened by the full costs of environmentally protective technology. Full costs are high where people are poor and low where the ability to pay is rather high. The ethical conflict created by conflicting goals of full cost and ability to pay pricing considerations could be resolved by a rule that requires that rich nations or individuals cover the gap between the poor persons' ability to pay and the full costs of water protection.

2.6.3. *Limits of Cost-Benefit Analysis Applied to Water Resource Projects*

Agenda 21 provides that water resources planning programs identify costs of environmental protection and operation and benefits of investment. Agenda 21 does not expressly prescribe that all water resource decisions should be made by comparing costs and benefits, nor does it discourage decisionmakers from making water resource decisions based on such comparisons. Because Agenda 21 does call for an analysis of costs and benefits, water resource decisionmakers should be aware of some of the ethical controversies embedded in cost-benefit analysis (CBA).

CBA is often used as a decision rule in environmental matters because decisionmakers assume that governments should choose options that maximize or at least improve human welfare or utility. Therefore, according to some economists, government should choose those sustainable development options for which benefits most exceed costs.

Like other welfare-maximizing techniques, CBA can be criticized for: (1)valuing environmental entities on the basis of willingness to pay and thereby ignoring other, nonmonetary values; (2)the failure to consider questions of intergenerational and intragenerational distributive justice; (3)the use of discounting to determine present value of future benefits in a way that unfairly reduces the value of environmental benefits that become manifest in the future; (4)the need to put costs and benefits on a single scale in a way that biases monetary values at the expense of nonmonetary values; (5)the difficulty in quantifying long-term environmental benefits because of pervasive scientific uncertainty in predicting impacts; and (6)in predicting consequences, often conflating wealth with happiness, a confusion not supportable by classical utilitarian ethics.

For these ethical reasons, CBA should not be used as a prescriptive rule in decisionmaking about water resource projects. CBA can, however, be helpful as a heuristic tool in analyzing alternatives, but the decisionmaker should understand the ethical limitations of the tool. One important limitation is that CBA, if used as a decision tool, can be objectionable from an ethical perspective in some circumstances.

3. Summary

Agenda 21 delineated seven interrelated program areas associated with freshwater resources: (1)integrated water resources development and management; (2)water resources assessment; (3)protection of water resources, water quality, and aquatic ecosystems; (4)drinking-water supply and sanitation; (5)water and sustainable urban development; (6)water for sustainable food production and rural development; and (7)impacts of climate change on water resources. These areas encompass many facets related to the provision of adequate quantities of water of appropriate quality for multiple uses, including water supplies and the maintenance of diverse aquatic ecosystems. Interrelationships between surface water and groundwater and freshwater and coastal resources are also incorporated in the seven program areas.

Numerous scientific and technical uncertainties are related to the seven program areas. Examples described in this chapter are related to the water resources planning process and necessary technical analyses. Planning uncertainties are typically related to the lack of baseline data and the need to consider alternative scenarios for future development. Many uncertainties and research needs can be delineated for surface water and groundwater hydrologic and quality modeling. Developed models range from empirical statistical approaches to sophisticated three-dimensional algorithms which incorporate multiple hydrodynamic, biotic, and abiotic processes. Finally, monitoring of flows and chemical and biological parameters of water quality is critical for effective planning and management of water resources. Monitoring data is particularly deficient in developing countries.

Scientific and technical uncertainties can be used to delineate both global and national research needs related to freshwater resources. Paramount in the needs list are topics such as effective water resources planning under varying conditions of uncertainty, establishment of cause-effect linkages for water resources development projects (including human health impacts), cost-effective monitoring and quantitative modeling of significant indicators of the status of freshwater resources, prioritization methods for pollution sources, and systematic decision analysis methods for comparisons of planning alternatives for resource management and development projects. Particular attention should be given to the development of valuation methods for analyzing the economic implications of protection and development of freshwater resources.

An effective and integrated water resources management program must be based on a holistic approach involving multiple elements related to both technical and policy issues. Such a program must address both surface water and groundwater resources, water demand and usage limitations, pollution sources and control, and predecision analysis associated with land-use planning and development activities. Twelve elements in an effective integrated water resources management program include: (1)the recognition and delineation of relationships between surface water and groundwater resources; (2)statutory authority and water usage and quality standards; (3)resource characterization studies (quantity and quality); (4)resource usage studies; (5)a water usage conservation program; (6)resource pollution studies; (7)a pollution source control and resource protection program (point and nonpoint sources); (8)technically based decisionmaking regarding usage allocations, wastewater (point and nonpoint) discharges, and permitting program; (9)a monitoring and enforcement program; (10)an emergency response program; (11)an institutional coordination program (related to element 2 above); and (12)a public awareness, public participation, and education program.

It is also critically important for water resource decisionmakers to understand that water resource projects raise important ethical questions and issues throughout project conceptualization, design, and implementation. These ethical questions include, but are not limited to, issues about: (1)the responsibility or obligations of humans to other humans, including future generations; (2)obligations to nonhuman animal species, plants, and ecosystems; (3)fairness among and within nations; (4)resolving matters of scientific uncertainty about consequences of actions; and (5)the ethical limits of economic tools.

Because science, economics, and law are likely to be the linguistic tools that structure debate about water resource projects, and because these disciplines are often understood to be ethically neutral but are in fact value-laden throughout, it is critically important for decisionmakers to understand the ethical controversies often hidden in these linguistic tools. Moreover, in democracies that implement water resource projects, it is not enough for decisionmakers to understand the ethical issues often hidden in technical analyses; decisionmakers should also expose the ethical questions and their proposed resolutions so that citizens may understand the ethical positions taken in their name.

4. References

Anderson, M.G., and T.P. Burt. 1985. Modelling Strategies. In *Hydrological Forecasting*, M.G. Anderson and T.P. Burt, eds. John Wiley and Sons, Ltd., New York, pp. 1-13.

Bouwer, H. 1978. *Groundwater Hydrology*. McGraw-Hill, New York.

Broderius, S.J. 1990. Modeling the Joint Toxicity of Xenobiotics to Aquatic Organisms: Basic Concepts and Approaches. EPA/600/D-90/063. U.S. Environmental Protection Agency, Duluth, MN.

Brookes, A. 1988. *Channelized Rivers—Perspectives for Environmental Management*. John Wiley and Sons, Ltd., Chichester, England, pp. 67-71.

Brown, P. 1994. *Restoring the Public Trust*. Beacon Press, Cambridge, MA, pp.53-54.

Canter, L.W. 1985. *River Water Quality Monitoring*. Lewis Publishers, Inc., Chelsea, MI.

Canter, L.W. 1986. Functions and Activities of Ground Water Protection: Implications for Institutional Coordination. *The Environmental Professional* 8(3): 219-224.

Canter, L.W. 1991. Water Resources Management: Fundamental Considerations and Linkages with Land Planning. Workshop on Environmental Planning, Milano, Italy, October 21-22.

Canter, L.W. 1993a. The Role of Environmental Monitoring in Responsible Project Management. *The Environmental Professional* 15(1): 76-87.

Canter, L.W. 1993b. Pragmatic Suggestions for Incorporating Risk Assessment Principles in EIA Studies. *The Environmental Professional* 15(1): 125-138.

Canter, L.W. 1995. *Environmental Impact Assessment*. McGraw-Hill, New York, Chapters 5, 7.

Canter, L.W., and D.M. Fairchild. 1986. Post-EIS Environmental Monitoring. In *Methods and Experiences in Impact Assessment*, H. Becker and A. Porter, eds. D. Reidel Publishing Company, The Hague, The Netherlands, pp. 265-285.

Canter, L.W., and D.A. Sabatini. 1990. Use of Aquifer Vulnerability Mapping in Ground Water Management. In *Proceedings of the International Conference on Groundwater Resources Management*, Bangkok, Thailand, November 5-7, pp. 227-251.

Canter, L.W., and B.E. Vieux. 1993. Environmental Impact Assessment in Transboundary Rivers. *Water Power and Dam Construction* 45(5): 26-29.

Canter, L.W., R.C. Knox, and D.M. Fairchild. 1987. *Ground Water Quality Protection*. Lewis Publishers, Inc., Chelsea, MI.

Chapman, D., ed. 1992. *Water Quality Assessments—A Guide to the Use of Biota, Sediments, and Water in Environmental Monitoring*. Chapman and Hall, Ltd., London, p. 9.

(COE) U.S. Army Corps of Engineers, Lower Mississippi Valley Division. 1980. A Habitat Evaluation System for Water Resources Planning. U.S. COE, Vicksburg, MS.

(COE) U.S. Army Corps of Engineers, Waterways Experiment Station. 1987. Water Quality Models Used by the Corps of Engineers. Information Exchange Bulletin E-87-1. U.S. COE, Vicksburg, MS.

Corbitt, R.A. 1990. *Standard Handbook of Environmental Engineering*. McGraw Hill, New York.

(CTAGWV) Committee on Techniques for Assessing Ground Water Vulnerability. 1993. *Ground Water Vulnerability Assessment—Contamination Potential Under Conditions of Uncertainty*. National Research Council, National Academy Press, Washington, DC.

da Cunha, L.V. 1988. Climatic Changes and Water Resources Development. In *Proceedings of Symposium on Climate and Geo-Sciences: A Challenge for Science and Society in the 21st Century*, Louvain-la-Neuve, Belgium, May 22-27.

Domenico, P.A., and F.W. Schwartz. 1990. *Physical and Chemical Hydrogeology*. John Wiley and Sons, Inc., New York.

Dworkin, R. 1978. *Taking Rights Seriously*. Harvard University Press, Cambridge, MA.

Dzurik, A.A. 1990. *Water Resources Planning*. Rowman and Littlefield Publishers, Inc., Savage, MD, pp. 83-92.

(ECE) Senior Advisors to ECE Governments on Environmental and Water Problems, Economic Commission for Europe, United Nations. 1992. Methods and Techniques for Prediction of Environmental Impact. ECE/ENVWA/21. United Nations, New York.

Elder, J.F. 1988. Metal Biogeochemistry in Surface-Water Systems—A Review of Principles and Concepts. USGS Circular 1013. U.S. Geological Survey, Denver, CO.

(EOP) Principles and Guidelines, Executive Office of the President. 1983. Economic and Environmental Principles and Guidelines for Water and Related Land Resources Implementation Studies. Water Resources Council, Washington, DC.

(EPA) U.S. Environmental Protection Agency, Office of Ground Water Protection. 1987. Guidelines for Delineation of Wellhead Protection Areas. U.S. EPA, Washington, DC.

(EPA) U.S. Environmental Protection Agency, Office of Science and Technology. 1991. Policy on the Use of Biological Assessments and Criteria in the Water Quality Program. U.S. EPA, Washington, DC.

(ERL) Environmental Resources Limited. 1982. *Environmental Impact Assessment—Techniques for Predicting Effects in EIA*, Vol. 2. ERL, London.

(ESCAP) Economic and Social Commission for Asia and the Pacific. 1990. Environmental Impact Assessment—Guidelines for Water Resources Development. ST/ESCAP/786. United Nations, New York.

Funtowicz, S.O., and J.R. Ravetz. 1991. A New Scientific Methodology for Global Environmental Issues. In *Ecological Economics*, R. Costsanza, ed. Columbia University Press, New York, pp. 137-152.

(FWS) U.S. Fish and Wildlife Service. 1980. Habitat Evaluation Procedures (HEP). ESM 102. U.S. FWS, Washington, DC.

Grigg, N.S. 1985. *Water Resources Planning*. McGraw-Hill, New York, pp. 27-34.

Heredia, R.C. 1994. The Ethical Implications of Global Climate Change: A Third World Perspective. In *Proceedings on Ethical Dimensions of the United Nations Program on Environment and Development Agenda 21*, D.A. Brown, ed. Earth Ethics Research Group, Harrisburg, PA, pp. 121-128.

Henderson-Sellers, B. 1991. *Water Quality Modeling—Vol. IV—Decision Support Techniques for Lakes and Reservoirs*. CRC Press, Boca Raton, FL.

(IHD-WHO) IHD-World Health Organization Working Group on the Quality of Water. 1978. Water Quality Surveys. Studies and Reports in Hydrology No. 23. WHO, Geneva, Switzerland.

(IMC) Interim Mekong Committee. 1982. Environmental Impact Assessment—Guidelines for Application to Tropical River Basin Development. Mekong Secretariat c/o ESCAP, Bangkok, Thailand.

(IWR) Institute for Water Resources, U.S. Army Corps of Engineers. 1992. Guidelines for Risk and Uncertainty Analysis in Water Resources Planning. 92-R-1. U.S. COE Fort Belvoir, VA, pp. 1-2, 17-21, and 38.

James, A., ed. 1993. *An Introduction to Water Quality Modeling*. John Wiley and Sons, Ltd., West Sussex, England.

Jonas, H. 1984. *The Imperative of Responsibility*. Unviversity of Chicago Press, Chicago.

Legg, D.L. 1978. Gas Supersaturation Problem in the Columbia River Basin. In *Environmental Effects of Large Dams*, W.L. Chadwick, ed. American Society of Civil Engineers, New York, pp. 149-164.

Lyon, W.A. et al. 1984. *Social and Environmental Objectives in Water Resources Planning and Management*. Committee on Social and Environmental Objectives, Water Resources Planning and Management Division. American Society of Civil Engineers, New York, pp. 2-10.

Metcalf and Eddy, Inc. 1991. *Wastewater Engineering—Treatment, Disposal and Reuse*, 3d ed. McGraw-Hill, New York.

(NAS) National Academy of Science. 1987. *Responding to Changes in Sea-Level: Engineering Applications*. National Academy Press, Washington, DC.

Nemerow, N.L. 1978. *Industrial Water Pollution: Origin, Characteristics and Treatment*. Addison-Wesley Publishing Company, Reading, MA.

Nemerow, N.L., and A. Dasgupta. 1991. *Industrial and Hazardous Waste Treatment*. Van Nostrand Reinhold Company, New York.

Novotny, V., and G. Chesters. 1981. *Handbook of Nonpoint Pollution*. Van Nostrand Reinhold Company, New York.

Ott, K. 1994. Ethical Questions Embedded in Water Resources Provisions of UN Agenda 21. In *Proceedings on Ethical Dimensions of the United Nations Program on Environment and Development Agenda 21*, D.A. Brown, ed. Earth Ethics Research Group, Harrisburg, PA, pp. 217-234.

Ott, W.R. 1978. *Environmental Indices: Theory and Practice*. Ann Arbor Science Publishers, Inc., Ann Arbor, MI.

Partridge, E. 1990. On the Rights of Future Generations. In *Upstream/Downstream—Issues in Environmental Ethics*, D. Scherer, ed. Temple, Philadelphia.

Paustenbach, D.J., and R.E. Keenan. 1989. Health Risk Assessment in the 1990s. *Hazmat World* 2(10): 44-54.

Petersen, M.S. 1984. *Water Resources Planning and Development*. Prentice-Hall, Englewood Cliffs, NJ, pp. 54-66.

Priscoli, J.D. 1994. A Perspective on Some Emerging Ethical Dilemmmas in Water Resources Management. In *Proceedings on Ethical Dimensions of the United Nations Program on Environment and Development Agenda 21*, D.A. Brown, ed. Earth Ethics Research Group, Harrisburg, PA, pp. 247-260.

Ray, C. 1990. Use of QUAL-I and QUAL-II Models in Evaluating Waste Loads to Streams and Rivers. In *Environmental Modeling for Developing Countries*, A.K. Biswas, T.N. Khoshoo, and A. Khosla, eds. Tycooly Publishing, London, pp. 65-79.

Sagoff, M. 1981. At the Shrine of Our Lady of Fatima or Why Political Questions Are Not All Economic. *Arizona Law Review* 23: 1283-1298.

Sanders, T.G., ed. 1980. *Principles of Network Design for Water Quality Monitoring*. Colorado State University, Ft. Collins, CO.

Schneider, S.H., P.H. Gleick, and L.O. Mearns. 1990. Prospects for Climate Change. In *Climate Change and U.S. Water Resources*, P.E. Waggoner, ed. John Wiley and Sons, Inc., New York, pp. 41-73.

Smalley, D.H., and J.K. Novak. 1978. Thermal Phenomena Associated with Reservoirs. In *Environmental Effects of Large Dams*, W.L. Chadwick, ed. American Society of Civil Engineers, New York, pp. 29-50.

Stakhiv, E.Z. 1986. Achieving Social and Environmental Objectives in Water Resources Planning: Theory and Practice. In *Proceedings of an Engineering Foundation Conference on Social and Environmental Objectives in Water Resources Planning and Management*, W. Viessman and K.E. Schilling, eds. American Society of Civil Engineers, New York, May, pp. 107-125.

Starfield, A.M., and A.L. Bleloch. 1986. *Building Models for Conservation and Wildlife Management*. MacMillan Publishing Company, New York.

United Nations. 1992. *The Earth Summit*. U.N., New York, pp. 334-360.

(USGS) U.S. Geological Survey. 1982. National Handbook of Recommended Methods for Water Data Acquisition. U.S. GS, Reston, VA.

Waggoner, P.E., ed. 1990. *Climate Change and U.S. Water Resources*. John Wiley and Sons, Inc., New York.

(WHO) World Health Organization, Panel of Experts on Environmental Management for Vector Control. 1983a. Report of the Third Meeting—September 1983. WHO, Geneva, Switerland, pp. 27-43.

(WHO) World Health Organization, Regional Office for Europe. 1983b. Environmental Health Impact Assessment of Irrigated Agricultural Development Projects — Guidelines and Recommendations. WHO, Copenhagen, Denmark.

(WRC) Water Resources Council. 1983. Economic and Environmental Guidelines for Water and Related Land Resources Implementation Studies—Environmental Quality (EQ) Procedures. WRC, Washington, DC.

(WSTB) Water Science and Technology Board. 1990. *Ground Water Models—Scientific and Regulatory Applications*. National Academy Press, Washington, DC.

Chapter 9
TOXIC SUBSTANCES AND AGENDA 21: ETHICAL AND POLICY ISSUES IN THE SCIENCE AND ITS IMPLEMENTATION

Carl F. Cranor[1]

1. Introduction

This chapter addresses some of the scientific tools for assessing risks from toxic substances, one of the major concerns of the United Nations Agenda 21 (Chapter 19). One cannot, however, treat this as merely a technical or scientific problem. As I argue throughout, the science itself is laden with normative considerations. An important policy question is the extent to which public health and environmental threats should be addressed as if they were merely scientific or technical questions. We need an ethic or policy decision about how much scientific evidence of harm to demand before taking preventive or remedial action. We should also understand the normative, indeed moral, commitments that are implicit in the scientific procedures themselves and in the practices of the professions that implement them. For example, implicit in typical scientific procedures are normative commitments, burdens of proof, and practices for addressing scientific uncertainty that can easily frustrate health-protective and environmentally protective courses of action of the kind envisioned in Agenda 21 for dealing with hazardous materials and toxic substances. Failure to revise present scientific practices will tend to favor the status quo of ignorance about potentially toxic substances and of slow governmental action, while improvements of both would mitigate these problems. Present scientific practices used in risk assessment need evaluation and revision in order not to enshrine the status quo ante.

In what follows, I briefly describe some of the Agenda 21 programs toward toxic chemicals, examine some of the scientific tools for assessing toxic risks, argue that these are laden with policy and normative considerations, recommend improvements in these tools and techniques, and show the implications of the recommendations of Agenda 21 on toxic substances.

2. The Unknown Threat of Unevaluated Substances

Both the industrialized and the developing world face a potentially serious problem of unknown magnitude: a threat to human health from toxic substances. This presentation uses carcinogens as the representative of that problem, but numerous other toxic substances, such as reproductive toxins and neurotoxins, as well as substances posing threats to the ecosystem may pose similar risks.

---

[1]College of Humanities and Social Services, College Building South, University of California at Riverside, Riverside, CA 92505, U.S.A.

J. Lemons and D. A. Brown (eds.), Sustainable Development: Science, Ethics, and Public Policy, 215–253.

In the United States, more than 400,000 deaths per year are attributed to cancer (U.S. Congress, OTA 1982: 70-71). Major sources of these deaths are tobacco (which causes about 25 percent to 40 percent of the deaths due to cancer, with a best estimate in the neighborhood of 30 percent) and alcohol (which causes about 3 to 5 percent of all cancers) (U.S. Congress, OTA 1982: 69-72). Cancer caused by both substances is for the most part self-inflicted harm (except for any effects from inhaling secondary smoke, which now appear substantial). However, a number of cancers may be not self-inflicted but caused by others. Conservative estimates suggest that at present about 5 to 15 percent of all cancers are caused by workplace exposure (with a best estimate of 10 percent), about 2 percent are caused by air pollution, and perhaps as many as 35 percent are caused by diet (U.S. Congress, OTA 1982: 86-88, 91, 108). Others make much higher estimates of environmental causes of cancer. The U.S. Public Health Service estimated that "as much as 20 percent or more" of cancers now and in the future may result from past exposure to six known carcinogens (Bridbord and French 1978). Even an industry report indicated that "the full range [of total cancer attributable to occupational exposure] using multiple classifications may be 10 to 33 percent or perhaps higher, if we had better information on some other potentially carcinogenic substances" (Epstein and Swartz 1981). This report indicates that asbestos exposure alone may account for 29,000 to 54,000 cancer deaths. Moreover, even cancer deaths attributed to smoking may hide other causes of cancer, because this is a multifactorial process (Epstein and Swartz 1981). Several authors have disagreed with the emphasis placed on dietary contributions to cancer (see, e.g., Schmahl et al. 1981).

Identifying the environmental causes of cancer is a difficult job with many complications, because cancer appears to be a multifactorial disease (Schmahl et al. 1981; Tomatis 1988; Tomatis et al. 1990). How much of the threat from cancer is created by people as opposed to being naturally caused is not clear. How much of it is in some sense self-inflicted is also not clear. There are at best rough estimates, but the estimates indicated here are sufficiently high to be of concern. If as few as 10 to 15 percent of cancer deaths (a very low estimate) are caused by the alterable behavior of others, the number is substantial—40,000 to 60,000 per year (Scott 1990). If the number is greater, as some suggest, the problem is much worse.

The task of assessing the substances that might cause cancer is daunting because of the large numbers involved. If we consider only commercial chemicals as a potential source of cancer, we face considerable uncertainty because we are largely ignorant about their potential toxicity (U.S. Congress, OTA 1982: 12). There are estimates of from 55,000 to 100,000 such substances in commerce. Only about 6,000 to 7,000 of this universe have been evaluated in animal tests, many inadequately, with about 10 to 16 percent of them testing carcinogenic (U.S. Congress, OTA 1982: 12, 130). Other estimates vary widely. General Electric scientists, examining a seven-volume list from the Public Health Service and using a "relaxed criterion" of carcinogenicity, found that about 80 percent were carcinogenic. Several federal agencies, reviewing shorter lists of substances about which they may have had some antecedent concerns, found 24 to 52 percent of the substances to be carcinogenic (U.S. Congress, OTA 1982).

Against this background it is important to identify carcinogens, to estimate the risks they pose, and then to determine the best policy toward regulating the substances that pose risks of concern. At present, however, little is known about the universe of chemical substances (California EPA 1991). Few carcinogens have been identified. Moreover, when there are clues to their toxicity, for example, from structural similarity to known carcinogens, these are frequently not followed up (Johnson and Parnes 1979). Even when they have been identified in animal tests, only 10 to 20 percent have been evaluated quantitatively for their potency in risk assessments (U.S. Congress, OTA 1987). When such assessments have been completed, the substances have not necessarily been regulated, even though that may be warranted (U.S. Congress: OTA 1987: 19).

One reason (but not the only one) for the delay in the identification, assessment, and regulation of risks from carcinogens is the way scientific evidence is currently used for these purposes. The burdens of proof and other procedures used in science to establish a theory or causal relationship tend to be much more demanding than those that might be adopted for legal or public health purposes. Thus, there can be a conflict between scientific assessments of substances in the research laboratory and the preventive strategies that might be appropriate for public health or environmental protection purposes. As I argue throughout, how much evidence we demand or require for public health and environmental protection is a normative or a policy question that is typically not even broached.

3. Agenda 21

Chapter 19 of the Earth Summit Agenda 21 notes two major problems facing countries in the environmentally sound management of toxic substances: "(a)lack of sufficient scientific information for the assessment of risks entailed by the use of a great number of chemicals and (b)lack of resources for assessment of chemicals for which data are at hand" (U.N. Agenda 21 1992: 186). Chapter 19 recommends six program areas to address the problems. Three of these are improvements in generating or using scientific information about potentially toxic substances: "(a)expanding and accelerating international assessment of chemical risks; (b)harmonization of classification and labeling of chemicals; [and] (c)information exchange on toxic chemicals and chemical risks" (U.N. Agenda 21 1992: 186). (The more general Rio Declaration on the Environment and Development [Principle 9] also supports the improvement in scientific understanding and exchange of information [U.N. Agenda 21 1992: 10].) These three programs are the main focus of this chapter. The fourth is the "establishment of risk reduction programs"; the fifth is "strengthening of national capabilities and capacities for management of chemicals," and the sixth is "prevention of illegal international traffic in toxic and dangerous products" (also endorsed by the Rio Declaration [Principle 14]) (U.N. Agenda 21 1992: 186). The analysis of this chapter has implications for the last three programs as well. The reconception of risk assessment suggested in this chapter and some alternative procedures made possible as a result facilitate risk reduction programs and facilitate the strengthening of national capabilities for risk assessment. Finally, the reconception of risk assessment facilitates national autonomy in risk assessment and risk management which a strong prohibition of the illegal international traffic in toxic substances helps to secure (a concern also embodied in the Rio Declaration's Principle 11) (U.N. Agenda 21 1992: 10).

Understanding some of the risk assessment tools for identifying, assessing, and managing the risks from chemical substances together with the policy assumptions implicit in these sciences can improve the policies and procedures for addressing problems in this area. Treating the identification and regulation of toxic substances mainly as a problem to be addressed by "objective" science will exacerbate current problems, I believe, and is to be avoided. By contrast, recognizing the ethical and policy implications implicit in risk assessment tools opens up the possibility of addressing the problems in innovative ways that will better achieve the goals of Agenda 21.

4. The Scientific Tools for Assessing the Risks From Carcinogens

Governmental administrative agencies use scientific or quasiscientific "risk assessment" procedures to provide evidence about risks from toxic substances. The aim of agency actions is to estimate the risks to human beings from exposure to toxins in order to prevent or reduce those risks. This chapter considers animal bioassays and human epidemiological studies two of the main pieces of evidence relied upon in these institutions to ascertain risks to human beings. The best evidence that a substance causes cancer to human beings is provided by well-done epidemiological studies with large samples and sufficient follow-up. However, I begin by considering animal studies. This is the evidence much more frequently relied upon by administrative agencies.

The uncertainties inherent in inferences from animal bioassays present one class of problems in their use for the predictions of risks to human beings. Both actual and possible scientific uncertainties are large enough that two different researchers using exactly the same data points from an animal study can come to quite different conclusions concerning risks to human beings. The uncertainties and the policies used to overcome them, together with scientific practices followed in their use, permeate regulatory science with public policy or moral considerations. As a result, regulatory scientific decisions are mixed judgments of science and value, and quite properly so. This makes regulatory science much more normative and much less like ordinary, core areas of science than we might suppose. This conclusion undermines the idea that risk assessment is objective and is independent of risk management decisions.

Moreover, if some of the practices of research science are insensitively transposed to regulatory contexts, they may determine and even beg the public policy outcomes in ways often unbeknownst to practitioners. Scientists unwisely demanding more and better data, withholding scientific judgment until there is sufficient research, and using too-demanding standards of evidence can impede public health aims. Taken individually, each of the foregoing may frustrate the discovery of risks of concern for the substance under consideration. Perhaps of greater importance, the combined effect of these tendencies is to slow the scientific evaluation of carcinogens, and thus regulatory efforts, to a snail's pace. This prevents agencies from assessing identified but unevaluated carcinogens or from devoting resources to identifying other toxins. While these factors individually and collectively present substantial problems in risk assessment and regulatory activities, they present opportunities as well. There are procedures for addressing some of the shortcomings of present methods that remedy some of the difficulties and that are in accord with many of the recommendations of Agenda 21.

Epidemiological studies provide an example of a different and additional set of problems. In many circumstances, unwitting commitment to traditional scientific procedures in the design and interpretation of epidemiological studies (and scientific data more generally) can beg the normative issues at stake. Even when there are no practical evidence-gathering problems or none of the uncertainties noted above, either scientists or the risk managers who use their data mathematically may be forced into a dilemma. They may have to choose between adhering to the evidentiary standards typical of research science and interpreting the data in "regulatory-sensitive" ways—providing results that are sensitive enough to detect the risks of concern and avoiding false negatives.

Risk assessments based upon animal and human studies suffer from shortcomings that can easily frustrate the aims of the institutions in which they are used and thus affect the regulation of toxic substances. Whether these problems arise and whether they are exacerbated or ameliorated depends upon the scientific and policy responses to them. A tempting response is to make risk assessments more nearly like normal science. In many circumstances this is very likely to hide the risks to human beings. Consequently, decisionmakers should balance the pursuit of the goals of research science with those of the institutions in which they are used.

5. Predicting Risks from Animal Bioassays

Risk assessment, "the characterization of the potential adverse effects of human exposures to environmental hazards," seeks to provide accurate information about risks to human beings or the environment so that agencies in fulfillment of their legal mandates can regulate exposure to potentially carcinogenic substances (NRC 1984: 18-19). After scientists in the technical, scientific part of the agencies have provided an estimate of risks to human beings from exposure to toxic substances, they then give this information to the risk managers. Risk management is concerned with managing the risks in accordance with statutory requirements and other economic, political, and normative considerations (NRC 1984: 19).

Risk assessment is typically divided into hazard identification, dose-response assessment, and environmental risk assessment. Hazard identification is "the process of determining whether exposure to an agent can cause an increase in the incidence of a health condition (cancer, birth defect, etc.)" (NRC 1984: 19). Dose-response assessment seeks to characterize quantitatively "the relation between the dose of an agent administered or received and the incidence of an adverse health effect in exposed populations and estimating the incidence of the effect as a function of human exposure to the agent" (NRC 1984: 19). Both hazard identification and dose-response assessment rely on inferences from animal bioassays and human epidemiological studies. Exposure assessment estimates risks to human beings from exposure to carcinogens when they are released into the environment, soil, groundwater, and air. I focus here mainly on hazard identification and dose-response assessment for illustrative purposes.

Risk assessment in the present state of knowledge, however, is a third-best solution to the problem of estimating harms to human beings from exposure to toxic substances. The ideal is a "harm assessment." If we had perfect information, we would accurately assess the harmful effects to people from exposure to toxic substances. This would provide us with exact numbers of deaths, diseases, and environmental harms, and we would not overestimate or underestimate the effects of toxic exposures, but we still might not be able to determine *which* persons would be affected.

If we distinguish between risk and uncertainties, a risk is the probability of an unfortunate or undesirable outcome, when such probabilities can be assigned to outcomes (Rescher 1983: 5). Thus, a risk assessment, properly speaking, aims to estimate the probabilities of harms from toxic exposures and is a second-best solution to a harm assessment. For a whole population and for accurate probabilities of harm, this alternative would very closely approximate the morbidity and mortality rates of a harm assessment.

At present, the task of regulators is more complicated than this, for substantial uncertainties can obtain in trying to predict such harms. Thus, we should think of risk assessments not as risk assessments properly speaking, but as "risk and uncertainty assessments." This is the third-best solution to a harm assessment. For my purposes, "risk assessment" will refer to this third possibility—the present state of the art.

Matters of considerable moment depend upon the products of risk assessment. On the one hand, in many cases one answer (a projection of high enough risks to require response) may impose substantial costs on the affected industry and perhaps the larger public. On the other hand, another answer (a projection of a risk low enough so that regulation is not required) may leave innocent people at risk from exposure to dangerous substances.

Because risk assessment is not a perfectly accurate procedure, regulatory agencies will make mistakes. Two kinds of mistakes can be made at the level of hazard identification: false positives and false negatives. A false positive occurs when one mistakenly identifies a substance as a carcinogen. A false negative is a failure to identify a substance as a carcinogen when it is one.

A second class of mistakes might occur at the level of dose-response assessment and regulation: overregulation or underregulation of the substances in question. Overregulation occurs when a substance is regulated in accordance with a particular statute too stringently for the kind and degree of harm that it causes. The substance might cause no harm of regulatory concern, or much less harm than an agency believed. By contrast, underregulation occurs when a substance is regulated under a particular statute to a much lesser degree than it should be. In what follows, I often use "false positives" to refer generically to the mistaken identification of a substance as harmful and to overregulation. Similarly, "false negatives" often refers generically to a failure to identify toxins and underregulation. (When it is necessary to distinguish between mistakes of identification and regulation, I do so.) Both kinds of mistakes are illustrated in Table 1. The possibility of scientific and regulatory mistakes raises the normative question of how to cope with the uncertainties in risk assessments and regulation. On whom should the costs of mistakes fall? How shall we err?

Once substances have been identified as carcinogens, dose-response information is then combined with actual exposure information at all doses on the appropriate population at risk in order to estimate the magnitude and extent of the risk to human beings or the environment. Finally, the risk information is combined with economic, policy, statutory, and technological feasibility information so that administrative agencies can then decide how best to manage properly the risks in question. Figure 1 is a schematic of the risk assessment-risk management relationship.

There are a number of advantages to using animal studies as evidence that a substance causes cancer in human beings. Many experts believe that "animal evidence alone should serve as the basis" for regulating carcinogens (DOL 1980: 5061). Most substances that induce cancer in one mammalian species also induce cancer in others. A finding of "carcinogenicity in rodents is proof that the chemical is carcinogenic in a mammalian species" (ISG 1986: 274). The pathological development of tumors in various species of animals in most cases is believed to resemble that in humans. Human and animal molecular, cellular, tissue, and organ functions are thought to be similar (Rall 1979: 179). "In the absence of adequate data on humans, it is reasonable, for practical purposes, to regard chemicals for which there is sufficient evidence of carcinogenicity in animals as if they presented a carcinogenic risk to humans" (ISG 1986: 234).

Animal studies also have several advantages over human epidemiological studies. For one thing, few industrial chemicals have been adequately tested by epidemiological studies

Table 1. False Positives and False Negatives.

	Possibilities in Causal Relationships	
Possible Test Results	Null hypothesis is true: Benzene is not positively associated with leukemia	Null hypothesis is false: Benzene exposure is positively associated with leukemia
Test does not show that benzene exposure is associated with leukemia	No error	False negative
Test shows that benzene exposure is associated with leukemia	False positive	No error

Figure 1. Relation Between Risk Assessment and Risk Management.

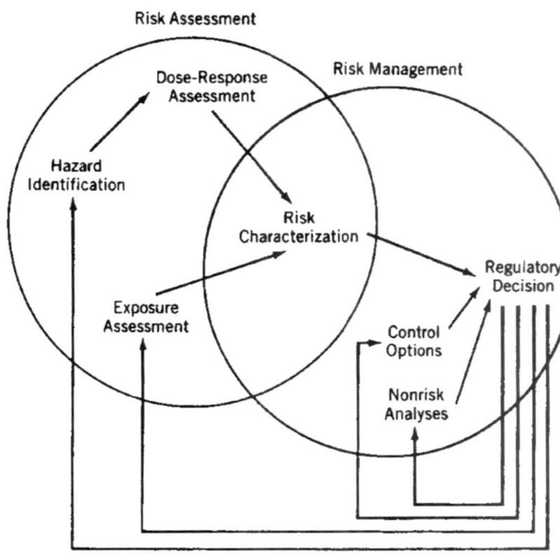

to discover whether they cause cancer in humans (as of 1988, researchers seeking to compare epidemiological with animal bioassay estimates of cancer potency considered epidemiological results for only 23 chemicals) (Allen et al. 1988). For another, epidemiological studies are frequently too insensitive to detect relative risks of concern. Further, an epidemiological survey that is not positive is of questionable merit for showing that a substance is ''safe'' because of sensitivity problems. (These last two points are considered later.) Moreover, it may still be too soon to observe the carcinogenic effects of many substances, since cancer typically has latency periods of up to 40 years (Tomatis 1988).

Even after a sufficient latency period has elapsed, it may be difficult to trace diseases to particular substances because of the insensitivity of epidemiological studies and because almost no toxic substances leave a unique "fingerprint" of their presence. Occasionally a substance leaves a trace behind (e.g., asbestos fibers) or is so rare it can more easily be detected. Animal studies are faster and cheaper than human studies. Moral considerations also provide reasons for using animal studies. There is no justification to wait for "evidence of harm in exposed people when risks can be established relatively quickly by animal experimentation" (Rall 1980: 5061).

In an ideal animal study, usually three or sometimes four groups of about 50 animals each are studied: (1)a control group; (2)an experimental group fed the maximum tolerated dose (MTD) of a toxic substance, "a dose as high as possible without shortening the animals' lives from noncarcinogenic toxic effects" (U.S. Congress, OTA 1987: 39); (3)an experimental group fed one half the MTD; and (4)sometimes an experimental group fed one fourth the MTD. Tumor data from each of the experimental groups then become fixed data points on a graph,.if the tumor results differ statistically from the control group. (This is schematically indicated in Figure 2.) To estimate low-dose responses in animals, researchers then use a computer model to fit a mathematical model through experimental data points and through those seen in control groups. (Figures 3 through 5 indicate the general problem.)

Figure 2. Dose-Response Evaluation Performed to Estimate the Incidence of the Adverse Effect as a Function of the Magnitude of Human Exposure to a Substance.

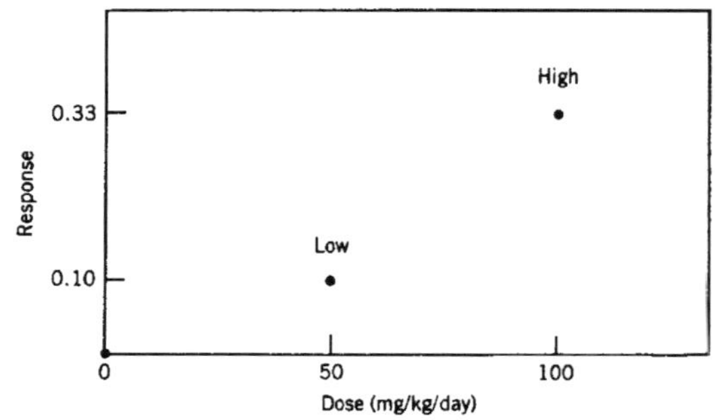

Figure 3. Dose-Response Curve.

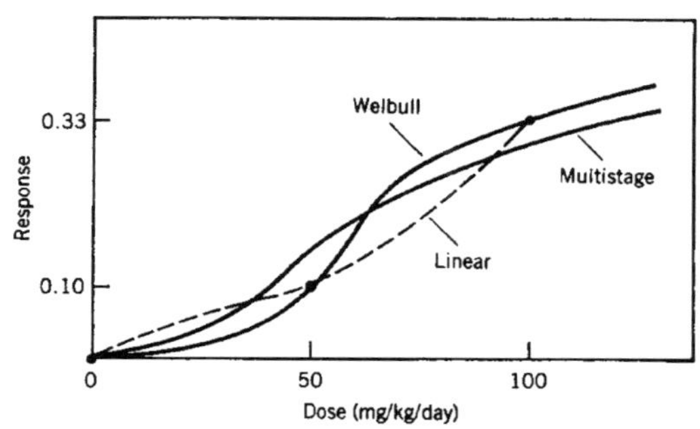

Figure 5 is misleading in some respects. This is an extrapolation from the same animal data using different mathematical models that can extrapolate to quite different values. However, there are biological constraints such that some of these extrapolations do not make sense. (These are discussed later.) Moreover, for all the models, if the background cancer in control animals (or the general population) works by the same mechanism as cancer induced by the administered carcinogen, resulting in an additive carcinogenic effect, all the models become linear at low doses, rather than curvilinear as some appear here. As long as the background cancer rate is 1 percent or higher, this linear effect results. In addition, all the curves would be shifted to the left because of the additivity effect (Cornfeld et al. 1980; Van Ryzin 1982: 130; Zeise et al. 1987).

Animal studies have some shortcomings, however, which we should understand in order to appreciate the foundation of regulatory standard-setting and some of the evidentiary limitations. Estimating risks from animal studies requires a number of inferences from the established experimental data from laboratory animals to the projection of end-point risks to human beings. These inferences have a number of uncertainties and inference gaps that must be bridged in order to produce the risk numbers. Inference gaps arise because there is insufficient information available to settle the scientific questions at issue. These gaps are distinguished from measurement uncertainties, which are features of scientific information

Figure 4. High-Dose to Low-Dose Problem.

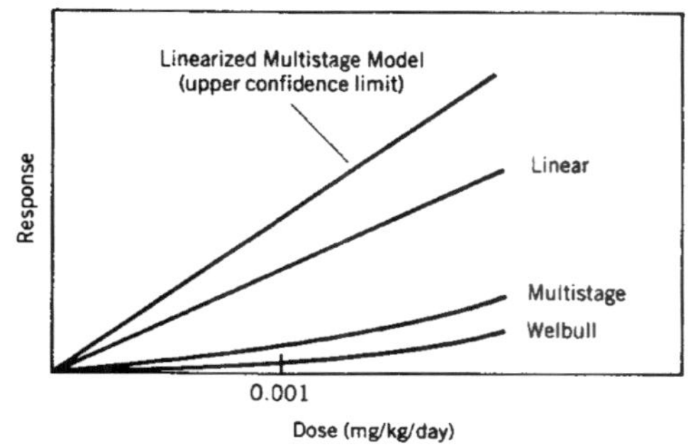

Figure 5. Extrapolated Risks Due to Benzene Exposure.

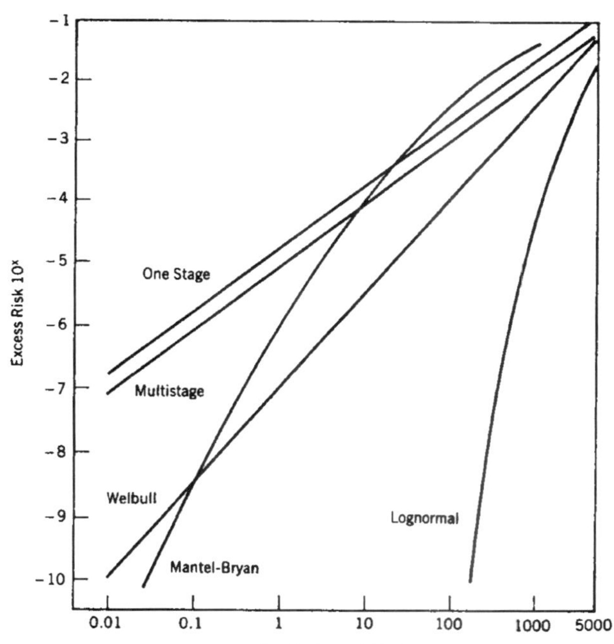

that cannot be measured precisely. Rather, in many cases the inference gaps result from insufficient understanding of the biological mechanisms involved or the relationships between biological effects on one species compared with another. Scientists typically use mathematical models or other generalizations to fill the gaps, and these might produce scientific predictions that differ substantially from one another depending on the model chosen. Thus the range of possible answers will produce a range of uncertainty. Such models introduce uncertainties because surrogates for the proper quantities (if they have even been identified) are used, because some possibly appropriate variables may be excluded, or because the proper model for representing and quantifying the data is not known (Finkel 1990: 12-15). Moreover, uncertainties can be introduced because of the inherent variability in biological response between individuals (Zeise 1991). Concerns about the uncertainties in arriving at dose-response (potency) estimates for substances lead to considerable controversy among critics about their use in predicting risks to human beings (HUSPH 1991). Nonetheless, there is a substantial consensus for continuing their use (ISG 1986).

One kind of uncertainty is introduced because the etiology of carcinogenesis is insufficiently understood. Scientists do not know which mathematical model is the correct one for representing the mechanism of carcinogenesis and for making the extrapolation from high-dose effects in animals (i.e., from the experimentally established data points) to low-dose effects in animals (i.e., to effects typically beyond the experimentally established data points) (DOL 1980: 5190). In addition, there are insufficient data points from experimental evidence to enable researchers to find a unique curve to fit the data produced by controlled experiments. (If mathematical formulas have enough variables, it is relatively easy to fit a large number of curves to the data points.) Thus, both insufficient theoretical understanding and too few data contribute to the uncertainties that plague dose-response risk assessment procedures.

After extrapolating from high-dose effects to low-dose effects (more typical of human exposure) in animals, risk assessors estimate the low-dose effects for human beings. The use of different rodent-to-human extrapolation models can produce differences in risk estimates of 6 to 13 to 35, depending on which of several methods is used (DOL 1980: 5190). In this case, the difference is due partly to differences in theoretical understanding and partly to lack of a sufficient experimental data base for toxic substances. Other uncertainties that plague toxicological risk assessments concern the relative weighting of positive and negative animal studies for the same substance, whether benign and malignant tumors should be weighted equally as evidence for carcinogenicity in animals (and thus in humans), whether and how data from two or more nonhuman species should be combined, and so on, but I do not pursue those here (NRC 1984: 29-33).

The foregoing are some of the main steps leading to an estimate of the carcinogenic potency of a substance based on animal data. The next step is to indicate how much of a substance will reach human beings through the environment—via soil, air, water, or food. Environmental fate models provide a way of estimating this. There are also uncertainties in these procedures, although I do not develop this in great detail. According to Lee and Chang (1992), models for predicting the transport of ethyl benzene from leaking gasoline tanks differ by as much as a factor of 1,500, and there appears to be no easy way to validate the correct model for making such assessments. They also point out that although theoretical models for predicting the environmental fate of substances such as benzene, xylene, toluene, and ethyl benzene are reasonably well understood, at least in laboratory settings, applying the models under actual field conditions could produce substantial differences in the estimates of the risks. The differences in this case are for the most part due not to a failure of theoretical understanding of the environmental fate of substances under controlled conditions, for this seems to be reasonably well understood. Instead there is typically insufficiently detailed site-specific information about the behavior of substances in the soil, groundwater, or air (Lee and

Chang 1992). Additional on-site field research could remove some of these uncertainties (this is not true for many of the uncertainties connected with toxicological risk assessment). But it is time-consuming, probably expensive, and of limited value, since the results could vary by individual substance, type of soil, underlying impermeable stratum, groundwater properties, and other factors specific to the location.

The cumulative theoretical uncertainties that can be introduced by uncertainties at each step of the risk assessment process could be substantial. For example, if, at each step of a chain of inferences, alternative inference guidelines or choice of models would introduce in a risk assessment difference of only magnitude 2, and if there were 10 such inference gaps, then the cumulative theoretical differences mathematically could be as large as 1,024 (2^{10} = 1,024). If there were 20 magnitude 2 such gaps, it is mathematically possible for the cumulative difference to be as large as 1,048,576 (2^{20} = 1,048,576) (Brown 1989). Even though these are mathematical possibilities, there are not typically so many gaps, and such large differences do not materialize (discussed later).

The potential quantitative differences among models to fill an inference gap in a carcinogen risk assessment, however, could easily be much greater than a factor of 2. In toxicological risk assessment, high-dose to low-dose extrapolation models can vary by several orders of magnitude (Cothern et al. 1986). The use of upper confidence limits versus maximum likelihood estimates (a very unstable point) (Zeise 1990) in estimating high-dose to low-dose extrapolations can vary from a factor of 2 to 5 where there are good dose-response data up to several orders of magnitude at the lowest doses where there are not. Interspecies scaling factors, used to account for the different toxicological effect in different mammalian species, can vary up to a factor of 35. Use of pharmacokinetic information, which enables a scientist to estimate the dose of a substance reaching an internal target organ, may change a risk assessment by a factor of 5 to 10 or even more compared with the dose of a substance reaching an external exchange boundary in an animal or human, such as the nose, skin, or mouth (Hattis 1988). An agency's choice of *de minimis* risk thresholds, which trigger regulatory action, might differ by one or two orders of magnitude from that of another agency (Zeise 1991).

However, the logically possible uncertainties that might exist in risk assessment have not tended to materialize in actual risk assessments among agencies or other groups performing them. For one thing, for many risk assessments, there may not be as many inference gaps as the possibilities sometimes suggest. For another, empirical data constrain some of the choices. For example, some high-dose to low-dose extrapolation models "seem impossible to interpret in terms of any biological description" (Zeise et al. 1987). Moreover, where there are good data on both animal and human response rates, many of the typical assumptions used to infer human risks from animal studies tend to predict risks to humans that agree fairly closely with epidemiological studies (Allen et al. 1988). Furthermore, different regulatory agencies frequently agree on the same models, even though scientific data and theories are not fully adequate to support such choices. And there has been some attempt at coordination between federal and state agencies in the United States.

While the differences between actual risk assessments have tended not to be as great as the mathematical possibilities, there can be substantial discrepancies in risk assessments for particular substances. Some of these results for two agencies are summarized in Figure 6. A comparison of potency estimates done by the California Department of Health Services and by the Carcinogen Assessment Group of the U.S. Environmental Protection Agency (EPA) shows that 80 percent of the time the two agencies were within a factor of 5 of each other, 17.5 percent of the time between a factor of 5 and a factor of 25 of each other, and 2.5 percent of the time off by a factor of 25 or more (Zeise et al. 1989). Although the extreme discrepancies are the exception to fairly close agreement between the agencies, they do show that there can be major disagreements nonetheless.

Figure 6. Frequency Diagram of Differences Between EPA and CDHS Potency Values (Zeise 1990).

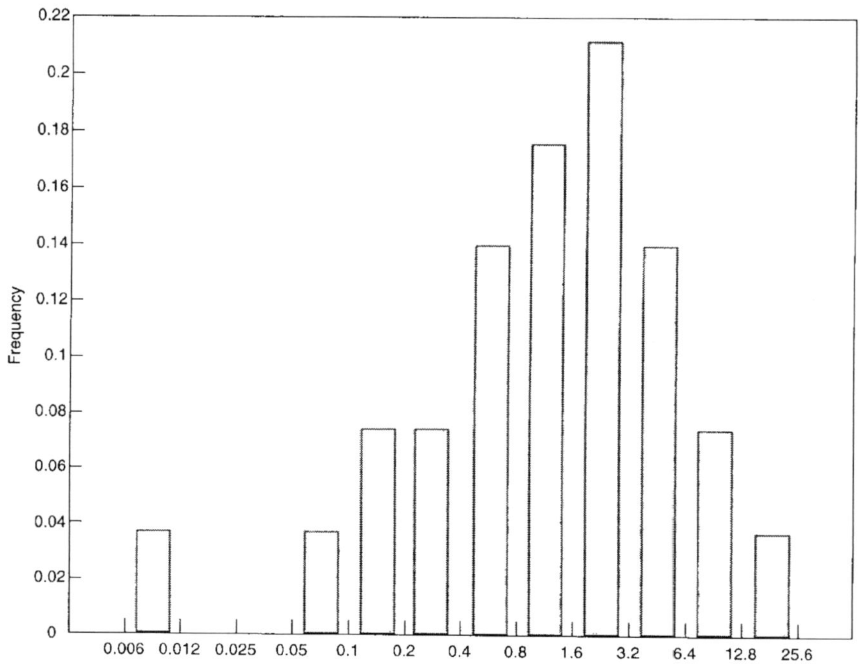

The preceding remarks indicate that carcinogen risk assessments: (1)can suffer from substantial uncertainties, (2)in fact sometimes can differ sufficiently to make a substantial difference in the actual risks that are predicted, and (3)may well in some cases make substantial differences in regulations that are issued.

In order to provide guidance for the approximately 50 inference gaps created by the absence of data and theoretical understanding, a National Academy of Sciences report suggested that regulatory agencies adopt policy (or inference) guidelines (NRC 1984: 1750) or assumptions to bridge the gaps in question (U.S. Congress, OTA 1987: 25). Agencies have adopted five kinds of assumptions to guide the choices of models mentioned earlier: (1)assumptions used when data are not available in a particular case, (2)assumptions potentially testable, but not yet tested, (3)assumptions that probably cannot be tested because of experimental limitations, (4)assumptions that cannot be tested because of ethical considerations (U.S. Congress, OTA 1987: 25), and (5)policy judgments not to underestimate the risks. These policies lead to considerable controversy, for they can make a difference in the estimation of risks to human beings. Frequently, little in the way of biological evidence in the present state of knowledge can determine the choice between the models.

I believe, however, that we cannot avoid the use of inference guidelines, since we need to make decisions concerning the effects of toxic substances on human beings in a timely manner. We cannot wait for these inference gaps to be closed by scientific research. In many cases this could take years, while in others they may never be eliminated. Part of the reason for this concern is that the scientific basis of risk assessment, because of the uncertainties involved, exhibits in a much more radical form a feature of all empirical inferences: the evidence available for the inference underdetermines the inference or the theory that aims to explain the evidence. This general point about our empirical theories and beliefs was well articulated by Quine (1969: 78) in *Word and Object*:

> To the same degree that the radical translation of sentences is under-determined by the totality of dispositions to verbal behavior, our own theories and beliefs in general are under-determined by the totality of possible sensory evidence....

There is ample support for the general point that evidence underdetermines theories or scientific inferences. First, at least since Descartes, philosophers have discussed the possibility of persons having mistaken beliefs based upon the evidence before them. The evidence does not guarantee the conclusions one might typically infer from it. Second, a paradigm of scientific inference, so-called inference to the best explanation, rests on the possibility that there may well be several plausible alternative explanations to account for the evidence available to an observer. The resulting problem is to characterize which explanation is the best one. Again, the possibility (and in some cases the likelihood) of alternative explanations in science is evidence for the general thesis of underdetermination.

Thus, the evidentiary underdetermination of an inference is both an old problem for all empirical beliefs (Descartes' point) and a problem common to all scientific inferences (Quine and the point about inference to the best explanation). So the evidentiary problems of risk assessment are not new. However, the underdetermination of models or theories in carcinogen risk assessment is so much more radical as to make risk assessment substantially different from the case of ordinary scientific inferences. The uncertainties resulting from lack of understanding are much greater than in more well-established areas of science.

The last remark, then, leads to the following additional points: inference guidelines, or at least some kind of choices among competing models, cannot be avoided, if risk assessors are going to provide needed evidence in a timely manner for the assessment and regulation of toxic substances. The inference guidelines (or choices) must be chosen (made) on the basis of some reason, but since scientific data radically underdetermine the choice and perhaps more radically underdetermine this choice than similar choices in other areas of science, some other consideration(s) must determine it.

These additional considerations are typically of several kinds. Some are scientific or other empirical generalizations, but ones not necessarily well supported in a particular case. Examples include the claim adopted by many agencies that carcinogens do not act by means of a threshold mechanism, or the claim that since a surface area extrapolation is used for other purposes of comparing the metabolic activity between species, a similar extrapolation model should be used to predict the toxic effects of a substance from one substance to another. Sometimes decisions are made on the basis of simplicity or ease of calculation, and agencies sometimes choose a middle-ground theory simply because it is midway between alternatives they might adopt (Latin 1988).

In addition, risk assessors acknowledge that some of the decisions of scientific models are chosen on nonscientific policy or moral grounds. Typically they reason that risk assessment procedures should not underestimate risks to human beings at the end of the process or that they must be prudent in protecting public health. For example, they might use a health-protective high-dose to low-dose extrapolation model, use the biological response in the most tumor-sensitive mammalian species as a basis of risk assessment, or count benign as well as malignant tumors as evidence of a substance causing a carcinogenic response in experimental animals. Even though agencies acknowledge the policy role in risk assessment and frequently make health protective choices, not all may be health-protective, for they ignore other choices that could reduce the underestimation of risks. For example, sensitive subpopulations, such as the elderly or nursing infants, are seldom considered. In some cases, humans are much more sensitive to substances than test animals (e.g., benzidine), and the timing of exposure is frequently not taken into account (Bailar et al. 1988, Finkel 1989, Zeise 1990).

The U.S. National Academy of Sciences (NAS) argued that important inference guidelines are based upon policy considerations:

> Guidelines unavoidably embody both scientific knowledge and risk assessment policy. In the past, regulatory agencies typically used a conservative [health-protective] approach in the development of risk assessment policy....(NRC 1984: 76)

The NAS further notes that risk assessments could avoid the use of policy guidelines, but argues that:

> ...guidelines very different from the kinds described could be designed to be devoid of risk assessment policy choices. They would state the scientifically plausible inference options for each risk assessment component without attempting to select or even suggest a preferred inference option. However, a risk assessment based on such guidelines (containing all the plausible options for perhaps 40 components) could result in such a wide range of risk estimates that the analysis would not be useful to a regulator or to the public. Furthermore, regulators could reach conclusions based on the *ad hoc* exercise of risk assessment policy decisions. (NRC 1984: 76)

Such *ad hoc* choices would still involve policy considerations, but they would be hidden, not identified as public choices on the part of an agency.

The above considerations do not deductively warrant the conclusion that risk assessments in the present state of knowledge are policy- or value- or morally laden, but given current practice and the radical underdetermination of the theory by the evidence, they strongly support such conclusions. Thus, we might somewhat oversimplify the point by saying that whether risk assessment indicates a risk to human beings and the seriousness of that risk are, in part, normative policy considerations. Thus, not just the notion of an acceptable risk (an obvious normative concept), but also (in the present state of knowledge of carcinogen risk assessment) the ideas of a risk and the severity of the risk are partly normative notions, for they are the product of combined scientific and normative judgments. Given the uncertainties in the quasiscientific field of risk assessment and given the typical procedures adopted by agencies for coping with such uncertainties (and the recommendations by the NAS for coping with them), the scientific neutrality of factual inquiries that appear to be typical of normal scientific inquiries (in subatomic physics, molecular biology, geology, etc.) does not obtain for risk assessment.

The upshot of the foregoing observations and arguments is the following: in fact actual risk assessments have relied heavily upon policy or moral considerations, in fact agencies at present rely upon such considerations, and the NAS recommends that agencies continue to rely upon such considerations. Finally, because of these arguments, it seems proper and plausible that agencies should continue to do so. Thus, it appears the quasiscientific process of risk assessment in its present state is substantially different from other areas of science; it is a process substantially permeated by policy considerations and seems destined to continue to be so permeated in the indefinite future. What is important for this chapter is the content of the policy considerations for the particular institutional setting in which they are used for risk assessment.

6. Normative Implications of the Scientific Uncertainties in Inferences from Animal Studies

A number of points emerge from the preceding discussion. First, because of substantial uncertainties, and because of the radical underestimation of the theories by the evidence, carcinogen risk assessment differs markedly from core areas of science.

Second, in the present state of knowledge, risk assessments are substantially influenced, and correctly so, by normative judgments. The notion of a risk and the extent of a risk which is the outcome of present carcinogen risk assessments is at least in part a normative notion.

Third, the lack of information and the treatment of uncertainties in traditional scientific practices, which are typical of and even essential to pursuit of scientific truth for its own sake, may well paralyze risk assessment and regulatory activity. One such practice is the demand for more and better data about the substance under consideration. A particular substance may have properties and operate by biological mechanisms different from others. Yet too much emphasis on additional data can slow regulatory evaluation of the substance and can divert agency efforts from considering other chemicals. Moreover, scientists' postponing judgment until sufficient facts are available and using demanding inference standards before ascertaining that a substance is toxic will have similar effects. The justification behind such inferential caution appears to be that by keeping the chances of scientific mistakes quite low, when one obtains a positive result, one can have considerable confidence that one's addition to scientific knowledge is not the result of random error or a mistake. In building the edifice of science, by keeping the odds of mistakes low, one ensures that each brick of knowledge added to the structure is solid and well cemented to existing bricks of knowledge. Were one to take greater chances in generating new knowledge, the edifice would be much less secure. This cautious attitude is considered important in keeping the scientists from chasing chimeras and wasting time, money, and human resources. This is not the whole story, however, for it may also be important to protect against false negatives—failing to discover an effect when it is there.

In risk assessments, the cautious scientific attitude might well have the consequence that harmful responses from exposures to toxic substances are not ascertained until scientists are certain (as measured by the standards of reputable scientific inquiry) that there is such a response. In addition, scientists may be reluctant to endorse a scientific theory that would justify a particular scientific model for risk assessment inferences, unless there were sufficient scientifically defensible support for the theory. Both instantiations of this attitude could paralyze risk assessment activity because of the many uncertainties and because the available evidence greatly underdetermines the inferences needed in risk assessment. This may not always be the case in the future, but in the present state of knowledge it is. The cautious attitude toward inferences and the demand for more and better data before judging that a substance is toxic is a moral or policy judgment that favors, even enshrines, the status quo. Thus, the demand for the perfect or the very good (evidence and inferences) can frustrate the good (health-protective policies).

Fourth, common to the practices just described is an effort to prevent false positives. However, in regulation, false negatives are of perhaps greater importance. The aim of regulation typically is to protect against toxic substances causing or contributing to human beings contracting diseases. Thus, what is typically of lesser concern in purely scientific inquiry is of much greater concern in regulatory inquiries and to the general public (Lawless 1977). But how much concern is justified in each case will be partly a function of the institutional setting and the aims of the activities in question. Moreover, while this might vary from country to country depending on its level of development and its view of public health and environmental protection, it appears that Agenda 21 in the Precautionary Principle and many of its programs envisions governmental action, based on less than the most demanding evidence. (This point is developed later.)

Fifth, in any case, as I argue later, some balance of the two kinds of mistakes should be achieved for risk assessment activities in their institutional settings. The proper balance is in part an institutional, legal, and moral issue depending upon the context of inquiry.

Sixth, the cumulative effect on regulation of many of the foregoing practices typical of research science is that risk assessment and regulation are slow. Animal bioassays are time-consuming. Since the inception of the animal bioassay program of the U.S. National Cancer Institute and the National Toxicology Program, as of July 1991 there had been 367 animal bioassays (and since some of the studies are duplicative, this represents fewer than 367 substances) (NTP 1991). Typically it takes about two years from nomination of a substance to its acceptance for testing, two more years for the actual experiment, two years for evaluating the tumors from experimental animals, and one year for writing up the results—approximately seven years total from nomination to final write-up (U.S. Congress, OTA 1987: 17). Five years is the very minimum for definitive test results from animal studies even without regulatory delay in nomination. However, this procedure is much too slow to evaluate all the suspected carcinogenic substances, for at present there are 55,000 to 100,000 chemical substances in existence (NRC 1984: 3), some of which will be toxic. Moreover, many more substances are created all the time; the EPA estimates that from 900 to 1,800 substances are introduced into commerce each year (U.S. Congress, OTA 1987).

Risk assessments appear to be as slow or slower than the animal bioassays, even though they should take less time. From the late 1970s to the fall of 1990, the federal EPA and risk assessors in the state of California using similar data bases completed about 50 risk assessments of carcinogens (Zeise 1990: 139-141). According to the Office of Technology Assessment, as of November 1987, the EPA Carcinogen Assessment Group (CAG) had performed risk assessments on only 22 of 144 substances (15 percent) testing positive in at least one National Toxicology Program animal bioassay. For 61 substances for which there was even better evidence for their potency because they tested positive for carcinogenesis in three or four animal experiments, the EPA's CAG performed only 9 risk assessments (15 percent) between 1978 and 1986 (U.S. Congress, OTA 1987: 20). Under California's Safe Drinking Water Act of 1986 (Proposition 65), 369 carcinogens had been identified as of April 1991. However, as of fall 1991, risk assessments had been performed on only about 74 of them, leaving 295 unevaluated (Prop 65 1990). Thus, even if animal bioassays could be done instantaneously, risk assessments lag far behind, taking from one-half to five person-years just for the detailed assessment (Zeise 1991).

At present, animal bioassays and the risk assessments based on them cannot be done quickly enough to survey the existing substances or to keep up with the introduction of new substances into the marketplace. If this is a concern, and if it points to a need to evaluate the toxicity of chemical substances more rapidly, as I think it does, then the rate of risk assessments becomes a relevant consideration in institutional and risk assessment design. Risk assessments should be done much more rapidly, and this would to some extent increase the number of substances subjected to regulatory evaluation. (It would not address the problem totally, because the underlying animal bioassays still take about five years.) If scientists or parties to the risk assessments insist on detailed evaluations of each substance to protect against false positives, this will perpetuate the slow pace.

Seventh, a point that emerges from several of the preceding remarks is that our scientific, institutional, and policy response to uncertainty and to the degree of certainty we require for regulatory action may promote or frustrate the many institutional and social goals served by both risk assessment and regulation. We implicitly or explicitly make policy decisions about how to respond to uncertainty, lack of data, and ignorance of underlying mechanisms. It thus is important to acknowledge that these polices are being made, to identify appropriate institutional goals, and to recognize how our responses to the uncertainties that plague risk assessment will affect the pursuit of such goals. Agenda 21 appears to acknowledge some of these concerns and responses to them (discussed below).

As we have discussed, an attempt to make risk assessment more carefully scientific may paralyze regulation, because the knowledge is not available and because almost all current regulatory laws tend to preserve the status quo until evidence for changing it is provided. On the other hand, if agencies expedite risk assessments and do not wait for answers to these scientific questions, then the "science" of risk assessment to some extent will rest on approximations and will be further permeated with either scientific or nonscientific policies. Thus, it will be quite different from ordinary science. (Of course, even in its present state, it already differs substantially from ordinary scientific inquiry.) Scientific data and theories clearly have a role in determining the health effects of human exposure to toxic substances. However, the science has to be good enough only for the balance of concerns faced by the institution.

How good is good enough depends upon the many aims of the institutions, legal constraints imposed upon them and, more broadly, matters of moral and political philosophy about the kind of world in which we want to live. The argument here is a cautionary note, a reminder, that there are substantial limitations to the extent to which risk assessments can measure up to present standards of good scientific evidence and continue to serve the aims of the regulatory institutions in which they are used. We should not confuse risk assessment for regulatory purposes with ordinary science, where the aim is pursuit of truth for its own sake, nor should we necessarily expect it always to measure up to standards of evidence required for peer-reviewed scientific journals. Instead, we should recognize it for what it can tell us, consistent with its evidentiary limitations, about the phenomena in question. In using these data, however, we should not lose sight of the many other aims of risk assessment, for either environmental or public health protection, which should properly modify the aims of risk assessment.

A larger point is that the many uncertainties pervading carcinogen risk assessment may make it difficult, if not impossible, for scientists to remain wholly faithful to their own scientific traditions (developed in circumstances in which the pursuit of truth for its own sake is the aim) while providing data that will permit timely and justifiable public health and environmental protections. More likely, fidelity to scientific tradition will produce agency paralysis. Thus, scientists may face both cognitive dissonance if asked to participate in risk assessments and peer pressure not to tailor risk assessments to serve regulatory aims. This should not deter scientific input into or participation in environmental regulation; instead scientists should recognize the nature of the enterprise together with its evidentiary limitations and the timely need for answers and do the best they can within that context.

Finally, however, even if the problems mentioned did not exist, there remain more fundamental problems with aspects of carcinogen risk assessment or of scientific inference more generally that should be of concern to scientists, philosophers of science, and policymakers in this area. These are considered next.

7. Problems in the Statistics of Human Epidemiological Studies and Animal Bioassays

Epidemiological studies of human beings exposed to a substance have the potential to provide the best scientific evidence that the substance is carcinogenic at specific levels of exposure. Whether they in fact provide such evidence depends upon whether they suffer from some possible practical and theoretical difficulties. Practical evidence-gathering problems such as poor record-keeping, job mobility (for workplace studies), and exposure to more than one toxin may frustrate good studies of the workplace. Lack of long-term monitoring, populace mobility, and confounding factors frustrate general environmental health studies. Long latency periods for diseases typically caused by carcinogens make it difficult to conduct well-done, reliable studies (Cranor 1993: 30-31). However, even if none of these problems exists, theoretical considerations indicate that in many circumstances the design and interpretation of such statistical studies may beg some of the normative concerns at issue. This last factor is of such importance that it should serve as a cautionary reminder of possible shortcomings from statistical scientific inferences more generally.

7.1. DISCOVERING RISKS

Human health risks at particular exposure levels can be detected either through cohort or case-control epidemiological studies. A cohort study compares the incidence of disease in a group exposed to a health hazard with the incidence of disease in a group representative of the general population (Mausner and Bahn 1974: 320; Rothman 1986: 62-74). In a case-control study, people diagnosed as having a disease (cases) are compared with persons who do not have the disease (controls) (Mausner and Bahn 1974: 312-13; Rothman 1986: 62-74). Fewer people are needed in a case-control than in a cohort study, for only those with the disease, not those exposed to a risk factor, are the objects of examination. In either case, in a good study, a positive correlation between a risk factor and the disease indicates that those exposed will tend to develop the disease and those not exposed will tend not to develop it.

Case-control studies are essentially retrospective. The researcher takes a group that has contracted a disease, compares the characteristics of that group and its environment with a properly representative control group, and tries to isolate factors that might have caused the disease. Cohort studies can be retrospective or prospective. In a prospective study, a sample population exposed to a potential disease-causing factor is followed forward in time. Its disease rate is then compared with the disease rate of a group not exposed to the potential disease-causing factor. In a retrospective study, the same method is employed, but historical data are used. The researcher studies the cold record of a group of people exposed to some suspected disease-causing factor over some period to establish their disease rate. That rate is then compared with the disease rate for nonexposed groups.

Each kind of study has its advantages and its problems. Case-control studies can provide estimates of relative risk, incur little expense because the sample sizes are small, and are especially suited to the study of rare diseases (Mausner and Bahn 1974: 316; Rothman 1986: 79-82). They have several disadvantages. Careful analysis is required to ensure a properly representative control group (Rothman 1986: 64-68). The incidence rate cannot be derived, for there are no appropriate denominators for the populations at risk (Mausner and Bahn 1974: 316). Case-control studies, like retrospective cohort studies, require historical information about their subjects, which creates problems of accuracy and documentation (discussed later). Sometimes it is difficult to separate and measure the effect of one risk factor compared with another (Mausner and Bahn 1974: 316). For example, rubber workers are exposed to vinyl chloride, polychlorinated biphenyls, chloroprene, selenium compounds, benzidine and its salts, aniline, carbon tetrachloride, and benzene, all of which are either suspected or federally regulated carcinogens (Schottenfeld and Haas 1979: 156-59). Case-control studies also run the risk of recall bias, since both the informant and the interviewer know the subject has the disease (Mausner and Bahn 1974: 320).

In contrast, a prospective cohort study is free from recall bias. And cohort studies yield incidence rates and attributable risk as well as relative risk. But cohort studies, particularly prospective ones, have their drawbacks too. They require much larger samples than case-control studies to detect the same risk, and they require a long follow-up period, which increases with the latency period of a disease. Such studies are thus costly (Mausner and Bahn 1974: 323-25). In a prospective study, subjects may drop out. In a retrospective study, they may be difficult to trace. Criteria and methods may change as the years progress. Finally, since most carcinogens have a latency period of 5 to 50 years, there are ethical problems in exposing people to suspected carcinogens for the period a prospective cohort study requires (Schottenfeld and Haas: 1979).

7.2. PRACTICAL EVIDENCE-GATHERING PROBLEMS

The cost and bioethical aspects of prospective cohort studies prompt most epidemiologists to rely on case control or retrospective cohort studies. However, there are several practical difficulties inherent in relying on the historical information required for such studies. Frequently, industry data on workplace exposure to potentially harmful substances are inadequate. When this is a problem, epidemiologists must resort to a worker's duration of employment as a surrogate measure of total exposure. The proper interpretation of these data, like any indirect measurement, is understandably a point of controversy, and in any event, companies often fail to keep the required information. Even if such data exist, they do not necessarily reveal which employees actually worked in the contaminated quarter (DOL 1980: 5040). Data on environmental exposure is likely to be even less reliable, since environmental exposures will be less well defined than workplace exposures.

As indicated, employees are often exposed to more than one chemical agent, which makes both case-control and retrospective cohort studies much more difficult, if not impossible, to conduct. In addition, the dosage of exposure frequently varies over time (DOL 1980: 5040).

Job mobility and population heterogeneity also pose problems for workplace epidemiology. Since there is considerable job mobility in American employment, the effect of a carcinogen can easily be overlooked. Typically, the briefer the exposure, the longer the latency period of the disease, unless the exposure was at a very high dose (DOL 1980: 5040). Even if an epidemiologist has data for one population and its set of characteristics for either a cohort or case control study, it is difficult to extrapolate to other populations and their characteristics (DOL 1980: 5040). Populations can vary in socioeconomic status, age at which exposure occurred, smoking history, and other factors that affect susceptibility and confound the studies (DOL 1980: 5042). Population mobility and lack of good monitoring are also problems for environmental epidemiological studies.

The long latency periods of diseases typically caused by carcinogens may be a more serious problem. Thus, even though a scientist has none of the practical problems mentioned previously and has sufficiently large samples to avoid insensitive studies (addressed later), if subjects are not followed for a long enough period, a disease effect may be missed. These practical problems make it difficult, perhaps nearly impossible in some cases, to obtain scientifically respectable results to quantify health risks and to provide even the most rudimentary dose-response curve for a substance. In fact, one researcher has suggested that the relevant data are missing for most chemical substances and industrial processes (DOL 1980: 5044). Thus, researchers might fail to detect a risk of concern, even when one exists, because of such practical problems.

One consequence of failing to have data about adverse health effects, even when in fact they exist, is that this favors the legal status quo. If employees or the public are not protected legally from toxic substances or if they are protected less than they should be, then as long as there is no evidence of risks, even when such risks exist, they remain unprotected. By contrast, if the legal status quo prevents the introduction of potentially toxic substances into commerce until they are proven safe enough for human exposure, lack of evidence of safety protects the status quo. There is a problem with studies that show no association between exposure and contraction of disease: frequently, studies are too insensitive to establish the requisite claim of safety even though they do not show an "adverse effect." We pursue this further below.

7.3. THEORETICAL DIFFICULTIES

To illustrate the theoretical problems, consider cohort observational epidemiological studies. Observational studies, typically relied upon to identify risks to human health from exposure to toxic substances, depend "on data derived from observations of individuals or relatively small groups of people" (U.S. Congress, OTA 1982: 137) and in which exposure is not assigned by the investigator. They are then analyzed with "generally accepted statistical methods to determine if an association exists between a factor and a disease and, if so, the strength of the association" (U.S. Congress, OTA 1982: 137).

A wise and conscientious epidemiologist with perfect evidence, but with constrained sample sizes for studying relatively rare diseases, faces potentially controversial moral and social policy decisions in order to design and use an epidemiological study. If scientists uncritically follow scientific conventions and practices used in pursuit of knowledge for its own sake and in the requirements for publishing in reputable scientific journals, contrary to what most scientists believe, they may unwittingly have dirty hands, contaminating their scientific results with implicit social policy outcomes and begging the policy issues at stake. In fact, the very attempt to make the science rigorous as required by the ideals of the profession may beg or frustrate the regulatory questions for which the studies are done (or used). Thus, one kind of scientific objectivity leads implicitly to dirty hands. This problem is not easily avoided, for while a more sophisticated presentation of scientific results (discussed later) leaves scientists with clean hands, this merely shifts the problem to someone else.

To see these points, we must review the theory of hypothesis acceptance and rejection, in order to introduce enough terminology to characterize the main risk and proof variables with which epidemiologists must work and to understand the logic of scientific proof available in this area. I focus on hypothesis rejection and acceptance because it is the traditional statistical approach used in much of science and statistics. However, epidemiology and perhaps the statistics of scientific reasoning are moving away from the model of hypothesis acceptance and rejection. Rather than explain alternatives to this model (although some are suggested toward the end of his chapter), I use it as an example to illustrate certain problems that can arise from the use of demanding standards of evidence. Thus, because the field is in flux, there will be some who will not accept the model on which I focus, but I believe many (even most) still accept it, and even for those who do not accept it, it can still usefully serve as an illustration of a general class of problems (Fleiss 1986, Walker 1986, Poole 1987, Thompson 1987, Goodman and Royall 1988, Swan 1991).

In trying to determine whether a substance such as benzene is a human carcinogen, a scientist considers two hypotheses. The first (the null hypothesis, H_0) predicates that exposure to benzene is not associated with greater incidence of a certain disease (e.g., leukemia or aplastic anemia) than that found in a nonexposed population. The second (the alternative hypothesis, H_1) indicates that exposure to benzene is associated with a greater incidence of such diseases.

Since epidemiology considers samples of both exposed and unexposed populations, by chance alone a researcher risks inferential errors from studying a sample instead of the whole population in question. A scientist runs the risk of false positives (the study shows that the null hypothesis should be rejected [and the alternative hypothesis accepted] when in fact the null hypothesis is true) or false negatives (the study shows that the null hypothesis should be accepted when in fact the null hypothesis is false [and the alternative hypothesis is true]). A false positive is designated a type I error, and a false negative is called a type II error

Table 2. False Positives and False Negatives in Epidemiology.

	Possibilities in Causal Relationships	
Possible Test Results	Null hypothesis is true: Benzene is not positively associated with leukemia	Null hypothesis is false: Benzene exposure is positively associated with leukemia
Test does not show that benzene exposure is associated with leukemia	No error	Type II error False negative β
Test shows that benzene exposure is associated with leukemia	Type I error False positive α	No error

(summarized in Table 2). Figure 7 illustrates the relation between false positives and false negatives for two distributions of a test measure. (In Figure 7, *C* is the decision cutoff such that "responses to the right of *C* are declared positive and those to the left declared negative" [Ades 1990].) Statistical theory provides estimates of the probability of committing such errors by chance alone. The probability of a type I error is normally designated a and the probability of a type II error is designated ß. Conventionally, a is set at .05 so that there is only a 1 in 20 chance of rejecting the null hypothesis when it is true (Walker 1977). The practice of setting $\alpha = .05$ I call the "95 percent rule," for researchers want to be 95 percent certain that when new knowledge is gained and the null hypothesis is rejected, it is correctly rejected.

Conventional practice also sets ß between .05 and .20 when a is .05, although conventions are less rigid on this than for values of a. When ß is .20, one takes 1 chance in 5 of accepting the null hypothesis as true when it is false—for example, the chance of saying benzene is not associated with leukemia when in fact it is (Feinstein 1977). When ß = .20, the power (1—ß) of one's statistical test is .80, which means scientists have an 80 percent chance of rejecting the null hypothesis as false when it is false. The low value for a probably reflects a philosophy about scientific progress and may constitute part of its justification (Giere 1979). It is an instantiation of the cautious scientific attitude described earlier. When the chances of false positives are kept low, a positive result can be added to scientific

Figure 7. Relation Between False Positives and False Negatives for Two Distributions of a Test Measure (Ades 1990).

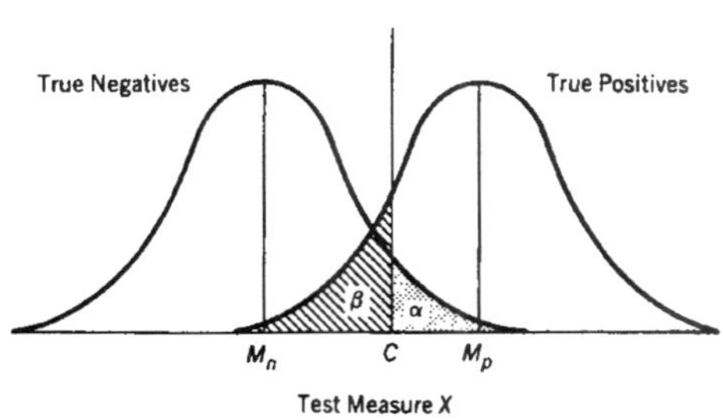

knowledge with considerable confidence that is not the result of random chance (Feinstein 1988; Greenland 1991). Were one to tolerate higher risks of false positives, take greater chances of new information being false by chance alone, the edifice would be much less secure. A secure edifice of science, however, is not the only important social value at stake.

One can think of α, β (the chances of type I and type II errors, respectively), and 1-β as measures of the "risk of error" or "standards of proof." What chance of error is a researcher willing to take? When employees in an industry or the general public may be contracting cancer (unbeknownst to all) even though a study (with high epistemic probability) shows they are not, is a risk to their good health worth a 20 percent gamble?

If we think of α, β, and 1-β as standards of proof, how much do we demand of researchers and for what purposes? Must researchers be more than 51 percent sure that benzene is a carcinogen presenting a risk to employees in the workplace before regulating it? Or, equivalently, should scientists in agencies be permitted to take a 49 percent chance (β = .49) that substances are not high-risk carcinogens to the populace when in fact they might be? Such questions only precede more complex matters, for the standards of proof demanded of statistical studies have implications for the costs of doing them and for the relative risks that can be detected. The mathematics of epidemiological studies together with small sample sizes and rare diseases typical of environmentally caused cancer force serious policy choices on researchers and regulators alike, when these studies are used in legal contexts to estimate risks to people.

In order to see some of the trade-offs we need two other variables: *N*, the total number of people studied in the exposed and unexposed samples, and *d* the relative risk one would like to be able to detect (Feinstein 1977). Relative risk is the ratio of the incidence rate of disease for those exposed to a disease-causing substance to incidence rate among those not exposed (Mausner and Bahn 1977: 322):

$$\text{Relative risk} = \frac{\text{incidence rate among exposed}}{\text{incidence rate among nonexposed}}$$

For instance, if the incidence rate of lung cancer in the nonexposed population is 7/100,000, and the incidence rate among heavy smokers is 166/100,000, the relative risk is 23.7. The value of concern depends upon many factors, including the seriousness of the disease, its incidence in the general population, and how great a risk, if any, the exposed group justifiably should be expected to run. (Relative risk can be misleading if the disease rate in the general population is quite low, for example, 1/10,000,000. Thus, one needs to take into account this and other factors in evaluating the overall seriousness of the risk.) With α and β fixed, the relative risk one can detect is inversely related to sample size: The smaller the risk to be detected, the larger the sample must be.

The variables α, β, *d*, and *N* are mathematically interrelated. If any of the three are known, the fourth can be determined. Typically, α is specified at the outset, although it need not be. Because the variables are interdependent, however, crucial trade-offs may be forced by the logic of the statistical relations, as the following examples indicate (summarized in Figures 8 and 9) (Cranor 1990).

Alternative 1: Suppose we want to discover whether a suspected carcinogen *C* is associated with a particular cancer *L*. Suppose the incidence of *L* in the general population is 8/10,000, and suppose we rely upon the 95 percent rule. We want to be 95 percent sure that when no association exists between *C* and *L*, our study shows that none does. Thus we set α at .05. Suppose we also wish to have very small odds of false negatives, so we specify that β should be .05. Thus, the chances of false positives and false negatives are low and equal. Suppose further that we regard a relative risk of 3 (d = 3) as a "serious" risk worth investigating for public health purposes. Given these antecedent desiderata, in order to achieve them, we would have to study at least 13,495 people exposed to C, and (I assume for the sake of simplicity) an equal number who are not exposed (or 26,990 people total) to obtain statistically significant results at a relative risk of 3 (Walter 1977; Cranor 1983). That very likely might be prohibitively expensive, and it would be practically very difficult to follow participants, thus threatening the feasibility of the study. Thus, a moral consideration, the value of the most accurate information for detecting potential harms (that is, tests with low and equal chances of type I and type II errors), can enter at the outset of a study.

Alternative 2: Next, assume everything is the same except ß and sample size. Suppose we would tolerate a lower ß = .20, so we have only 1 chance in 5 of committing a type II error. Given these values, we would have to study at least 7,695 people exposed to C and the same number who are not exposed (or 15,390 people total) to obtain statistically significant results with a power of .80 to detect a relative risk of 3. This study would also likely be prohibitively expensive, and it might be difficult to find such large groups of exposed individuals. Scarcity or the impracticality of following large groups of people would still prevent the most accurate results even if type I and type II errors are not equal.

Alternative 3: Suppose we could not study the large numbers required in alternatives 1 and 2, but we could study only 2,150 in the exposed and nonexposed groups. Suppose we also want to be 95 percent confident (α = .05) of results favoring the null hypothesis and 80 percent confident (1-ß = .80) of detecting an elevated relative risk should it exist, when the prevalence of the underlying disease is 8/10,000. What relative risk can we hope to detect? At best, we only could detect a relative risk of 6, or two times higher than the risk we thought was "serious" enough to warrant social attention. Put differently, given the values for α, ß and N, our study could not even detect the relative risk of concern with 80 percent confidence, when it exists. Thus, small samples, forced by cost considerations or impracticalities and a demand for accuracy, mean that our test cannot even detect the risks of concern that by hypothesis motivated the study.

Alternative 3 suggests some interesting results for "negative" or "no effect" studies. If a study were negative or showed no effect between the chemical C and the disease L, the most that we could infer would be that the relative risk to people in the exposed group is not as high as the relative risk tested for in the study. Thus, negative studies show nothing about relative risks smaller than the test can detect. Regulatory agencies regard such results as useful mainly for setting upper bounds on risks to people.

Alternative 4: The mathematical interrelations between α, ß, d, and N are flexible enough, however, to enable us to detect a lower relative risk, say d = 3.8, by making some trade-offs. If we kept N and α constant (N = 2,150; α = .05), ß would have to be correspondingly raised to .49, lowering the power of the test 1-ß to .51 (Walter 1977; Cranor 1993: 467-88). Because ß =.49, there is now, however, a 49 percent probability of mistaking a toxic substance for a benign substance by chance alone, when in fact the substance is toxic. The study now faces two problems. The smallest relative risk we could detect among the 2,150 exposed population would be 3.8 (still slightly higher than the relative risk of concern). And we could detect that only if we were willing to take 49 percent odds of leaving that group exposed to a possibly harmful carcinogen. This is a morally dubious alternative, for our false negative rate is no better than the toss of a fair coin.

Alternative 5: The mathematical relations permit another alternative. Holding sample size constant, if we want to be able to detect a relative risk as low as 3.0 with 80 percent confidence when it exists, we could increase α instead of ß. With a commitment to ß = .20, the resultant α would have to be about .33 to enable us to detect a relative risk of 3.0. Now we could be only 67 percent confident of not incurring false positives (Rimm 1980). Thus, even though we can reach statistically significant results for a relative risk of 3 by increasing α to .33, one third of the time we run the risk of mistakenly adding to the stock of scientific knowledge. According to typical conventions in science, results from such studies would not likely be published in reputable scientific journals, for the chances of type I errors are too large (U.S. federal agencies once suggested a continuum of α values, but now are more rigid) (IRLG 1979: 241-69; ISG 1986: 281-82). Thus, we would be tolerating results that might be considered somewhat less accurate in order to increase our chances of detecting risks of concern. Therefore, such a study would sacrifice one kind of accuracy important to the scientific community in order to try to improve the accuracy of the study for public health purposes.

These examples constitute the decision tree displayed in Figure 8. It is not immediately evident which alternative is the most attractive. Alternatives 1 and 2, although the most accurate, are excluded for reasons of cost or impracticality. Were scientists forced to adopt alternatives 3 or 4, either might put those exposed to toxic substances at considerable risk because the scientists could not detect the risk of concern when it existed (on alternative 3) or because of high false negative rates (on alternative 4). On alternative 5, scientists risk undermining the credibility of their research and increasing the risk of making a mistake in scientific research (although the odds of this are still not as high as the false negative rates in alternative 4). The logic of epidemiology and study costs, together with small sample sizes and rare background disease force these difficult moral choices on "scientific" research.

As striking as the preceding examples are, they only suggest the statistical problems a

Figure 8. Some Choices in Conducting a Cohort Study of a Relatively Rare Disease (Cranor 1993).

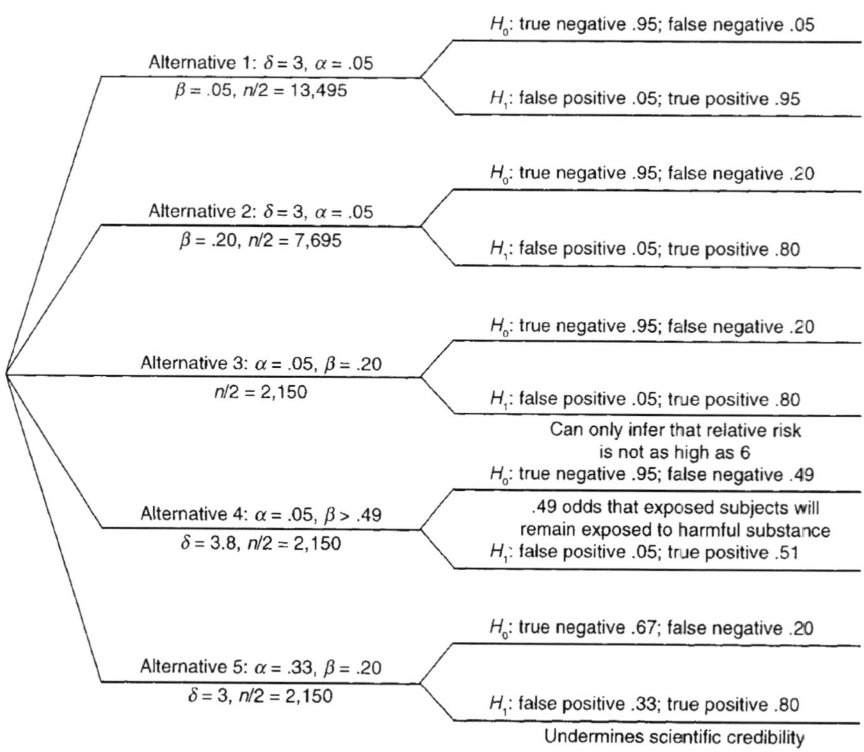

cohort study of a typical environmentally caused disease (e.g., benzene-induced leukemia) might pose. Alternatives 1 through 5 in Figure 8 assume that the prevalence of the hypothetical disease *L* in the general population is 8/10,000. If the background disease rate were rarer by a factor of 10, which is more realistic because it is the rate, for example, of leukemia, then our decision tree would exhibit the even more surprising results displayed in Figure 9. The sample sizes required for analogues of alternatives 1 and 2 increase tenfold, the smallest relative risk that could be detected in the analog of alternative 3 is 39, the false negative rate in the analog of 4 greatly exceeds .60, and even increasing a to .33 in alternative 5 with $\beta = .20$ does not lower the smallest detectable relative risk below 12 (Cranor 1993: 227-228).

The preceding discussion is slightly misleading in one respect. Cohort studies require much larger samples than case-control studies to be equally sensitive. Case-control studies tend to use much smaller samples and are thus good for detecting rare diseases. The costs of using them would be much lower (Schlesselman 1974). However, I have focused on cohort studies because the mathematics is easier to explain and the mathematical trade-offs between sample size and relative risk for both kinds of study are similar. In addition, cohort studies are frequently used in occupational and environmental studies. Case-control studies also suffer from special methodological difficulties that may preclude their use for illustrating the

Figure 9. Some Choices in Conducting a Cohort Study (Cranor 1983).

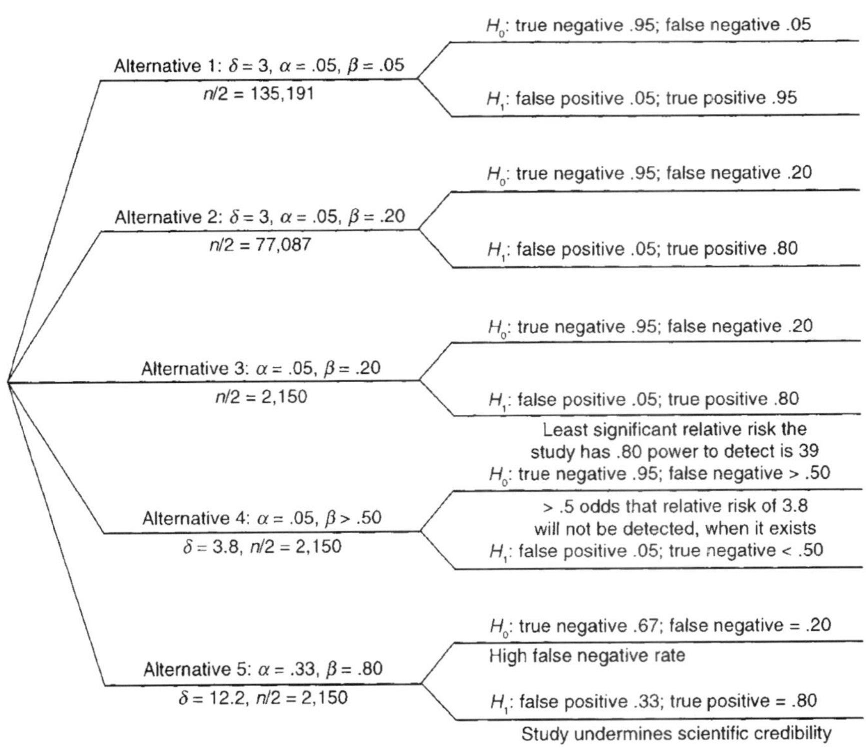

burden-of-proof problems with which I am concerned (Horowitz and Feinstein 1979).

The statistics of animal bioassays exhibit behavior similar to that of epidemiology, although the numbers are not quite as dramatic. Thus, the same trade-offs may be forced in animal studies; the rarer the disease rate is in control animals and the fewer treated animals in the sample with tumors, the less likely researchers would be to detect risks of concern, if they remain committed to the 95 percent rule. In such cases, scientists should consider using higher α values to ensure that elevated disease rates in the treated groups do not go undetected because of the scientific conventions of the statistics of the studies (Cranor 1988).

An additional problem can arise from scientists' or risk managers' implicit commitment to certain statistical variables. Consider the effect of the α-β asymmetry in testing large numbers of substances. As long as $\alpha>\beta$ and α is in the neighborhood of .05, we are doing "better" science conventionally conceived, but we may also be protecting possibly harmful chemicals better than we are protecting human health. Suppose that we have 2,400 substances to test and that 40 percent of them are carcinogens and 36 percent of them are not, with the remainder unclear. (These percentages are similar to the results obtained from testing by the National Toxicology Program.) With α at .05 and β at .20, assuming there are large enough study samples, epidemiological studies will result in 192 false negatives and 43 false positives. Thus, 192 substances will pose some risk of cancer to the populace (and how large a risk this is will depend upon the prevalence of the disease, the relative risk associated with the substance, its potency, and the number of people exposed), and 43 substances will be wrongly regulated or possibly banned altogether (depending upon the statutory authority in question).

Moreover, although I focused on what I call the 95 percent rule, the burden-of-proof problems resulting from administrative practices may be worse than I have indicated. It is probably rare when a federal agency or an organization such as the International Agency for Research on Cancer (IARC) using epidemiological studies for identifying or regulating carcinogens relies on a single study. Several positive studies, each at 95 percent confidence level, may be required before an agency is prepared to identify a substance as a carcinogen. With two such studies (assumed to be independent) at $\alpha = .05$, the chances of two such rare events occurring is .0025 ($.05^2$). To demand multiple studies, the first one or two of which are positive, then, is to be exceedingly cautious. If an agency refuses to identify a substance as a carcinogen or to regulate in the meantime, this imposes costs on the potential victims while the agency gathers more information. In addition, if the agency does not address the identification or regulation of other substances, there are further opportunity costs from inaction.

Finally, both cohort and case-control studies can suffer from confounding, the mixing of an effect of the exposure of interest with the effect of an extraneous factor. Confounding can lead to overestimation or underestimation of the causal effect, depending upon the direction of the effect (Rothman 1986: 89). This is one of the most serious problems facing epidemiology, but because the confounding effect is not consistent, there is no one remedy.

However, there are contexts in which a continued search for confounders frustrates public health protections. If researchers have evidence that a substance harms human health, e.g., cigarette smoke or asbestos, but continue to search for possible confounders to explain away observed associations between exposure to the substance and contraction of disease, this can delay action and possibly frustrate health protection. As Greenland (1991) notes, "One can *always* invoke unmeasured confounders to explain away observational associations. Thus, actions should not depend on the absence of such explanations, for otherwise action would never be taken."

The motivation of some researchers to search for confounders is similar to the motivation to require demanding standards of scientific evidence: sufficient proof to justify a scientific inference of a casual connection. Advocates of a careful search for confounders in such circumstances, such as Eysenck (1991), seek to establish such casual connections with "proof in the sense usually accepted in science" or possibly proof "beyond a reasonable doubt," because such facts if discovered will slay "a beautiful hypothesis." However, for the reasons previously discussed and for reasons that follow, such an approach poses problems in protecting the public health and the environment. In my judgment, Greenland's views are closer than Eysenck's to the correct approach.

7.4. INTERPRETING EPIDEMIOLOGICAL STUDIES

The preceding discussion of scientific standards of evidence indicates only some of the abstract logical trade-offs that exist between different variables that are used in an epidemiological study. Research practices could address some of these issues. In designing a survey, a scientist must decide which variables are to be independent and which dependent. At least one variable will be fixed: the disease rate in the general population for the disease that is the object of study. Three of the remaining four variables must be specified: sample size, α, β, and δ.

As a matter of present practice, scientists appear to specify α at some low value, typically $\alpha = .05$. The sample size is also likely to be fixed antecedently, because only a certain group of people is available for study (e.g., workers in a factory) or because costs limit the sample. If sample size is not fixed for one of these reasons, it may be chosen in light of other goals of the study that influence choice of the statistical variables, for example, α and β and the relative risk one believes is a matter of public concern. A study will then yield certain morbidity and mortality rates in both exposed and control groups, which as a matter of fact establishes an experimental-to-control group relative risk. Thus, in traditional practices, scientists specify $\alpha = .05$, and δ is in effect fixed as an outcome of the study. Finally, scientists may use a predetermined β (Simpson and Hill 1986) or solve for β (or 1-β, the power of the study). I assume they solve β, because that is the more flexible and charitable interpretation of the procedure.

However, when α is set antecedently (or specified as a matter of routine) and typically at a small value (<.05) because the 95 percent rule is being used, as a matter of experimental design this creates the possibility that risks of concern may go undetected, because the power of the test may be quite low (for the reasons indicated previously).

Scientific practices could be different, for scientists could be more flexible in evaluating the data and consider alternatives such as 5 above, which permits both α and β to vary in value. This flexibility would permit departures from the 95 percent rule, in order to interpret studies to detect risks of concern. Suppose there is a study in which the fixed data consist of the disease rate (8/10,000), sample size (2,100), and mortality rate (5), compared with 1.72 expected deaths. For purposes of interpreting this information, epidemiologists could vary the values of α and β. By adopting this procedure, however, there is a problem of what the fixed data show. Any pairwise combinations of α and β in the left-hand column of Table 3

Table 3. Pairwise Combinations of δ and β and Study Outcomes.

Positive Results			Negative Results
α	β	δ	
.10	.49	3.0	When $\alpha < .10$ (with β constant) or $\beta < .49$ (with α constant)
.15	.40	3.0	When $\alpha < .15$ (with β constant) or $\beta < .40$ (with α constant)
.20	.30	3.1	When $\alpha < .20$ (with β constant) or $\beta < .30$ (with α constant)
.25	.25	3.1	When $\alpha < .25$ (with β constant) or $\beta < .25$ (with α constant)

will show that the study outcome is positive, for all would permit researchers to detect a relative risk of about 3. Changing the variables slightly as indicated in the right-hand column will produce a negative study.

Note that δ is the least significant relative risk that can be detected when it exists. This example, together with Figures 8 and 9, shows there is considerable flexibility in interpreting the data of a study. How they are interpreted and used in certain regulatory and legal contexts will have important consequences for protecting human health.

7.5. PUBLIC POLICY ISSUES

We have seen that there can be a mathematical tension between the use of the 95 percent (or higher) confidence rule and other public policy and moral concerns we might have. However, there is no necessity to use the cautious scientific practice. Whether epidemiologists should be committed to the 95 percent rule in certain contexts is a normative, or policy, question. Scientists, moral philosophers, philosophers of science, lawyers, and those in public institutions with the authority to protect our health should explicitly acknowledge and address this question. Those charged with regulating our exposure to substances should consider the foregoing policy problems in the design of the study.

In addition, the reporting and use of epidemiological data given traditional scientific practices may not be as neutral and objective a project as scientists might believe. In the example just considered, what inferences one makes from the fixed data for subsequent regulatory or legal proceedings depends upon the choice of statistical variables and may have important consequences for our health and national wealth. Current practices provide conventional procedures for interpreting data. They are just conventions; these could be changed.

More important, there is not an obvious correct interpretation of the data, for the choice of values for α and β (and the inferences drawn from the study) depend in part upon wider uses to which the data will be put and the context in which they will be used. But this feature of the situation commits the decisionmaker implicitly, if not explicitly, to making judgments that are the equivalent of moral or social policy considerations. These equivalents of moral considerations must be relied upon in order to perform and interpret the studies in question.

Alternatives are open to scientists, however, which preserve the objectivity of the science and insulate them from the "dirty" policy decisions that infect some traditional practices (when they are used uncritically). Scientists could present their results as in Figure 8 or 9 (alternatives 3 to 5) or in a set of power function curves (Figure 10) (Lehman 1959), or with appropriate confidence intervals (Mayo 1988). Recent discussions suggest there is a strong movement among epidemiologists away from tests of significance toward such presentations of data (Fleiss 1986; Walker 1986; Poole 1987; Thompson 1987; Goodman and Royall 1988). All three kinds of presentations indicate some of the objective limits to the data. A power function curve indicates lower bounds on the relative risk that could be detected (if it existed) relative to particular α values. Confidence intervals indicate the "actual magnitude of the effect as well as the precision of the estimate" (Goodman and Royall 1988). Common to such presentations is that they show how the data of a test place limits on what one can infer from them. In addition, the concern to display the objective limits of data serves the scientific goal of understanding, but this information is then used in regulatory contexts for decisionmaking (Poole 1987), and those who must make the decisions must be able to understand the information and use it.

However, one cannot escape the fact that making inferences from data derived from small samples for rare diseases is like trying to eliminate the pucker in a wall-to-wall carpet that is too large for a room—a problem in one area can be eliminated only by creating a problem in another area. Thus, if scientists use α and β values uncritically in the design of their

Figure 10. Power Function Curves for Sample Sizes 2,150 and for Disease Rate of 8/10,000. (See Appendix C in Cranor 1993.)

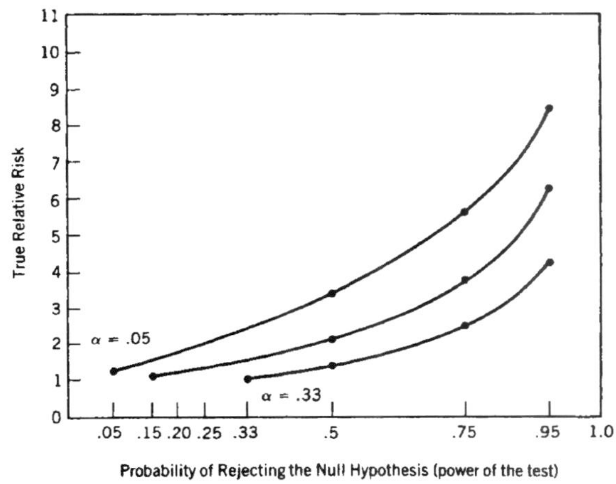

studies and the reporting of their data, they implicitly will make some of the important social policy decisions. If, on the other hand, they present the results of their studies in the most objective manner that they can, their hands remain clean, but someone else must face precisely the same social policy trade-offs. This problem raises both public policy issues, already sketched and issues of professional ethics (Cranor 1993: 40-47).

Since the reporting, interpretation, and use of epidemiological data are not normatively neutral, and we could change traditional scientific practices, we should face the use of the 95 percent rule in these contexts as a normative question, no matter who makes the interpretive decisions. The 95 percent rule in the preceding discussion serves as a surrogate or as an exemplar for the use of demanding standards of scientific evidence for regulatory and public health purposes. We have seen how this can adversely affect health protective goals. Thus, the reader should keep in mind plausible generalizations of the 95 percent rule as examples of how demanding standards of evidence or other scientific practices could frustrate pursuit of public health or environmental goals when more sensitive and more flexible interpretations of data would better serve these aims.

In other institutional contexts we have clearly faced such issues. In the criminal law, for example, avoiding wrongful damage to someone's reputation and well-being is so important that we spend considerable sums of money and deliberately impose difficulties on proving guilt in order to avoid wrongly inflicting harsh treatment and condemnation on the defendant. We could save money and have more numerous unjust outcomes if we thought it worth the human costs, but we do not. Clearly a number of moral and cost considerations have influenced the institution of the criminal law. We have been quite self-conscious in debating the considerations that bear on the design and workings of the criminal law. Somewhat analogous problems arise in environmental health institutions concerning the interpretation and use of scientific studies; I suggest that similar debates should address these issues. This point is especially important to some of the main goals of Agenda 21 (discussed later).

Thus, the problem is that the evidentiary standards of science as illustrated by the 95 percent rule may be much more demanding than the standards of evidence that would be required where the scientific evidence might be used. If this is correct, then, if government or public health agencies use conventional scientific standards, by default their science in many cases runs the risk of not discovering risks of cancer or of begging the normative questions at issue. Under postmarket regulatory statutes (which predominate in governing the regulation of carcinogens), a commitment to demanding scientific standards may well prevent the discovery of risks and lead to lowered protections for the public.

Given the wider aims of both public health and environmental law and the evidentiary standards that typically must be met in the law compared with science, there may not be good reasons in these legal contexts to require risk assessment science to meet the same evidentiary burdens as required in normal scientific pursuit of the truth for its own sake. Thus, I would urge that for regulatory science, agencies adopt evidentiary standards much closer to those of the institutions it is meant to serve.

The following reasons might be offered for retaining the 95 percent rule (and other analogous demanding standards of evidence) even in regulatory and other legal contexts:

1. It is the prevailing tradition.
2. Scientists should be cautious about additions to scientific knowledge so that additional bricks of knowledge are well made and well cemented to the existing scientific structure.
3. The 95 percent rule provides a useful standard, a benchmark against which all epidemiological studies (and scientific studies more generally) can be compared. If studies were conducted with substantial departures from the 95 percent rule, they would no longer have a kind of automatic credibility (represented by the 95 percent rule) and perhaps would have to be scrutinized much more closely. It also provides a neutral standard in debates that are otherwise politically and morally charged.
4. The aim in an epidemiological study is to provide evidence of a causal relationship between a substance and a disease, and not using the 95 percent rule undermines this aim.
5. The 95 percent rule, when used in certain regulatory contexts, protects the status quo.

None of these arguments provides an overriding reason for always using the 95 percent rule, and several do not constitute even a prima facie reason for using it. Reason 1, although correct, begs the question whether the rule should be followed in all contexts, especially in nonresearch regulatory and legal settings. Reason 2, although correct for basic research that aims to add to the stock of fundamental knowledge, is less appropriate for regulatory contexts. For regulation, epidemiological and other statistical studies are aimed not at discovering new scientific results, but at trying to discover whether risks to health exist. And sometimes the aim is merely to confirm or deny the carcinogenicity of a substance that is in a chemical class with other substances known to be carcinogenic. Neither case presents a good reason for always adhering to the 95 percent rule.

The third reason is important for consistency, but only that. Consistency is not an overriding reason for following a certain practice, if the practice otherwise would produce bad results in a particular area. In addition, such studies are not automatically accepted at present—they receive considerable scrutiny. If the 95 percent rule were abandoned in some regulatory contexts, this would add only marginally to the usual controversies. In particular, not using this rule would merely add to the existing complaint by some interest groups seeking to undermine risk assessments that risk assessments are not "scientific."

A related point is that a commitment to scientific standards of evidence may be the only stable and neutral reference point in debates that otherwise seem driven by political interests, policy considerations, and a good deal of uncertainty. There is much to this concern. However, it is not clear that a commitment to the 95 percent rule, which can beg regulatory questions, is the best way to address it. Thus, the use of this rule is considerably less neutral than it appears on the surface and than members of the scientific community may sometimes believe. Presenting data in the most objective manner possible is a meritorious consideration. But how the data is used, whether to infer a risk of concern or to infer no such risk or to infer there is no evidence about risks, is clearly an important policy decision to be settled by broader moral and political philosophic concerns, or a matter to be settled by the institution in question. How much of a "clue" to carcinogenicity is provided by studies that fall short of the 95 percent is an important issue to be settled by the evidentiary standards appropriate to the institutional context.

Furthermore, it is a good thing to have questions of health risks decided in large part on normative policy grounds. Some degree of accuracy in estimating risks is important, but risk assessment is an inexact "science" (Cranor 1988). As I indicated earlier, perhaps it is much better to treat both the kind and amount of evidence needed to estimate a risk and the acceptability of the risk in part as social decisions, rather than treating risk estimation as a purely scientific decision and the acceptability of the risk as the only policy decision, especially when the scientific part of the decision may beg the normative issues. In addition, several researchers have found that scientists' attitudes toward their research results and toward public policy issues are substantially influenced by their place of employment. Industry scientists are more skeptical that substances pose risks of harm than are academic or government scientists (Lynn 1987). Given the possibility of normative slants to scientists' work, it seems that a better approach is to choose openly and deliberately the normative concerns we want to influence the choice of models in risk assessments and the interpretation of statistical studies. A public, community decision about these matters through the mechanism of a regulatory agency seems the appropriate approach in a democracy.

Fourth, if indeed the aim is to provide evidence of a causal relationship between a substance and a disease, then we surely want to understand this. However, how definitively this must be established before we decide to take action as a matter of public health and environmental policy is another matter. Different kinds and amounts of evidence may be needed before one asserts for purposes of understanding the definitive existence of causal claims versus deciding for public health and environmental purposes what to do. For regulatory purposes, one might well accept the results of epidemiological studies not based on the 95 percent rule in order better to detect potentially harmful substances. For more fundamental research purposes, for example, discovering whether a whole class of substances—say, the arsenicals—appear to be carcinogenic, one might want to have at least some of the studies established by the 95 percent rule.

Finally, although in fact the 95 percent rule used in evaluating carcinogens may preserve the status quo, it is for precisely this reason that we should reexamine the use of the rule. The status quo may result in harm to humans and the environment; burdens of proof implicit in science combined with the demanding standards of evidence enshrine the status quo. If that is the case, or if we have good reasons to think that it might be, then we should adopt different policies toward the use of scientific evidence. As we saw above, for small samples and relatively rare diseases, use of the 95 percent rule under a post-market regulatory statute in many circumstances will protect commercial manufacturers and sellers of a substance better than potential victims. In weighing the balance between risks of wrongly regulating commercial substances and wrongly leaving people exposed to potentially carcinogenic substances, the latter seems the more important concern, although this depends upon the facts of the case and one's larger legal and moral policy decisions.

In addition to the preceding rebuttals, there are some more positive reasons for adopting more context-sensitive procedures for various legal purposes. When sample sizes are small and the background rate of disease is relatively rare (<8/10,000), departures from the 95 percent rule make it possible for epidemiologists better to detect harmful substances at a certain relative risk. This is especially important for detecting low but possibly substantial relative risks. In regulatory contexts, where a major aim of the enterprise is to predict risks to human health and to prevent them, if possible or if feasible (depending upon the statute), departures from the 95 percent rule may better serve this preventive aim.

The 95 percent rule and analogous demanding standards of evidence need not be abandoned in all scientific contexts, nor should they necessarily be abandoned in all regulatory contexts. Instead, scientists and policymakers should be more discriminating in their use of such standards and carefully consider the consequences of their use. In clinical trials of a drug in which the goal is to discover if a drug has therapeutic effects, the 95 percent rule might be relied upon, for research endeavoring to add to our fundamental knowledge about biochemical and therapeutic mechanisms should not be conducted if chances of incurring false positives are significant. (Although in contexts where life-saving drugs might be developed, evidentiary standards that are too strict can also have adverse social consequences.) Similarly, when one is conducting epidemiological research to establish knowledge as a foundation for further research, one might well want to retain the 95 percent rule. On moral and legal grounds it is likely there will be reasons for departing from the 95 percent rule in several contexts: in screening substances to try to discover those that pose harms to health and in preventive regulatory proceedings where the major concern is the forward-looking prevention of health harm and there is little fundamental research to be gained or upon which to build.

There is a generalization of the concerns raised in this section. Any specialist (at least in academic disciplines) is concerned about the validity and the defensibility of his or her inferences. We tend to be cautious in drawing inferences in order to avoid mistakes. Frequently, we are hyperskeptical in order to protect the field and prevent pursuit of false leads. The 95 percent rule is an exemplar of a minimal standard for good statistical inferences in scientific inquiries. By analogy with the arguments about epidemiology, to the extent that scientists are reluctant to conclude that suspect substances do not cause disease or death because the inferences cannot be justified on the best inference standards for discipline, a debate whether to regulate or not may be begged in favor of nonidentification of a harmful substance and nonregulation. By analogy with the recommendations made previously, scientists should scrutinize other demanding scientific inferences used in risk assessment and other public policy debates to see whether regulatory outcomes are biased by evidentiary practices used in a discipline.

Similarly, the use of strict scientific inferences in regulatory contexts should be addressed as moral or social policy questions. In many cases, evidentiary practices in science will beg policy questions; thus they should be examined for this possibility.

8. Implications for Agenda 21

Enough background on the science of risk assessment and some of its policy implications has been provided to suggest some policy consequences for Agenda 21 and some of the principles of the Rio Declaration. The first thing to note is that because of the probabilistic nature of and uncertainties in risk assessment, there will be mistakes. Moreover, because of the mathematical trade-offs necessitated by small sample sizes and rare diseases between false positives and false negatives, it is unlikely that without great cost we will be able to reduce the numbers of both false positives and false negatives. Since assessment procedures are not perfect and we cannot ensure that the numbers of false positives and false negatives will be zero, the next best alternative is to design processes that will minimize the total costs of mistakes. Putting the point symbolically, we should aim for min $[(N_{FN} \times SC_{FN}) + (N_{FP} \times SC_{FP}) + SC_T)]$, where N_{FN} is the number of false negatives, N_{FP} is the number of false positives, SC_{FN} is the social costs of false negatives, SC_{FP} is the social costs of false positives, and SC_T is the costs of the testing or assessment procedure itself. Thus, our response to the uncertainties present in scientific studies and to the mathematical trade-offs that are forced by statistical studies will have substantial consequences for promoting or frustrating the social goals for which the risk assessments are done. This suggests that there may well be alternative strategies for assessing and regulating toxic substances. In the remainder of this chapter, I suggest that we be prepared to act on the basis of something short of the very best scientific evidence of risks of harm. Instead, we should act on the best scientific information that is reasonably available or that can be obtained in an expeditious manner to address some of the problems identified in U.N. Agenda 21.

To that end, the precautionary principle that was articulated at the Earth Summit is quite consistent with the analysis of this chapter. Principle 15 of the Rio Declaration on Environment and Development states: "Where there are threats of serious or irreversible damage, lack of scientific certainty shall not be used as a reason for postponing cost-effective measures to prevent environmental degradation" (U.N. Agenda 21 1992: 10). Among other things, this principle suggests that where one has evidence of serious threats to health or the environment, lack of full scientific certainty should not be a reason for postponing the action. Thus, the precautionary principle seems to accord with many of the arguments of this chapter that we should be prepared to act in the environmental and public health area on the basis of good, but not necessarily the best, scientific evidence of risks of harm, even though the science does not measure up to the most rigorous standards. Some of the reasons for this, of course, are the pervasive presence of uncertainty on the one hand and the incompatibility of removing both false positives and false negatives from statistical studies on the other hand, as well as the normative concern about the importance of avoiding false negatives. Moreover, as noted above, one of the major problems with current approaches to risk assessment, and a temptation that scientists frequently face, is the demand to remove uncertainty by developing more and better scientific data and better theoretical understandings, as well as better mechanistic understanding of the harm from the risk in question. But slowness perforce, as result of such demands for detailed understanding and case-by-case assessment of risks, prevents scientists and policymakers from addressing other risks of concern. Consequently, an additional desiderata in environmental health science is the rate of toxic risk identification, potency assessments, and regulation, if it is merited.

Program areas A through D proposed in Chapter 19 of Agenda 21 appear to embody many of the recommendations made in this chapter. These include ((U.N. Agenda 21 1992: 187-191): (a)expanding and accelerating the international assessment of chemical risks, (b)harmonizing the classification and labeling of chemicals, (c)fostering information exchange on toxic chemicals and chemical risks, and (d)establishing risk reduction programs. If current risk assessments even in the most developed countries with the most advanced environmental science are slow, then there is much to be gained by other countries cooperating with those that have already assessed the risks of many substances and taken steps to regulate them. There is no need to reinvent the environmental science and accompanying regulations. Thus, for example, if the U.S. EPA or the California EPA have identified, evaluated, and regulated large numbers of carcinogenic, reproductive, or neurotoxins, then there is little need for other countries to duplicate this effort when they can achieve rapid identification and assessment of those substances by learning from the efforts of others.

Moreover, certain procedures would expedite current potency assessment practices even more, and some possible procedures that would expedite the identification of hazards are still in their scientific infancy. Both could contribute to the generation of usable knowledge for other countries, and both can be adapted for circumstances in which few resources are available for assessment purposes. Current carcinogen potency assessment procedures, as we have seen, are slow, taking from 1/2 to 5 person-years per substance. While these are as accurate as they can be at present, they are costly to conduct and have substantial opportunity costs, because known carcinogens remain unanalyzed as a result of resources committed to more science-intensive case-by-case analyses. Conceiving of risk assessment as a procedure whose total costs we should minimize creates the opportunity of tolerating slightly less science-intensive case-by-case procedures, while decreasing the number of known but unassessed carcinogens, thus resulting in lower overall social and agency costs (Cranor 1993). Moreover, certain technical assessment procedures that are quite accurate and very fast can be used for many assessment and regulatory purposes; some have been adopted into regulation by the State of California (California Code of Regulations, Title 22, Section 12705).

California, for example, provided potency assessments of 140 previously unassessed carcinogens plus 75 previously assessed carcinogens in an eight-month period in 1991-92. The approximation procedures used for these purposes provided nearly as accurate results as the time-consuming science-intensive potency assessment procedures. Moreover, these were done with very few agency personnel at a cost of approximately $40,000 instead of a cost of $7 to $70 million that would be spent doing detailed case-by-case science-intensive analysis (Zeise 1991).

Thus, not only can governments expedite the identification, assessment, and regulation of toxic substances by utilizing existing information, but also they can increase the rate of identification and assessment of toxins by adopting procedures that are nearly as accurate as more time-consuming and costly methods that have proven useful to some of the administrative agencies within the United States. Thus, I strongly endorse Agenda 21 items 19.11 (noting the large number of unassessed substances that need to be addressed), 19.12 (noting that because risk assessment is resource-intensive, duplication of effort among countries should be avoided), and 19.13 (recommending the assessment of several hundred chemicals before the year 2000). In the international context, assessing toxic substances can be made more efficient by strengthening international cooperation (19.33-19.35, 19.40), by making the best use of available resources (19.14, 19.15, 19.40), and by avoiding unnecessary duplication of effort. In addition, however, although Agenda 21 does not speak to this point, scientific agencies could further improve risk assessment by adopting some of the approximation procedures that have been used successfully by the California EPA that provide faster but still accurate ways of assessing the risks from carcinogens. Adopting programs A through C, as well as adopting some of the expedited assessment procedures described above, also supports risk reduction program D (19.44). Once countries have identified toxins and have some sense of their potency, these nations are in a much better position to introduce risk reduction programs, because they have better information. This can lead to product and industrial process reformulations as well as to regulation. A key component, however, is knowledge of the nature and extent of present risks. Thus, the analysis of this chapter supports many of the important recommendations of Chapter 19 and some of the more general principles but also suggests further recommendations in the spirit of Chapter 19. Finally, if expedited hazard identification procedures prove to be accurate, this will lessen reliance on time-consuming animal bioassays, which are not only slow but also morally controversial because of the necessity of killing the experimental animals.

The recommendations in the above analysis have implications for the two remaining programs of Chapter 19 as well: "[S]trengthening of national capabilities and capacities for management of chemicals and...prevention of illegal international traffic in toxic and dangerous products." To the extent that there are relatively inexpensive expedited and approximation procedures for identifying and assessing toxic substances, this makes the task of strengthening the information-generating capabilities of governments much easier. Insofar as such methods might be easier and cheaper to use, requiring relatively fewer resources and relatively less training to implement, they would enable countries to have some capacity for acquiring information about risks at relatively low cost. This is the case, for example, for two expedited carcinogen potency assessment procedures evaluated by this researcher and the California EPA. One procedure to assess the potency of carcinogens requires only a computerized or hard-copy data base and a calculator for determining the potency of several hundred carcinogens. A second procedure requires the same data base, a more experienced toxicologist's eye for the data, and a relatively inexpensive computer program to produce very accurate potency numbers. The state of California has assessed simultaneously some 200 substances in less than one year by these two methods. The first procedure relies upon the most sensitive site in the most sensitive species reported in the data base, and thus with very little training one could identify such information. The second procedure requires more sophisticated judgment about what the data do and do not show, as well as limitations to the data, and thus it requires considerably more training of the individual(s) in question. The second procedure also necessitates a computer and a particular software program. Neither procedure requires expensive investment in large numbers of toxicologists or extensive libraries of information—both are kinds of barefoot potency assessments. Similarly, if inexpensive identification procedures, e.g., by use of short-term tests, can be found for carcinogens, this would also facilitate the strengthening of capacities for assessing the assessment and management of chemicals.

Both kinds of potency assessments described above and the possibility of expedited identification procedures are possible because of reconceiving the nature of risk assessment and what we should expect from it. Thus, recognizing the normative components of risk assessment and the importance of minimizing the total costs of false positives and false negatives conceptually opens up these possibilities and suggests the possibility of alternative designs for risk assessment procedures.

Finally, while no one is likely to disagree with the prevention of illegal international traffic in toxic and dangerous products, there is an important implication for this subject from the above discussion. Different countries may evaluate the importance of avoiding false positives and false negatives differently depending upon their different stages of development, the need for economic development, and the national judgment about the fragility of ecological and public health protections. Such different evaluations can easily be accommodated by the risk assessment and implicit risk management procedures articulated above. I have not proposed a univocal standard for all peoples and all places. A more rigid scientific approach to evidence and testing procedures would be less likely to permit this degree of flexibility; in particular, it would tend to frustrate more health-protective approaches. The analysis of this chapter thus serves national autonomy in assessing and managing toxic substances. If one country endorses assessment and management procedures that tolerate fewer false negatives and more false positives, a strong international program prohibiting illegal trafficking in toxic substances helps to secure that country's regulatory autonomy. If another country is willing to permit fewer false positives and more false negatives, strong international control programs for traffic in toxic substances also serves its national autonomy in this area. The unauthorized shipment of toxic substances across borders frustrates a country's regulatory autonomy unless it explicitly invites the dumping of toxic substances within its borders. (Even a country that invited the dumping might not want unauthorized dumping, for it might want to know what was being dumped within its borders.) Thus, the approach to risk assessment and risk management discussed in this chapter permits national autonomy in both risk assessment and risk management which an international program preventing the illegal trafficking in toxic substances would reinforce.

There is a caveat, however, to the preceding paragraph. While Principle 11 of the Rio Declaration and the analysis of this chapter envision the possibility of different environmental and public health standards for different countries, this may appear to be inconsistent with program area B (19.4[b] and 19.24), which recommends the harmonization of the classification and labeling of chemicals. To the extent that risk assessments and hazard identifications are not standardized (that is, identical), this will make harmonization of classifications more difficult. This is not an insuperable problem, however. If there are different potency assessments from one country to another, for purposes of classifying the carcinogenic potency of a substance, classification and labeling schemes could include the most health-protective potency and the least health-protective potency (or alternatively provide a range of risks) so that those handling toxic substances had a sense of the range of carcinogenic potencies. Of course, the classification and labeling of substances would indeed be easier if there were identical potency numbers for each substance. However, the attempt to achieve identical numbers among countries is both unlikely to be realized because of the uncertainties in risk assessment and likely to bring the present snail's pace of risk assessment nearly to a halt. It seems much better for international purposes for countries to expedite the identification and assessment of toxins for now, even if there are disagreements between countries, in order to develop some reasonably accurate information about the large universe of unidentified and unassessed substances. The classification and labeling of toxins can accommodate different assessments as an interim measure. Over a much longer term, disagreements among countries about particular assessments can be resolved as the science improves. A serious mistake, if the analysis of this chapter is correct, would be for countries to wait to develop standardized classification schemes until science-intensive, case-by-case risk assessments had been concluded. This would ensure near-paralysis in the assessment and control of toxic substances.

Despite my endorsement of many features of Agenda 21 on toxic substances, some provisions of Chapter 19 could exacerbate current problems, depending on how the recommendations are implemented. Thus, for example, section 19.14 (as well as 19.20) indicates that there should be an "agreed-upon approach to quality assurance, application of assessment criteria, peer review, and linkages to risk management activities, taking account of the precautionary principle." To the extent such activities are dominated by a science-intensive effort to evaluate the toxicity of substances on a case-by-case basis with an emphasis on being certain about the risks from the substances, this will frustrate many of the goals stated at the outset of Chapter 19 of Agenda 21. Moreover, to the extent that peer review, scientific quality assurance and using common frameworks are catchwords for insuring scientific rigor, this is also likely to frustrate the very activities otherwise fostered by Chapter 19. Conversely, to the extent that the precautionary principle is endorsed, this may lead administrative agencies in the direction of utilizing existing data and information, sharing expertise, and perhaps even expediting processes that at present appear to be much slower than they need to be.

We have choices in designing risk assessment and risk management procedures, as the analysis of this chapter has tried to show. Such choices can reinforce the status quo, or they can promote modifications of it in accordance with Agenda 21 and in accordance with national priorities in protecting public health and the environment consistent with sustainable development. The important questions to be answered in guiding these choices are "How demanding shall we make the science?" and "How shall we err?" We can exercise our scientific skepticism responsibly or irresponsibly about whether substances pose risks to human health and the environment. A significant aspect of responsible risk assessment and risk management is normative in nature, as this chapter has tried to show.

9. References

Ades, E.A. 1990. Evaluating Screening Tests and Screening Programs. *Archives of Disease in Childhood* 65: 792-795.

Allen, B.C., K.S. Crump, and A.M. Shipp. 1988. Correlations Between Carcinogenic Potency of Chemicals in Animals and Humans. *Risk Analysis*, Dec: 531-544.

Bailar, J.C., E.A.C. Crouch, R. Shaiklr, and D. Spiegelman. 1988. One-Hit Models of Carcinogenesis: Conservative or Not? *Risk Analysis* 8: 485-497.

Bridbord, K., and T. French. 1978. *Estimates of the Fraction of Cancer Related to Occupational Exposures.* National Cancer Institute, National Institute of Environmental Health Sciences, National Institute of Occupational Safety and Health. Washington, D.C.

Brown, R. 1989. Presented at *Pesticides and Other Toxics: Assessing Their Risks.* University of California, Riverside.

California (EPA) Environmental Protection Agency, Health Hazard Assessment Division. 1991. Expedited Method for Estimating Cancer Potency: Use of Tabulated TD_{50} Values of Gold, et.al. Presented to the Proposition 65 Science Advisory Panel. April 26.

Cornfield, J., K. Roi, and J. Van Ryzin. 1980. Procedures for Assessing Risk at Low Levels of Exposure. *Archives of Toxicology* Supplement 3: 295-303.

Cothern, R.C., W.L. Coniglio and W.L. Marcus. 1986. Estimating Risks to Health. *Environmental Science & Technology* 20(2): 111-116.

Cranor, C. 1983. Epidemiology and Procedural Safeguards in Workplace Health Protections in the Aftermath of the Benzene Case. *Industrial Relations Law Journal* 5: 372-389.

Cranor, C. 1990. Some Moral Problems with Risk Assessment. *Ethics* 101: 123-143.

Cranor, C. 1988. Some Public Policy Problems With the Science of Carcinogen Risk Assessment. *Philosophy of Science* 2: 467-488.

Cranor, C. 1993. *Regulating Toxic Substances: A Philosophy of Science and the Law.* Oxford University Press, New York.

(DOL) United States Department of Labor, OSHA. 1980. Cancer Policy. Fed. Reg., 45: 5002, at 5190.

Epstein, S.S. and J.B. Swartz. 1981. Fallacies of Lifestyle Theories. *Nature* 289: 130.

Eysenck, H.J. 1991. Were We Really Wrong? *American Journal of Public Health* 133(5):429-32.

Feinstein, A.R. 1977. *Clinical Biostatistics.* C.V. Mosby, St. Louis, pp. 320-25.

Feinstein, A.R. 1988. Scientific Standards in Epidemiologic Studies of the Menace of Daily Life. *Science* 242: 1257-1263.

Finkel, A.M. 1990. *Confronting Uncertainty in Risk Management: A Report.* Center for Risk Management, Resources for the Future, Washington, DC. pp. 12-15.

Finkel, A.M. 1989. Is Risk Assessment Too Conservative: Revising the Revisionists. *Columbia Journal of Environmental Law* 14: 427-467.

Fleiss, J.L. 1986. Significance Tests Have a Role in Epidemiologic Research: Reactions to A.M. Walker. *American Journal of Public Health* 76: 559-560.

Giere, R. 1979. *Understanding Scientific Reasoning.* Holt, Rinehart and Winston, New York.

Goodman, S.N., and R. Royall. 1988. Commentary: Evidence and Scientific Research. *American Journal of Public Health* 78.

Greenland, S. 1991. Invited Commentary: Science Versus Public Health Actions: Those Who Were Wrong Are Still Wrong. *American Journal of Public Health* 133(5):435-36.

Hattis, D. 1988. Presentation at University of California, Riverside. April 19.

Horowitz, RI., and A.R. Feinstein. 1979. Methodologic Standards and Contradictory Results in Case Control Research. *The American Journal of Medicine* 66: 556-563.

(HUSPH) Harvard University School of Public Health. Center for Risk Analysis. 1991. *OMB v. the Agencies: The Future of Cancer Risk Assessment* (conference). Cambridge, MA.

(IGS) Interagency Staff Group. 1986. *Chemical Carcinogens.* International Agency on Cancer. p. 234.

(ISG) United States Interagency Staff Group on Carcinogens. 1986. Chemical Carcinogens: A Review of the Science and its Associated Principles. *Environmental Health Perspectives* 67: 201-282.

(IRLG) United States Interagency Regulatory Liaison Group. 1979. Scientific Bases for Identification of Potential Carcinogens and Estimation of Risks. *Journal of the National Cancer Institute* 63(1): 241-268.

Johnson, W.M., and W.D. Parnes. 1979. Beta-Naphthalamine and Benzidine: Identification of Groups at High Risk of Cancer. In *Annals of the New York Academy of Sciences: Public Control of Environmental Health Hazards,* E.C. Hammond and I.J. Selikoff, eds. New York Academy of Sciences, New York. pp. 277-284.

Lawless, E.W. 1977. *Technology and Social Shock.* Rutgers University Press, New Brunswick, NJ.

Lee, D.Y., and A. Chang. 1992. Evaluating Transport of Organic Chemicals in Soil Resulting from Underground Fuel Tank Leaks. In *46 Purdue Industrial Waste Proceedings,* Lewis Publishers, Chelsa, MI. pp. 131-140.

Lehman, E.L. 1959. *Testing Statistical Hypotheses.* John Wiley and Sons, New York.

Lynn, F. 1987. OSHA's Carcinogen Standard: Round One On Risk Assessment Models and Assumptions. *The Social and Cultural Construction of Risk,* B.B. Johnson and V.T. Covello, eds. D. Reidel, Bingham, MA. pp. 345-358.

Mausner, J.S., and A.K. Bahn. 1974. *Epidemiology: An Introductory Text.* W.B. Sunders, Philadelphia.

Mayo, D. 1988 . Toward a More Objective Understanding of the Evidence of Carcinogenic Risk. *Philosophy of Science Association* 2: 489-503.

(NRC) National Research Council. 1984. *Toxicity Testing: Strategies to Determine Needs and Priorities.* National Academy Press, Washington, DC.

(NTP) National Toxicology Program. 1991. *Chemical Status Report.* Division of Toxicology Research and Testing, July 7.

Poole, C. 1987. Beyond the Confidence Interval. *American Journal of Public Health* 77: 195-199.

(Prop 65) Proposition 65 Science Advisory Panel. 1990.

Quine, W.V.O. 1960. *Word and Object.* The MIT Press, Cambridge, MA. p. 78.

Rall, D.P 1980. Testimony. In *Federal Register* 45: 5061.

Rall, D.P. 1979. The Role of Laboratory Animal Studies in Estimating Carcinogenic Risks For Man. In *Carcinogenic Risks: Strategies for Intervention.* International Agency for Research on Cancer, Lyon, France, p. 179.

Rescher, N. 1983. *Risk: A Philosophical Introduction to the Theory of Risk Evaluation and Management..* University Press of America, Washington, D.C., p. 5.

Rimm, A. 1980. *Basic Biostatistics in Medicine and Epidemiology.* Appleton Century Crofts, New York.

Rothman, K.J. 1986. *Modern Epidemiology.* Little, Brown and Co., Boston, MA.

Schlesselman, J.J. 1974. Sample Size Requirements in Cohort and Care-Control Studies of Disease. *American Journal of Epidemiology* 99: 381-84.

Schmahl, D., R. Preussman, and M.R. Berger. 1981. Causes of Cancer—An Alternative View to Doll and Peto, *Klinische Wochen-Schrift* 67: 1169-1173.

Schottenfeld and Haas 1979. Carcinogens In the Workplace. *CA-Cancer Journal for Clinicians* 29:144-159.

Scott, J. 1990. Job-Related Illness Called "America's Invisible Killer." *Los Angeles Times* August 31: A4.

Simpson, M. and C. Hill. 1986. Congressional Research Service.

Swan, S. 1991. Chief, Reproductive Epidemiology Section, California Environmental Protection Agency. Personal communication.

Tomatis, L. 1988. Environmental Cancer Risk Factors: A Review. *Acta Oncologica* 27: 465-472.

Tomatis, L., et.al. 1990. *Cancer: Causes, Occurrence and Control.* International Agency for Research on Cancer, Lyon, France.

Thompson, W.D. 1987. Statistical Criteria in the Interpretation of Epidemiologic Data. *American Journal of Public Health* 77: 191-194.

(U.N.) United Nations Agenda 21: The United Nations Programme of Action From Rio. 1992. The United Nations Publication, New York.

U.S. Congress, Office of Technology Assessment. 1982. *Cancer Risk: Assessing and Reducing the Dangers in Our Society.* Westview Press, Boulder, CO.

U.S. Congress, Office of Technology Assessment. 1987. *Identifying and Regulating Carcinogens.* GOP, Washington, D.C. OTA-BP- H-42.

van Ryzin, J. 1982. Current Topics in Biostatistics and Epidemiology. *Biometrics* 28: 130-138.

Walker, A.M. 1986. Reporting the Results of Epidemiologic Studies. *American Journal of Public Health* 76: 556-558.

Walter. 1977. Determination of Significant Relevant Risks and Optimal Sampling Procedures in Prospective and Retrospective Comparative Studies of Various Sizes. *American Journal of Epidemiology* 105: 387-397.

Wong, O. 1990. A Cohort Mortality Study and a Case Central Study of Workers Potentially Exposed to Styrene in Reinforced Plastics and Composite Industry. *British Journal of Industrial Medicine* 47: 753-762.

Zeise, L. 1990 . Issues In State Risk Assessment: California Department of Health Service. In *Proceedings: Pesticides and Other Toxics: Assessing Their Risks,* J.L. White, ed. University of California, College of Natural and Agricultural Sciences, Riverside, CA, pp. 135-144.

Zeise, L., W. Pease, and A. Kelter. 1989. Risk Assessment for Carcinogens Under California's Proposition 65. In *The Western Meetings of the American Association for the Advancement of Science.* San Francisco, CA.

Zeise, L., R. Wilson, and E.A.C. Crouch. 1987. Dose-Response Relationships for Carcinogens. *Environmental Health Perspectives* 73: 295-308.

Zeise, L. 1991. Chief, Reproductive and Cancer Hazard Assessment Section, Office of Environmental Health Hazard Assessment, California Environmental Protection Agency. Personal communication.

Note: Portions of this chapter are taken and modified from the introduction, "Assessing Toxic Substances Through a Glass Darkly," and Chapter 1, "The Scientific Background," of *Regulating Toxic Substances: A Philosophy of Science and the Law* by Carl F. Cranor.

Chapter 10
NUCLEAR WASTE AND AGENDA 21

Kristin Shrader-Frechette[1]

1. Introduction

Nuclear power and its attendant waste are similar to the Ring in J.R.R. Tolkien's *The Lord of the Rings*. The Ring gave mastery over every living creature, but it was created by an evil power. As a result, it inevitably corrupted anyone who attempted to use it. The hobbits, who held the Ring, had to decide how to handle it. Erestor formulated the problem: "There are but two courses, as Glorfindel already has declared: to hide the Ring forever, or to unmake it. But both are beyond our power. Who will read this riddle for us?" (Tolkien 1965: 349-350; Blowers et al. 1991: xvii).

One important attempt to read the riddle of nuclear waste emerged as a result of the 1992 United Nations (U.N.) Conference on Environment and Development (UNCED) held in Rio de Janeiro. UNCED produced a variety of recommendations on environmental protection and sustainable development and also published Agenda 21, proposals to guide international action regarding environment and development. Included in Agenda 21 is an entire chapter with proposals for how to deal with nuclear waste. In this essay, we shall (1)review the U.N. mandates regarding nuclear waste and outline the scientific context for understanding the problems associated with it, (2)evaluate the main ethical rationale for these mandates, and (3)explore the most important policy implications of these mandates. Our conclusion is that following the directives of Agenda 21 arguably requires us both to cease generation of nuclear waste and to avoid permanent disposal of it, at least for the time being (Shrader-Frechette 1991a, 1993a).

2. The U.N. Mandates, Their Scientific Context, and the Appeal to Ignorance

Chapter 22 of Agenda 21, dealing with radioactive waste, provides a number of mandates for dealing with the riddle of nuclear waste. These include: (1)minimizing the volume of nuclear wastes, (2)handling and disposing of them safely, (3)ensuring military conformity with environmental norms for dealing with the wastes, and (4)employing domestic facilities for the waste, rather than exporting them to nations less able to deal with their hazards. All of these mandates are designed both to reduce the risk from the waste and to protect potential victims of it.

Protecting people and reducing the risk from the waste, as the U.N. mandates, will be

[1]Distinguished Research Professor, Environmental Sciences and Policy Program and Department of Philosophy, University of South Florida, Tampa, FL 33620-5550, U.S.A.

J. Lemons and D. A. Brown (eds.), Sustainable Development: Science, Ethics, and Public Policy, 254–274.

difficult, as this section of the chapter argues, because even minute amounts of radioactive waste are extremely harmful to living things. Yet more than 400 commercial reactors in 26 nations throughout the world produce radioactive waste and spent fuel as a result of their generation of electricity. Military reactors, and the development of nuclear weapons, produce nearly as much radioactive waste, collectively, as the commercial processes used by nuclear utilities throughout the world (Blowers et al. 1991: 1,4; Lenssen 1992: 50; Shrader-Frechette 1993a: Ch. 2).

Commercial and military reactors are the main producers of radioactive waste that results from fissioning or splitting uranium or plutonium atoms. Fissioning creates new radioactive elements or radionuclides, usually by splitting uranium-235; the breeder reactors used in France convert nonfissionable uranium-238 into plutonium-239 after irradiation by neutrons. Because some of the radionuclides created by neutron absorption and fissioning have higher atomic numbers than uranium, they are called transuranic elements. Other radionuclides are lighter than uranium, and they are called fission products. Each of the more than 80 products created in the fission reactor is capable of releasing ionizing radiation. Radiation, the energy transferred as particles or waves that move through space or from one body to another, is ionizing when it is able to remove orbital electrons from other atoms or molecules and hence able to change their structure and molecular charge. Because ionizing radiation can change the structure of atoms or molecules, it can cause cancer, genetic defects, and birth defects (Lipschutz, 1980; EPA 1982; Bertell 1985: 19; Murray 1989; Blowers et al. 1991: 1ff.; Shrader-Frechette 1993a: 13).

Whenever objects or persons absorb energy and change structure as a result of ionizing radiation, they become radioactive and emit radiation in the form of alpha particles, beta particles, or gamma rays. Alpha particles are positively charged helium atoms; they contain two protons and two neutrons. Because alpha radiation is the most densely ionizing, it is potentially the most dangerous of the three forms of radiation, even though it is the least penetrating. If an alpha-emitter like plutonium, a transuranic, enters the body, it can cause significant cellular damage or even death. Beta particles have the same charge and mass as electrons. Streams of beta particles (beta radiation), such as those from strontium-90 and cesium-137, can cause skin burns. Usually accompanying alpha and beta emissions, gamma rays are high-energy, short-wave-length radiation. Like X-rays, gamma rays are a type of electromagnetic radiation. Because gamma rays (like X-rays) are so highly penetrating, several feet of concrete or rock are required for adequate shielding. Iodine-131 is an example of a gamma emitter. Even though they are naturally present, at low levels, throughout our environment, all three types of ionizing radiation are harmful if they are inhaled, ingested, or incorporated into the body. Gamma rays and X-rays are especially dangerous because they can damage or destroy living tissue, even from a distance of several feet or more. Alpha emitters, on the other hand, must be in touch with live tissue to cause harm. Because they can lodge in the body, they can deposit a very large quantity of energy in tissue (Shrader-Frechette 1993a: 14).

The curie (Ci), the most common measure of the intensity of radioactivity, is named after Marie Curie. With her husband Pierre, she discovered radium in 1898. (A curie is the quantity of a radioactive isotope that decays at the rate of 3.7×10^{10} disintegrations per second. Originally it was the amount of radioactivity given off by one gram of radium.) One nuclear reactor may contain over 10,000 megacuries of radioactivity. (A megacurie is one million curies.) In general, the greater the intensity of the radioactivity, the greater the damage it can inflict. Short-term effects of chronic exposure to radiation are radiation sickness—nausea, vomiting, dizziness, headache, and so on. Long-term effects of chronic exposure to radiation are cancer, reproductive failure, birth defects, genetic effects, and death. One reason that the effects of radiation are so serious is that there is no threshold for increased risk as a result of exposure to any radiation. Effects of radiation also are cumulative: successive exposures increase one's risk of harm (Shrader-Frechette 1983: Ch. 2; Shrader-Frechette 1993a: 14).

Each year, each 1,000-MWe reactor discharges about 25.4 metric tons of high-level waste, spent fuel. For 400 commercial reactors, worldwide, the annual high-level radwaste would be 10,160 metric tons per year. Yet only 10 micrograms of plutonium is almost certain to induce cancer. Several grams of plutonium, dispersed in a ventilation system, are enough to cause thousands of deaths (Grossman and Cassedy 1985: 48). Moreover, each of the 7,620 metric tons of high-level waste produced annually has the potential to cause hundreds of millions of cancers for at least the first 300 years of storage, and then tens of millions of cancers for the next million years (Murray et al. 1982: 586). Potential cancers caused by nuclear waste could be prevented, of course, with perfect isolation of the wastes for a million years. That is why most plans for high-level radwaste storage call for defense in depth, for burying it deep underground. Despite these plans, the U.S. Environmental Protection Agency (EPA) has warned that we cannot count on institutional safeguards for the waste beyond 100 years (Hawkins 1978: 27-29).

Moreover, the scientific feasibility of dry storage in geologic repositories, deep in salt beds or hard rock, has not been demonstrated. As a result, granite storage sites in Sweden have been vetoed as unsafe, Kansas salt beds have been rejected because they are riddled with holes, and the first model US repository for high-level radwaste, proposed for Yucca Mountain, Nevada, will not be ready before the first or second decade of the next century, if indeed it is ever ready (Shrader-Frechette 1993a).

In the absence of proof that we can successfully store radwaste, the technical problems associated with it are forcing us to take a great gamble. This is a gamble that our descendants will not breach the repositories through war, terrorism, or drilling for minerals; a gamble that water and heat will not combine to create nuclear reactors in underground waste, as has already happened in the former Soviet Union; and a gamble that ice sheets, volcanism, seismic activity, and geological folding will not uncover the wastes. Scientists have not been worried by these gambles because, for the first 35 years of commercial nuclear fission, they were saying that safely isolating the wastes would be easy, once they set their minds to the task. They have been like contractors who built houses without toilets (Flavin 1983: 31) and then alleged that constructing the toilets would be easy. Because no country has a permanent high-level disposal facility (Polluck 1986: 13), perhaps the task is not so easy as has been alleged.

Regardless of the technology used, anyone who favors a particular method of radwaste management must use some form of the fallacy of the appeal to ignorance—namely, "I know of no way in which containment could be breached; therefore, containment will probably not be breached." Without falling into logical fallacy, one cannot draw a positive conclusion about radwaste safety when the premises upon which it is based have so little empirical foundation. Indeed, in their site studies for the proposed Yucca Mountain facility—proposed as the world's first permanent repository for high-level nuclear waste—U.S. Department of Energy (DOE) assessors have repeatedly appealed to ignorance. They have explicitly affirmed that their inability to prove the site unsuitable is alone grounds for claiming it is suitable (Shrader-Frechette 1993a: 114-121). Of course, scientists are often forced to draw conclusions without empirical tests, and much progress in science consists of theorizing on the basis of slim empirical foundations. Because this highly theoretical approach is legitimate in science, however, it is not clear that it is applicable to a technology capable of putting millions of lives at risk. We know that, in the past, we were wrong to use an appeal to ignorance when we dumped radwaste into the sea, when we treated mastitis with radiation, when we used X-rays to determine shoe fit, and when we subjected U.S. soldiers to nuclear-test fallout during peacetime. We may likewise be wrong to use an argument from ignorance to conclude that we can isolate radwastes in perpetuity.

A second technical problem is how we can completely isolate the radwaste from the biosphere and yet monitor it to insure that containment has not been breached. Complete isolation appears to preclude adequate monitoring, and adequate monitoring appears to preclude isolation. At the proposed Yucca Mountain facility in the United States, the waste will not be monitored beyond the first 50 years (Shrader-Frechette 1993a). This raises the possibility, of course, of leaks that go undetected, posing potential for great harm.

Another technical problem is how to guarantee the so-called neutrality criterion. This widely used government criterion specifies that the levels of risk to which future generations will be subjected because of the waste must be no greater than those of present persons (Shrader-Frechette 1993a: Ch. 9). Because of the absence of good inductive evidence, any suggestion that this criterion can be met amounts to another argument from ignorance. Other assumptions—that the future will be geologically, politically, and institutionally stable enough to insure that current plans for radwaste management are not disrupted—likewise amount to little more than arguments from ignorance (Polluck 1986: 15; Shrader-Frechette 1993a).

Appeals to ignorance are especially evident in the scientific studies of Yucca Mountain, Nevada, the proposed site for the world's first permanent geological repository for high-level nuclear waste and spent fuel. Hydrological, geological, social, and ethical uncertainties remain, however, although the government has spent $3 billion on site studies (Shrader-Frechette 1993a). Current plans for future U.S. storage of high-level radioactive waste at Yucca Mountain require the steel canisters to resist corrosion for as little as 300 years. Nevertheless, the DOE admits that the waste will remain dangerous for longer than 10,000 years. Government experts agree that, at best, they can merely limit the radioactivity that reaches the environment and that "there is no doubt that the repository will leak over the course of the next 10,000 years" (ERDA 1975: X-28). The U.S. government has extrapolated, on the basis of past leaks at its nuclear waste facilities, and has said that future leaks should occur at a rate of two to three per year. Using (U.S.) government-estimated exposure levels (580 person rem) at each radwaste site, each existing facility could cause approximately 12 cancers and 116 genetic deaths per century (AEC 1974: 3-83; ERDA 1975: X-74, II. 1-57; Amato 1986: 221). These numbers appear relatively small until one realizes that, for the period of storage required, the cancer deaths alone could be in the tens of thousands per storage site. Admittedly there are new technologies that might reduce these projected problems. Any new technology, however, is unavoidably dependent upon fragile and short-lived human institutions and human capabilities. It wasn't faulty science or technology, after all, that caused Three Mile Island or Chernobyl. It was human error. Likewise it could well be human error that is the insoluble problem with managing nuclear waste.

3. Nuclear Waste and Hydrogeological Uncertainty

Even if it were possible to avoid human error in managing radioactive waste, there would still be a problem with human uncertainties in the areas of hydrology and geology. Major uncertainties associated with hydrogeological predictions fly in the face of safely managing nuclear materials for the long time periods during which they are most dangerous (see Shrader-Frechette 1993a: 42-50).

One reason the predictions are so uncertain is that they are based on extrapolations from short-term studies. Indeed, geologists who are asked to make predictions typically are forced to extrapolate either to the past or to the future on the basis of what they observe in the present. Often geologists use morphologic expressions of rocks, for example, or general principles describing dynamic geological processes operating through time, when they make inferences and methodological value judgements about past geologic events (Watson 1969: 488-494). Frequently they use the absence of certain phenomena in the past and present as evidence for denying the likelihood that a specific geologic process will occur in the future. However, there must be other evidence when geologists extrapolate to the past or to the future on the basis of present data. Geologists know that, in general, the less we know about the processes and evidence supporting such extrapolations, the greater is the likelihood that we are wrong. Moreover, the smaller the empirical base used for such extrapolation, all things being equal, the greater the chance that it is not, or will not be, representative of past and future events and processes.

One problematic judgment made in many studies about permanent disposal of nuclear waste is that short-term tests (for several years or less) provide adequate information for very precise predictions of long-term behavior, for example, isolation of the waste for 10,000 years and container integrity for 1,000 years (Perry 1988; Beavers and Thompson 1990; Dobson et al. 1990; Halsey 1990). This judgment is highly controversial, in part because the extended time horizon for any high-level repository is several times longer than recorded human history (Sawyer 1990: 73). As M.I.T. geologist K. V. Hodges, a peer reviewer for the proposed Yucca Mountain nuclear waste repository, put it, the congressional mandate for siting a high-level repository is for predictive information 10,000 years into the future. Geology, however, as he points out, is an explanatory and not a predictive science (Hodges 1992: 362). Or, as Dartmouth geologist N. Oreskes noted: "The extrapolation of short-term to long-term studies at Yucca Mountain flies in the face of 300 years of geological practice" (Shrader-Frechette 1993a: 43). "Predictive geology," according to Hodges, "is predicated on the assumption of a sort of inverse uniformitarianism: the past geologic record is the key to future geologic activity" (Hodges 1992: 362-363). Yet at least since the recent revolution in plate tectonics, uniformitarianism—in any precise, predictive sense—has been largely abandoned by geologists. As Hodges affirms, "[T]ectonic predictions, when stripped of statistical sound and fury, are not much better than educated guesses" (Hodges 1992: 363). It is likewise "patently absurd" that we attempt to predict the probability of volcanic disruptions over 10,000 years. In asking for such predictions, claims Hodges, we are "asking the impossible" (Hodges 1992: 363).

Indeed, the entire 14-person peer reviewer group for the Yucca Mountain Evaluation confirmed the conclusions of Hodges about the impossibility of reliable 10,000-year predictions for any waste repository site. They said:

> Many aspects of site suitability are not well suited for quantitative risk assessment. In particular are predictions involving future geological activity, future value of mineral deposits and mineral occurrence models. Any projections of the rates of tectonic activity and volcanism, as well as natural resource occurrence and value, will be fraught with substantial uncertainties that cannot be quantified using standard statistical methods. (Younker et al. 1992: B-2)

If the Yucca Mountain peer reviewers are correct, then obviously making precise long-term geological predictions on the basis of short-term data is questionable.

Although long-term prediction is a problem in any area of science, nuclear repository predictions are particularly problematic as compared with those in other areas of science, not only because they deal with thousands of years but also because the potential dose commitments of radioactive isotopes (such as C-14, Pu-239, and I-129)—all with possibly serious health effects—extend from hundreds of thousands to millions of years (Hawkins 1978). Long-term predictions are more problematic in cases where their being wrong could lead to a human or environmental disaster. Such a disaster is especially worrisome because the EPA has explicitly warned that it is "impossible" to predict anything regarding the success of radioactive waste management beyond 100 years. The agency has noted that institutional controls are particularly problematic beyond a period of 100 years (EPA 1978: 10, 26).

One of the main worries about the judgment that a long-term repository risk is acceptable—given only short-term, incomplete data—is that some unforeseen catastrophic event could occur centuries from now that would compromise the integrity of the long-term facility. Because catastrophes have occurred in the past, there are grounds for similar worries about disruption of the waste-disposal facility. The U.S. landscape has a number of craters created by meteor hits, for instance, yet the annual probability of a meteor strike is quite low, just as the probability of repository flooding is likely small (NNWPO 1989). Even meteor hits, volcanism, or seismic activity, all of which have a low annual probability, assume major proportions for a long-term repository. In the case of volcanism or seismic activity, for example, it is not necessary to assume that the disruptive event would unearth the waste canisters. Even a small seismic dislocation of some geological features might be sufficient to flood the repository and leach the waste. Moreover, even if the per-year probability of dangerous seismic or volcanic activity is quite low, for example, 10^{-6}, this figure means that during the lifetime of the repository such an event would be virtually certain. An annual probability of 10^{-6} converts to a 10^{-3} likelihood over a thousand years.

Obviously, it is questionable whether one can make an inductive prediction that guarantees the radwaste permanent isolation from the biosphere on the basis of inductive data obtained during only one or several decades (Stein and Collyer 1984). Moreover, most of the Yucca Mountain experiments have been done and data have been obtained over periods of far less than a decade. Tests on migration of spent fuel and groundwater transport of radionuclides, for example, are typically only months in duration, for example, 2, 6, and 12 months (Smith 1988a, 1988b).

In the case of waste canisters, current experiments of three years' duration are particularly problematic, because the future temperatures in the proposed Yucca Mountain repository are expected to be as great as 200 degrees Celsius in the immediate vicinity of the waste (Hadlock 1980: 49), causing changes in the surrounding rock (Blacic et al. 1986). Moreover, in some experiments, all of the canisters made of the nuclear waste-package reference material have failed and showed stress-corrosion cracking when they were exposed to a one-year test in groundwater and tuff at the expected temperature of 200 degrees Celsius (Pitman et al. 1986). All these problems suggest that long-term experiments are essential. The shorter the time of the experiment, all things being equal, the more questionable the inductive value judgment that the data support precise predictions about repository behavior thousands of years in the future.

Loan companies find it difficult to predict mortgage rates for more than 30 or 40 years given nearly a century of information. Yet risk assessors have predicted confidently that "meteorological conditions...for over 40 years" provide a firm basis for concluding "that any radiological emissions would be effectively dispersed before they reach highly populated areas" near a proposed permanent repository (DOE 1986a: Vol. 2, pp. 6-24, 6-25). How could one know about dispersion 10,000 years hence? And how could one predict population centers so far into the future? Many of the radioactive isotopes that would be stored at sites like Yucca Mountain—such as I-129, Np-237, Cs-135, U-238, and Zr-93—have half-lives that are in the millions of years (Smith et al. 1982: 10, 51). During such long time periods of radiotoxicity, changes in climate, groundwater, precipitation, and volcanic activity could occur. Risk assessors, for example, need to predict precise phenomena associated with future climate, weather, mineralogy, and water composition, even though climate and weather are the most variable and rapid natural processes influencing a repository (DOE 1990: vii-xvii).

Ultimately, value judgments that short-term data provide an adequate basis for inferring extremely precise long-term behavior rely on an inductive inference, on the assumption that the future will be like the past. This is the basic assumption of all historical geology. The DOE has made exactly this judgment: "Yucca Mountain...would meet the U.S. Environmental Protection Agency standards...if present hydrologic, geologic, and geochemical conditions (as presently understood) persist for the next 10,000 years" (DOE 1986b: Vol. 3, p. C.5-55). How could one guarantee, however, that such conditions will persist? Long-term judgments about the suitability of Yucca Mountain for a waste repository are especially questionable because of the existence of lakes in the Great Basin of Nevada during the Wisconsin period two million years ago and because moderate variations in climate are sufficient to cause large changes in the hydrological budget of some of the closed basin systems in Nevada. An evaluation of the Quaternary history of the Yucca Mountain area reveals that, like other areas proposed for radwaste sites, it has undergone geomorphic change in the last million years, and it may undergo catastrophic landslides in the future (Mara 1980). Other assessors have calculated the probability of a volcanic disruption hazard given the natural historic seismicity of the Yucca Mountain region as 10^{-6} per year (Metcalf 1983; Crowe 1986). Even a U.S. National Academy of Sciences panel warned that its modeling results involve very "large uncertainty," that there are few data to constrain the complex hydrologic system at Yucca Mountain, and that its predictions "depend heavily on expert judgment" because of the "unprecedented" exactitude required for 10,000-year predictions (Raleigh et al. 1992). Because high-level radwaste requires essentially "permanent isolation from the biosphere" (Stein and Collyer 1984), some geologists have said that any planned repository must be built under the assumption that groundwater will eventually come in contact with the high-level waste (Roxburgh 1987: 183).

Assuming that short-term hydrogeological tests provide accurate predictions for long-term behavior has been one of the reasons for the erroneous underestimation of the potential for off-site radwaste migration at the low-level radioactive facility at Maxey Flats, Kentucky. Although the Maxey Flats site is disanalogous in many ways to the proposed Yucca Mountain site, several problematic risk-assessment methods appear to be similar at the two locations. The geologist from the New Jersey Geological Survey who did the original studies at the Kentucky site drilled and studied the wells for only ten days. As a result, he concluded that they were dry and that hydraulic conductivity was very low at the site (Walker 1962: 3). On the basis of his analyses, the Kentucky facility was opened the year he did his studies. Years later, other geologists and risk assessors observed the wells for a year and concluded that because some of them were filled with water, therefore hydraulic conductivity was quite high (Papadopulos and Winograd 1974: 29-30). Just as the longer-term studies gave the more accurate results—and a less optimistic picture of the Maxey Flats radioactive waste site—so also there is reason to question the methodological value judgment that short-term studies at proposed high-level sites, like Yucca Mountain, provide accurate data for precise long-term predictions. Moreover, as we have already argued, given the longevity of any site for permanent disposal of nuclear waste and given difficulties with accurate geological predictions, there are difficulties with any proposal to site a permanent geological repository on the supposition that precise long-term predictions about hydrogeology are reliable.

4. The U.N. Mandates and the Historical/Legal Context

Perhaps because they are worried about human error and about scientists' claims to store waste safely in perpetuity, citizens, environmental groups, nations, and regions have been in turmoil over the issue of radioactive waste storage. If society is to follow the U.N. mandates for nuclear waste, it is important to understand the historical and legal context surrounding the problem of radioactive waste. Virtually no one wants this waste in his or her backyard. In the United States, for example, numerous groups have charged in lawsuits that radiation exposure standards for proposed sites are too lax and violate the Safe Drinking Water Act. Over the last several years, the U.S. Congress has been besieged with more than 30 bills proposing to delay, abandon, or change the repository program established under the 1983 U.S. Nuclear Waste Policy Act (Raloff and Peterson 1987: 73).

In the absence of scientific certainty about our ability to store nuclear wastes, we need ethical criteria for accepting or rejecting the technological risks these wastes impose and for carrying out the U.N. mandates. Before developing some ethical criteria, however, we need to look at the historical and legal context in which the problem of nuclear waste has arisen. Standard accounts of nuclear history are often misleading regarding the source and the magnitude of our difficulties with radioactive waste. Some persons have claimed that we face the problem of nuclear waste both because of military activities and important hospital uses of nuclear medicine and because a number of utilities were eager to provide inexpensive electricity. Both claims are untrue. High-level radwaste is, for the most part, spent fuel rods from fission reactors and residues from fuel reprocessing (Murray et al. 1982: 569). Less than one percent of high-level radwaste is from medical activities (Sierra Club 1984: 2). Moreover, at least in the United States, approximately half of the high-level radwaste now needing storage is from commercial nuclear fission, not military activities (Deutch 1978: D-11, D-12, D-14). Nor do we have the problem of nuclear waste because industry was eager to generate electricity and fission was an economical means of doing so. Initially, industry was reluctant, both on economic and on safety grounds, to use fission to generate electricity. Worried about safety, every major U.S. corporation with nuclear interests refused to generate nuclear electricity unless some indemnity legislation was passed to protect them in the event of a major accident (Caldwell et al. 1974: 379; Shrader-Frechette 1983: 10-11).

The top lobbyist for the nuclear industry, the president of the Atomic Industrial Forum, has confirmed what numerous government committee reports from the 1940s and 1950s revealed. Commercial nuclear fission began, and was pursued, only because government hoped to justify continuing military expenditures in nuclear-related areas and to obtain weapons-grade plutonium (Novick 1976: 32-33; Shrader-Frechette 1983: 8-9). Moreover, at least in the United States, nuclear fission began only because government provided more than $100 billion in subsidies (for research, development, waste storage, and insurance) to the nuclear industry. Government also gave the utilities a liability limit (in the Price-Anderson Act) that protects licensees from 99 percent of all public losses and claims in the event of a catastrophic nuclear accident (Shrader-Frechette 1993a, pp. 96-99).

Twenty years after commercial fission reactors began operating, in 1976, the *Wall Street Journal* proclaimed them an economic disaster. Nuclear electricity has proved so costly that year 2000 projections for commercial fission reactors in the United States, for example, are now approximately one eighth of what they were in the mid-1970s. No new reactors have been ordered in the United States since 1974 (Flavin 1983: 33). The few U.S. nuclear manufacturers that are still in business have remained by selling reactors to other nations, often developing countries. Yet many of the commercial reactors going to these nations may not be in the best interests of the countries receiving them. Indeed, commercial nuclear fission may be the current version of infant formula. In the infant-formula controversy of the last two decades, U.S. and multinational corporations made great profits by exporting infant formula to developing nations. They were able to do so only by coercive sales tactics and by misleading foreign consumers about the merits of their products.

Some diplomats also have charged that developing nations are seeking fission-generated electricity as a subterfuge for obtaining nuclear-weapons capability (Shrader-Frechette 1985), through the plutonium byproduct. India exploded its first nuclear bomb, for example, by using plutonium produced by a reactor exported by Canada. Whether or not this is the reason for the survival of nuclear fission, the few countries still developing atomic energy do not have uncontroversial grounds for subscribing to the myth of the safe, economical atom.

Regardless of whether atomic energy is safe or economical, however, we still need to store the waste. Most governments and nuclear-industry representatives prefer permanent geological disposal of high-level wastes and spent reactor fuel. Currently, sites are under investigation for permanent disposal in Argentina, Finland, France, Germany, South Africa, Sweden, Switzerland, and the United States. Although possible sites exist in Argentina, Germany, South Africa, and the United States, no permanent repository has yet been chosen anywhere in the world. Geological disposal is also the selected method of managing intermediate wastes, although so far, only the (nuclear) Waste Isolation Pilot Project (WIPP) site in New Mexico has been built for possible permanent disposal of military transuranic wastes; other sites are under consideration in several countries (AEC 1974: 3-83; ERDA 1975: X-74, II. 1-57; Amato 1986: 221; Raloff and Peterson 1987: 73; U.S. Congress 1987, 1988, 1990; Carter 1989; Monastersky 1991: 228; Dunlap et al. 1992; Shrader-Frechette 1993a: 20).

Short-lived intermediate-level wastes and low-level wastes likewise are placed in shallow land burial trenches in France and in the United States, or in subseabed repositories in Sweden. Some low-level radioactive wastes are also incinerated or emitted into the air, rivers, or seas, as from reprocessing plants in Britain and France. Currently, the greatest source of radioactive pollution of the world's seas and oceans is the Windscale reprocessing facility in Sellafield, U.K. Until 1970, the United States dumped intermediate-level and low-level wastes into the oceans, and until 1983, the British dumped them (on behalf of the United Kingdom, Belgium, Switzerland, and the Netherlands) into the oceans. Most radioactive wastes, however, currently remain in storage, presumably waiting until governments come up with a final disposal site or management plan (Eid 1985; Carter 1987: 251ff.; Lowry 1990: 10-11; Shrader-Frechette 1993a: 21).

Nuclear wastes are subject to a number of laws and regulations throughout the world. Many of these restrictions have been developed in cooperation with the International Atomic Energy Agency (IAEA). Although the IAEA was founded in 1957, only recently has the agency developed several guidelines to govern the management of nuclear waste. These are: (1)For effluents containing radionuclides in amounts below authorized radiological-protection limits (based on the recommendations of the International Commission for Radiological Protection, the ICRP), one can follow the strategy of "dilute and disperse" to the environment. ("The solution to pollution is dilution.") (2)For waste containing only short-lived radionuclides, one can follow the strategy of "delay and decay." (3)For waste containing significant amounts of long-lived radionuclides, one can follow the strategy of "concentrate and confine." Only in late 1988, however, did the IAEA establish its first formal radioactive-waste policy body, the International Radioactive Waste Management Advisory Committee (Scheinman 1987; Blowers et al. 1991: 41ff.; Shrader-Frechette 1993a: 21).

Through United Nations agreements, member states have also accomplished some regulation of nuclear energy and radioactive waste. In the U.N. Convention on the High Seas, agreed to in Geneva in 1958, for example, each nation agreed to take measures "to prevent pollution of the seas from the dumping of nuclear waste, taking into account any standards and regulations which may be formulated by the competent international organization" (Blowers et al. 1991: 44; Shrader-Frechette 1993a: 22). Sea-dumping of radioactive wastes by some member states, however, continued until 1983. This and other examples suggest that, despite the international guidelines and information exchanges regarding radioactive waste accomplished through the IAEA and related agencies, it was not until the Chernobyl accident in 1986 that countries actually began to accept transnational regulation of nuclear technology. Even so, policy and execution regarding radioactive waste remains largely a matter of national law.

5. U.N. Mandates and the Ethical Context

Because nuclear waste policies are largely a matter of national law and because there are significant scientific obstacles to safe storage, it could be quite difficult to fulfill the ethical mandates for nuclear waste, as described in Agenda 21. What are some of the ethical problems the mandates address? Most of the major ethical problems associated with radioactive waste focus on the issue of equity. Kasperson, in his famous study of radwaste management, places them in three main groups: locus problems, legacy problems, and labor-laity problems. The locus issues have to do with where and how to site radwaste facilities. The labor-laity problems focus on whether to maximize either the safety of the public or that of nuclear waste workers, because both cannot be accomplished at once. The legacy problems have to do with exporting the risks and costs of radioactive waste to future generations (Kasperson 1983).

The key question raised by legacy concerns is whether one can justify intergenerational inequity by mortgaging the future, by imposing our debts of nuclear waste on subsequent generations. For example, U.S. citizens saddle their descendants with the health and financial risks of waste, then taxpayers in later centuries could be forced to pay an annual tab for radwaste storage between $1.9 and $3.8 million per reactor per year (Shrader-Frechette 1983: 57-58). This expenditure is obviously questionable because future generations should not be saddled with other persons' debts. Also, there is very little public funding, especially by taxpayers, for decentralized energy technologies, like solar, which are much less likely to burden the future (Berger 1976: 71).

A second legacy question is what sort of criteria might justify environmentally irreversible damage to the environment, like that caused by deep-well storage of high-level nuclear waste. Radwaste management schemes that are irreversible theoretically impose fewer management burdens on later generations, but they also preempt future choices about how to deal with the waste. On the other hand, schemes that are reversible allow for greater choices for future generations, but they also impose greater management burdens. If we can't do both, is it ethically desirable to maximize future freedom or to minimize future burdens (Shrader-Frechette 1993a)?

Perhaps the most important legacy question concerns the contribution of radwaste production and storage to the "plutonium economy" which is necessary for building nuclear weapons. Would our continuing to generate and store radwaste maximize or minimize future freedom and security? Would it add to the probability of a nuclear holocaust by providing materials necessary for arms (Shrader-Frechette 1985; Kipnis and Meyers 1987; Cohen 1989)? If the waste is not retrievable, presumably it would pose less of a threat regarding nuclear weapons.

Still other legacy questions have to do with use of social discount rates. Any alleged economies or safety associated with high-level radwaste storage are in large part questionable because of their dependence on a particular discount rate. Using a discount rate amounts to discounting future costs of radwaste storage, like radiation-related deaths or injuries, at some rate of *x* percent per year. Thus, at a discount rate of 10 percent, effects on people's welfare 20 years from now count only for one tenth of what effects on people's welfare count for now. With a discount rate of 5 percent, effects next year count for 1,000 times more than effects 200 years from now. Or, more graphically, with a discount rate of 5 percent, a billion deaths in 400 years counts the same as one death next year. A number of famous moral philosophers, like Parfit, have argued that use of a discount rate is unethical, because the moral importance of future events, like the death of a person, does not decline at some *x* percent per year (Burness 1981; Parfit 1983; Grossman and Cassedy 1985). Yet economists claim that radwaste disposal is economical only because they discount costs borne by future persons. Without discounting, it would be virtually impossible to justify the dangers and costs of storing radwaste for thousands of centuries.

Imposing nuclear waste on future generations might also be questionable from a practical point of view. Uranium will not be available much beyond the year 2000, even to supply existing fission reactors. Hence, after having generated tons of dangerous waste, nuclear energy (without the breeder and without fusion) may not have provided a long-term technological fix for our energy problems. From a practical point of view, it is not clear that the temporary benefits of nuclear fission are worth the permanent costs of radioactive waste (Joskow 1977).

In addition to the legacy issues, a number of locus or siting problems are associated with managing radioactive waste. One of the key difficulties here is vesting, allowing a company to obtain a return on its initial capital investment in a radwaste site. According to traditional vesting doctrine, if a company has invested large amounts in a particular location for managing nuclear waste, then this investment is considered to be a sufficient ethical condition for continuing the facility, even though the technology and the site may be discovered to be less than adequate. The obvious question this raises is whether there are ethical grounds for land-use controls. Are there ethical grounds for limiting property rights, even when such limitations fly in the face of current vesting doctrine (Sagoff 1988: 9-12)?

Still other, and even more far-reaching, locus questions arise because of the emergent field of land ethics. Consider, following Locke, Blackstone, Nozick, and others, that it may be impossible to justify acquisition of property rights in land, especially original property rights. If one has rights only to what is produced by one's labor, and if one's labor has not produced land, then it is questionable whether one might be said to have full property rights to land (Caldwell and Shrader-Frechette 1993: Ch. 4; Shrader-Frechette 1993b). But if so, then how does one justify siting any land for nuclear waste storage? This is a use that amounts to exercising the most extreme form of property rights, because it preempts both present and future choices about land use at that site.

Another important siting or locus issue is geographical equity. It is a foregone conclusion that radwaste repositories will be located in rural areas, away from major population centers (Leahy 1980: 2). One of the questions raised by considerations of geographical equity is whether it is fair to impose a risk on a person just because she lives in a rural community rather than a large city. Is it just to impose a higher risk of death/injury on a person simply because she lives in a waste-storing rather than a waste-producing community? Likewise, is it ethical for one geographical subset of persons to receive the benefits of nuclear-generated electricity, while a much larger set of persons bears the costs? Cost comparisons of alternative energy technologies typically ignore externalities like the $100 billion in nuclear subsidies already spent by U.S. taxpayers, part of which is for nuclear waste storage (Gofman and Tamplin 1971: 177, 199; Muchnicki 1973: 45; Primack and Von Hippel 1974: 7; Berger 1976: 94-97, 106-112, 144-147). In the past decade, for example, government subsidies to the nuclear industry, in the form of write-off for capital invested in plants not completed, was $4 billion in the United States alone (Flavin 1983: 41). It is arguably inequitable to ignore subsidies paid by all the taxpayers to an industry that primarily benefits only a small set of taxpayers.

Once taxpayer subsidies for costs like waste storage and decommissioning are included in the calculations, nuclear power can be shown to be more expensive than every other energy alternative (Miller 1976: 123; Flavin 1983: 1-33). The only way to make it viable is to remove it entirely from the discipline of the market and to increase taxpayer subsidies (Flavin 1983: 42). Already this removal from the market has resulted in members of future generations, and taxpayers generally, picking up the tab for billions of dollars of expenses that arguably should be borne by the nuclear industry and its beneficiaries. Hence radwaste funding raises a whole host of equity questions that affect both locus and legacy.

6. U.N. Mandates and the Equity Rationale

If one examines the Agenda 21 mandates for nuclear waste (see the first section of this essay), it is clear that their rationale is to provide equal treatment and equitable distribution of the risks and benefits associated with the wastes. Preventing innocent persons from becoming victims of radioactive wastes is an explicit rationale for the mandates expressed in Chapter 22 of Agenda 21. Avoiding this victimization is important, in large part, because all persons have rights both to equal treatment and to free informed consent to the risks imposed on them by others.

Persons have rights to equal treatment with respect to the risks of nuclear wastes, indeed with respect to all socially imposed risks, because all persons have an equal capacity for a happy life and therefore all persons, in principle, deserve equal treatment (Blackstone 1969: 121). They also have rights to equal treatment as evidenced by the fact that free, informed, rational persons typically agree to principles of equal rights and equal protection (Rawls 1971: 3-46). Moreover, principles of equal treatment are presupposed by all ethical schemes involving consistency, justice, fairness, and autonomy; it is impossible to have any consistent ethics without principles of equal treatment (Rawls 1971: 3-46). "Law itself embodies an ideal of equal treatment for persons similarly situated" (Pennock 1974: 2-6). If there were no duty to equalize the burden of societal risk imposed on one generation or segment of society for the benefit of another, then there could be neither authentic bodily security nor legal rights, and one group could victimize whatever generation or group that it wished to (Rawls 1971: 3-53; Shrader-Frechette 1991b: 112-117; Shrader-Frechette 1993a: Ch. 8).

Indeed, there is some evidence that current policies for dealing with nuclear wastes victimize both present persons who live near waste facilities and future persons likely to be harmed by stored wastes that migrate. For example, scientists studying the first proposed permanent facility for high-level nuclear waste, at Yucca Mountain, Nevada, have made several admissions. Because they have admitted that complete, perpetual containment of the waste is impossible (Williams 1980) and that the waste canisters will remain intact, at best, only for several hundred years (Berusch and Gause 1987), it is clear that permanent disposal of long-lived hazardous and nuclear wastes places the greatest risks on members of future generations. Because of this, permanent disposal implicitly sanctions an inequitable risk distribution according to which present persons receive most of the benefits of production of hazardous and nuclear wastes, whereas future persons bear most of the risks.

Proponents of present policies regarding wastes, however, are likely to argue that production of nuclear wastes benefits everyone and therefore that economics, efficiency, or the greater good justifies treating present and future generations differently (Lave and Leonard 1989: 1068-1069). There are at least two problems with this attempted defense of current policy. One difficulty is that, even on narrow economic criteria, the costs of generating long-lived hazardous wastes typically exceed the benefits, because future persons do not benefit directly from the wastes and because waste generation can be shown to be cost effective only if economists discount future deaths caused by waste storage. Hence, production of long-lived nuclear wastes can be shown to be cost-effective only if present and future persons are not given equal treatment in terms of economic accounting (Kneese et al. 1983: 219; Parfit 1983: 31-37).

Even if present and future benefits actually outweighed the risks, and even if future people received as many benefits from waste production as present people, there would still be an ethical problem with current policies of permanent disposal of hazardous and radioactive wastes. This second problem is that the risks for future people are much greater than the risks for present ones. Hence, any policy for managing long-lived hazardous and nuclear wastes ought to minimize the inequity between present and future people. Several ways of minimizing this inequity include funding a special trust whose monies are sufficient to insure both perpetual care for monitored surface-storage waste facilities and full compensation for present and future victims of hazardous and nuclear wastes. Inequity could also be minimized by establishing a "public defender for the future," someone who could act as a proxy for future persons in consenting to current policymaking regarding waste (Kasperson et al. 1983: 363-368; Shrader-Frechette 1993a: Ch. 9).

Although in practice future people obviously are unable to give or withhold consent to the risks that present individuals impose on them through long-lived hazardous wastes, there are several ways in which it is possible to determine whether we are protecting future people and their implicit rights to free, informed consent. One step is to employ policies for managing nuclear wastes that are likely to be sanctioned by an authentic and informed majority of people, across time (Rawls 1971: 356ff.). Current policies for dealing with high-level nuclear waste, for example, probably would not meet this first condition because the two or three generations that benefit from production of the waste are not a majority, historically speaking. Indeed, members of future generations potentially victimized by geological disposal of hazardous wastes likely represent a "silent majority" on whom a minority decision (based on short-term economic interests) has been imposed (Cochran 1983: 116). And if so, then it is unlikely that future people would theoretically or implicitly consent to current practices regarding long-lived hazardous wastes.

A second indicator that future people likely would not consent to our present policies regarding managing hazardous wastes is that even a majority of present people cannot be said to have given any form of consent to geological disposal. Polls indicate that a majority of people believe that radioactive waste disposal, for example, is not safe (Montange 1987: 408). Indeed, the NIMBY (Not In My Back Yard) syndrome is pervasive. In Nevada, for example (the site of the world's first proposed permanent, high-level nuclear waste facility), 80 percent of the population is opposed to a permanent repository in the state (Slovic et al. 1991: 1604). If one makes the reasonable assumption that the preferences of present people indicate something about the preferences of future people, then the opposition of this generation to permanent disposal provides grounds for arguing that subsequent generations would not be likely to consent either, because they would face an even greater risk than would present people from facilities built now (Shrader-Frechette 1993a: Ch. 8).

7. Policy Implications of the U.N. Mandates

If our ethical reasoning about equity and consent is correct, then the objection—that permanent disposal is cheaper or safer than other methods of handling long-lived hazardous wastes—is not compelling. The appropriate ethical response to such arguments is: "Cheaper for whom?" "Safer for whom?" It is certainly not cheaper or safer for members of future generations who comprise the majority of people likely to be negatively affected by long-lived hazardous wastes. In fact, even on utilitarian grounds, geological repositories represent perhaps the greatest good for the present number of people, but not the greatest good for the greatest number of all people. Hence, if one follows the mandates of Agenda 21, then building permanent repositories is justified, if at all, primarily on the grounds of the narrow self-interest of us in the 20th and 21st centuries.

Analogous to racism and sexism, the narrow self-interest of this generation might be called generationism. The power of whites over blacks does not give whites the right to do to blacks whatever they wish. The power of men over women does not give men the right to do to women whatever they wish. Likewise, our power over future persons does not give us the right to do to them whatever we wish. Might does not make right.

If might does not make right, then how do we develop policies to implement the Agenda 21 mandates? How do we prevent the might of present generations from dictating the right for future generations? One solution might be to develop policies similar to those already used in patient-doctor relationships. We all probably believe that it would be unethical for a doctor to impose a risky treatment on a patient without her free, informed consent. Yet most of us need to develop our moral sensibilities so as to see radwaste workers or radwaste siting in the same light. Until or unless a risk imposition receives the consent of those who are its potential victims, it cannot be justified. This means that those who wish to impose societal risks on others need to do whatever is necessary to compensate them to a degree adequate to obtain their free, informed consent. If no forms of compensation are adequate to obtain free, informed consent, then it is questionable whether the risk imposition can be justified.

In some cases in which it was not possible to meet the compensatory demands of a community proposed as a radwaste site, the facility was not located. In the cases in which compensation was able to be negotiated, however, often this was sufficient to insure community consent. Typically, industry-supplied compensation in cases involving nuclear waste includes tax breaks or funding community projects such as schools. According to social scientists studying waste siting, however, one form of compensation dominates all negotiation. The one factor almost always essential to achieving local consent is giving citizens/workers funding to control health and safety monitoring themselves at the facility. By forcing the waste managers to pay for outside monitors, citizens and workers are freed from relying on company monitoring. Moreover, once citizens and workers have greater control of their health and safety, they appear ready to give informed consent to the risk. By providing full compensating benefits for all unavoidable radwaste risks, industry can offset the inequities generated by some of the locus and legacy problems (Shrader-Frechette 1991b: 84-88, 116-121, 197-218).

Admittedly, compensation of future generations who bear the brunt of the legacy problems might be difficult. Obtaining free, informed consent and guaranteeing compensation in such cases requires that the consent of future people be obtained by means of representatives acting in their best interests. It also requires that those who store radioactive waste actually set up a fund for compensation of future persons possibly harmed by the waste. Requiring full compensation and consent in the case of future generations entails, in part, that we use an equal-opportunity criterion for intergenerational equity. To the degree that we cannot guarantee that persons a million years from now will have equal opportunity to protect themselves from our radwaste hazards, then to that same degree we cannot justify generating radioactive waste or imposing it upon others (Bodde and Cochran 1981: 7).

The rationale for requiring full compensation and consent, even regarding future people, is that any technology ought to pay its own way. Considerations of fairness dictate that a technology ought not to be subsidized by the deaths and cancers of people who are its unwilling victims. If nuclear power cannot pay its own way regarding compensation and consent, then it has no right to claim that it ought to be accepted. To the degree that nuclear power does not pay its own way, especially in regard to waste storage, it will continue to reinforce all the old income distributions and inequities. It will reaffirm a situation in which the poor and the uneducated bear the social costs of contemporary technology.

Moreover, if someone can impose a bodily risk of harm on others without their consent, then there is no recognition of a right to life, a right defended repeatedly in U.N. documents. If someone can profit by imposing a threat of physical harm on others without compensating them, then there is no right to due process. These considerations suggest that, if we refrain from requiring genuine informed consent and from guaranteeing complete compensation for radwaste risks, then nuclear waste will put more at risk than our health and that of future generations. It will put at risk our most basic rights to justice and equity and our most basic national and international institutions that guarantee these rights.

8. Achieving Environmental Protection Through NMRS

Perhaps the best way to guarantee basic rights and environmental protection in managing nuclear waste is to avoid permanent disposal and negotiate with host communities for temporary, monitored, retrievable storage of the waste for at least 100 years (Shrader-Frechette 1993a: 213-253). This wait-and-see position makes a great deal of scientific sense.

Wait and see if we can develop more resistant containers. Wait and see if we can devise a way to render radioactive materials less harmful. Wait and see if we can resolve some of the uncertainties regarding long-term safety at a permanent repository.

Establishing a system of geographically scattered, retrievable storage sites is an idea that has been around for many years. For temporary storage to be successful, however, government would need to negotiate site selection with potential host communities rather than impose facilities on them. A negotiated, monitored, retrievable system (NMRS) also would require the government to monitor the waste rather than simply bury it and leave it unmonitored, as planned at Yucca Mountain. Most important, it would enable the scientific and regulatory community to learn—in stages—how best to store high-level waste (HLW) safely. If some dangerous technologies—like those for waste disposal—are unforgiving, then it makes sense to lengthen the scientific and regulatory learning curves. Retrievability buys time and increases our scientific and ethical options.

After 100 years of experience with NMRS facilities, we would be in a better position to make decisions regarding permanent disposal. Transmutation—showering the waste with neutrons to convert fission products to stable or short-lived radioactive isotopes—might prove workable by then, thus enabling us to store the waste for a shorter time.

In 1972, Weinberg described the problem of nuclear wastes as a "Faustian bargain" (Shrader-Frechette 1993a: 182). In return for the present benefits of atomic energy, we must export the risks of its wastes to future generations. Because we have already made the bargain, we cannot avoid dealing with the radioactive wastes we have generated. We can, however, choose better or worse ways to live out the consequences of our pact with Mephistopheles. A better way to live out these consequences is to distribute the risks of nuclear waste as equitably as possible. Opposition to the proposed Yucca Mountain repository is bitter, in part, because Nevadans believe that their state is being treated inequitably; it must bear the HLW burden for the entire nation. The country as a whole made the Faustian bargain, but only Nevadans are supposed to pay the price. And because the Yucca Mountain waste would not be retrievable, future generations would also pay the price. They would not have the option of freely consenting to continued storage. Regional NMRS facilities thus are desirable because they would help spread both the temporal and spatial risk of HLW disposal. Multiple NMRS facilities would minimize the geographical and temporal inequities borne by people in particular locales and in future generations. A permanent, unmonitored, nonretrievable facility would place the greatest HLW risks on one region and on future generations.

The willingness of some communities to accept NMRS also suggests that temporary storage might be more politically acceptable than a permanent facility. Of course, the communities that might accept an NMRS facility are likely to be impoverished and looking for an economic boost. Morgan County, Tennessee, for instance, volunteered to host an NMRS facility. In the county, however, the unemployment rate is far above the national average, and per-capita incomes are low. Schools are poor and other services meager. About half of the local tax collected is needed merely to service the county's bonded debt. In offering to host a temporary waste facility, the representatives of Morgan County made it clear that the operators of the facility would have to underwrite a substantial portion of the county's operating expenses, including servicing its debt. Because poverty can raise serious questions about the voluntariness of a host community's informed consent, the negotiation process between governments and communities would have to maximize consent, equity, and due process. Nevertheless, 100-year use of NMRS facilities seems preferable to a central permanent repository because it would require the imposition of fewer, and shorter-term, burdens on the most vulnerable groups: host communities and future generations. Minimal fairness requires the current generation to clean up its own mess or to somehow pay its descendants—in full—to do it.

There are, of course, objections to NMRS. Perhaps the most obvious objections are that the facilities would be unsafe, that they would be targets for terrorist attacks, and that they would contribute to the proliferation problem. Although the DOE argues that a permanent geological repository would be safer than temporary, monitored facilities, the question is: "Safer for whom?" Certainly not safer for members of future generations who might be harmed by leakage from an unmonitored facility. Numerous members of Congress, study groups at sites wishing to host NMRS facilities, the Nuclear Regulatory Commission, and the U.S. Monitored Retrievable Storage Review Commission have affirmed the safety of 100-year storage (Shrader-Frechette 1993a: 213-253). The commission supported its arguments with detailed calculations. It estimated, for example, that the total radiation doses, both to the public and to workers, would be less in the case of an NMRS facility not linked to a repository than for a permanent facility handling the same amount of waste. The review commission indicated, however, that it did not believe the safety differences were great between the two options. It likewise emphasized that the NMRS option was safer than on-site storage at reactors, in part because the NMRS facility would employ experienced fuel handlers and would have a full staff available (see Shrader-Frechette 1993a: 213-253).

Admittedly, temporary storage facilities would be more susceptible to terrorism and sabotage than a permanent geological repository. But the monitoring and management of NMRS sites might make them better able to resist such attacks, once they occurred. Also, given current building technology, it should be possible to construct surface structures that are extremely protective.

Another objection to deferring the decision about a permanent repository and instead using NMRS facilities for a century is that, over the long term, such storage would be more expensive than permanent disposal. A host of analysts have looked at the cost question, but it is not clear that permanent geological disposal would be cheaper. The economics of permanent disposal depend on when and how much the facility leaks. The sooner and bigger the leaks, the higher the cost. But dollar costs are not the central consideration. The main reason many people claim NMRS sites are more expensive is that permanent geological disposal does not achieve the same level of pollution control. As just mentioned, a permanent site would leak. The scientific argument is largely over how quickly it would leak, how much it would leak, and how rapidly the leakage might reach the water table. Permanent disposal is premised on a philosophy of "dilute and disperse." In contrast, NMRS storage is based on containment. Dilution and dispersal of hazardous substances is always cheaper than containment.

The main problem with the economic objection to NMRS sites is that cost considerations, although important policy determinants, should not be the only or the primary determinants of waste policy. After all, if cost were the sole criterion for a reasonable choice, one might be able to argue for dumping radioactive materials into the sea or for using shallow land burial. The use of narrow economic criteria for waste management is also undesirable because we have produced the radioactive materials and thus we have an ethical obligation to do as much as is necessary and possible to protect subsequent generations. To argue that economics ought to be the principal determinant of waste policy would be to use an expediency criterion for recognition or denial of a basic human right to equal protection of the laws. The U.N. Universal Declaration of Human Rights does not say that we have rights to life, liberty, and the pursuit of happiness "provided that it is economical to recognize them." If nuclear waste policy is to be consistent with existing philosophical, legal, and political doctrines about human rights, then expediency should not be our primary guide.

Humankind has been civilized for only about 10,000 years, yet the United States faces the task of storing radionuclides such as plutonium, which remains dangerous for more than 250,000 years. Given our short experience in handling such materials, how can we deal adequately with long-lived radioactive waste? The short answer is: "We can't." We do not yet know how to do the job right. That's why permanent disposal of nuclear waste is a profoundly bad idea. Although he did not intend it, J.R.R. Tolkien, in *The Lord of the Rings*, suggested an answer to the riddle of nuclear waste. Humankind will eventually read the riddle. Agenda 21 has given us some clues. The rest is up to us.

9. References

(AEC) U.S. Atomic Energy Commission. 1974. Comparative Risk-Cost-Benefit Study of Alternative Sources of Electrical Energy. WASH-1224. U.S. Government Printing Office, Washington, DC.

Amato, I. 1986. Dangerous Dirt: An Eye on DOE. *Science News* 130(14): 221.

Beavers, J., and N. Thompson. 1990. Environmental Effects on Corrosion in the Tuff Repository. U.S. Nuclear Regulatory Commission, Washington, DC.

Berger, J.J. 1976. *Nuclear Power—The Unviable Option*. Ramparts Press, Palo Alto, CA.

Bertell, R. 1985. *No Immediate Danger: Prognosis for a Radioactive Earth*. The Book Publishing Company, Summertown, TN.

Berusch, A., and E. Gause. 1987. DOE Progress in Assessing the Long Term Performance of Waste Materials. In *Scientific Basis for Nuclear Waste Management*, J. Bates and W. Seefeldt, eds. Materials Research Society, Boston, pp. 13-28.

Blacic, J., et al. 1986. Effects of Long-Term Exposure of Tuffs to High-Level Nuclear Waste Repository Conditions, Final Report. Los Alamos National Laboratory, Los Alamos, NM.

Blackstone, W.T. 1969. On the Meaning and Justification of the Equality Principle. In *The Concept of Equality*, W.T. Blackstone, ed. Burgess, Minneapolis, pp. 119-123.

Blowers, A., D. Lowry, and B. Solomon. 1991. *The International Politics of Nuclear Waste*. St. Martin's Press, New York.

Bodde, D., and T. Cochran. 1981. *Conflicting Views on a Neutrality Criterion for Radioactive Waste Management*. Center for Philosophy and Public Policy, University of Maryland, College Park, MD.

Burness, H.S. 1981. Risk: Accounting for the Future. *Natural Resources Journal* 21(4): 723-734.

Caldwell, L.K., and K.S. Shrader-Frechette. 1993. *Policy for Land: Law and Ethics*. Rowman and Littlefield, Savage, MD.

Caldwell, W.S., et al. 1974. The "Extraordinary Nuclear Occurrence" Threshold and Uncompensated Injury Under the Price-Anderson Act. *Rutgers Camden Law Journal* 6(2): 360-383.

Carter, L. 1987. *Nuclear Imperatives and Public Trust*. Resources for the Future, Washington, DC.

Carter, L. 1989. Nuclear Waste Policy and Politics. *Forum for Applied Research and Public Policy* 4(3): 5-18.

Cochran, T. 1983. Conflicting Views on a Neutrality Criterion for Radioactive Waste Management. In *Energy and the Future*, D. MacLean and P. Brown, eds. Rowman and Littlefield, Totowa, NJ, pp. 110-116.

Cohen, S.M. 1989. *Arms and Judgment*. Westview Press, Boulder, CO.

Crowe, B. 1986. Volcanic Hazard Assessment for Disposal of High-Level Radioactive Waste. Los Alamos National Laboratory, Los Alamos, NM.

Deutch, J.M., and the Interagency Review Group on Nuclear Waste Management. 1978. Report to the President. TID-2817. National Technical Information Service, Springfield, VA.

Dobson, D., et al. 1990. Plans for Characterization of the Potential Geologic Repository Site at Yucca Mountain, Nevada. In *International Conference for High-Level Radioactive Waste Management*. U.S. Department of Energy, Las Vegas, NV.

(DOE) U.S. Department of Energy. 1986a. Nuclear Waste Policy Act, Environmental Assessment, Reference Repository Location, Hanford Site, Washington. DOE/RW-0070, 3 Vols. U.S. DOE, Washington, DC.

(DOE) U.S. Department of Energy. 1986b. Nuclear Waste Policy Act, Environmental Assessment, Yucca Mountain Site, Nevada Research and Development Area, Nevada. DOE/RW-0073, 3 Vols. U.S. DOE, Washington, DC.

(DOE) U.S. Department of Energy, Office of Scientific and Technical Information. 1990. Project History. In *Yucca Mountain Project Bibliography, 1988-1989*. DOE/OSTI-3406 (Suppl. 2) (DE90006793) U.S. DOE, Oak Ridge, TN, November.

Dunlap, R.E., M.E. Draft, and E.A. Rosa, eds. 1992. *The Public and Nuclear Waste: Citizens' Views of Repository Siting*. Duke University Press, Durham, NC.

Eid, C., ed. 1985. *Incineration of Radioactive Waste*. Graham and Trotman, London.

(EPA) U.S. Environmental Protection Agency. 1978. Considerations of Environmental Protection Criteria for Radioactive Waste. U.S. EPA, Washington, DC.

(EPA) U.S. Environmental Protection Agency, Office of Radiation Programs. 1982. Draft Environmental Impact Statement for 40 CFR 191: Environmental Standards for Management and Disposal of Spent Nuclear Fuel, High-Level and Transuranic Radioactive Wastes. EPA 5201102025. U.S. EPA, Washington, DC.

(ERDA) U.S. Energy Research and Development Administration. 1975. Final Environmental Statement: Waste Management Operations, Hanford Reservation, Richland, Washington. ERDA-1538, Vol. 1. National Technical Information Service, Springfield, VA, October.

Flavin, C. 1983. *Nuclear Power: The Market Test*. Worldwatch Institute, Washington, DC.

Gofman, J.W., and A. Tamplin. 1971. *Poisoned Power*. Rodale Press, Emmaus, PA.

Grossman, P.Z., and E.S. Cassedy. 1985. Cost Benefit Analysis of Nuclear Waste Disposal. *Science, Technology, and Human Values* 10(4): 46-50.

Hadlock, C. 1980. Technical Support of Standards for High-Level Radioactive Waste Management, Vol. D: Release Mechanisms. EPA 520/4-79-007D. U.S. Environmental Protection Agency, Washington, DC.

Halsey, W. 1990. Selection Criteria for Container Materials at the Proposed Yucca Mountain High Level Nuclear Waste Repository. In *NACE Corrosion '90*. U.S. Department of Energy, Las Vegas, NV.

Hawkins, D. 1978. *Considerations of Environmental Protection Criteria for Radioactive Waste*. U.S. Environmental Protection Agency, Washington, DC.

Hodges, K.V. 1992. Comment. In Report of the Peer Review Panel on the Early Site Suitability Evaluation of the Potential Repository Site at Yucca Mountain, Nevada. SAIC-91/8001. J.L. Younker et al., eds. U.S. Department of Energy, Washington, DC, pp. 362-363.

Joskow, P.L. 1977. Commercial Impossibility, The Uranium Market, and the Westinghouse Case. *Journal of Legal Studies* 6(1): 119-176.

Kasperson, R.E., ed. 1983. *Equity Issues in Radioactive Waste Management*. Oelgelschlager, Gunn & Hain, Cambridge, MA.

Kasperson, R.E., P. Derr, and R. Kates. 1983. Confronting Equity in Radioactive Waste Management. In *Equity Issues in Radioactive Waste Management*, R. Kasperson, ed. Oelgelschlager, Gunn & Hain, Cambridge, MA, pp. 363-368.

Kipnis, K., and D. Meyers, eds. 1987. *Political Realism and International Morality*. Westview Press, Boulder, CO.

Kneese, A., et al. 1983. Economic Issues in the Legacy Problem. In *Equity Issues in Radioactive Waste Management*, R. Kasperson, ed. Oelgelschlager, Gunn & Hain, Cambridge, MA.

Lave, L., and B. Leonard. 1989. Regulating Coke Oven Emissions. In *The Risk Assessment of Environmental and Human Health Hazards*, D. Paustenbach, ed. John Wiley, New York, pp. 1068-1069.

Leahy, P.J., U.S. Senator, Vermont. 1980. The Socioeconomic Effects of a Nuclear Waste Storage Site on Rural Areas and Small Communities. Statement before the Subcommittee on Rural Development, Committee on Agriculture, Nutrition, and Forestry, U.S. Senate, August 26, 1980. U.S. Government Printing Office, Washington, DC.

Lenssen, N. 1992. Confronting Nuclear Waste. In *State of the World 1992*, L. Brown, ed. Norton, New York, pp. 49-53.

Lipschutz, R.D. 1980. *Radioactive Waste*. Ballinger, Cambridge, MA.

Lowry, A. 1990. Radwaste Incineration in the U.S. *World Information Service on Energy (WISE) News Communique* 326(7): 10-11.

Mara, S. 1980. Assessment of Effectiveness of Geologic Isolation Systems. Geologic Factors in the Isolation of Nuclear Waste: Evaluation of Long-Term Geomorphic Processes and Catastrophic Events. Battelle Pacific Northwest Laboratory, Seattle, WA.

Metcalf, L. 1983. Preliminary Review and Summary of the Potential for Tectonic, Seismic, and Volcanic Activity at the Nevada Test Site Defense Waste Disposal Site. Desert Research Institute, Reno, NV.

Miller, S. 1976. *The Economics of Nuclear and Coal Power*. Praeger, New York.

Monastersky, R. 1991. First Nuclear Waste Dump Finally Ready. *Science News* 140(15): 228.

Montange, C. 1987. Federal Nuclear Waste Disposal Policy. *Natural Resources Journal* 27(Spring): 408.

Muchnicki, E. 1973. The Proper Role of the Public in Nuclear Power Plant Licensing Decisions. *Atomic Energy Law Journal* 15(1): 39-59.

Murray, J.P., J.J. Harrington, and R. Wilson. 1982. Chemical and Nuclear Waste Disposal. *The Cato Journal* 2(2): 569-586.

Murray, R. 1989. *Understanding Radioactive Waste*. Batelle Press, Columbus, OH.

(NNWPO) Nevada Nuclear Waste Projects Office. 1989. State of Nevada Comments on the U.S. Department of Energy Consultation Draft Site Characterization Plan, Yucca Mountain Site, Nevada, Vol. 1. U.S. Department of Energy, Carson City, NV.

Novick, S. 1976. *The Electric War*. Sierra, San Francisco.

Papadopulos, S., and I. Winograd. 1974. Storage of Low-Level Radioactive Wastes in the Ground: Hydrogeologic and Hydrochemical Factors. EPA-520/3-74-009. U.S. Environmental Protection Agency, Office of Radiation Programs, Washington, DC.

Parfit, D. 1983. Energy Policy and the Further Future: The Social Discount Rate. In *Energy and the Future*, D. MacLean and P. Brown, eds. Rowman and Littlefield, Totowa, NJ, pp. 31-37.

Pennock, P. 1974. Introduction. In *The Limits of Law* Nomos 15, Yearbook of the American Society for Political and Legal Philosophy, J. Pennock and J. Chapman, eds. Lieber-Atherton, New York, pp. 2-6.

Perry, J. 1988. A Lineament Analysis of Yucca Mountain, Nevada: The Proposed High-Level Nuclear Waste Repository. In *6th Thematic Conference on Remote Sensing for Exploration Geology: Applications, Technology, Economics*. Environmental Research Institute of Michigan, Ann Arbor, MI.

Pitman, S., et al. 1986. Corrosion and Slow-Strain-Rate Testing of Type 304L Stainless in Tuff Groundwater Environments. In *Corrosion '87*. Pacific Northwest Laboratory, San Francisco.

Polluck, C. 1986. *Decommissioning: Nuclear Power's Missing Link*. Worldwatch Institute, Washington, DC.

Primack, J., and F. Von Hippel. 1974. Nuclear Reactor Safety. *The Bulletin of the Atomic Scientists* 30(8): 5-12.

Raleigh, C.B., and the Panel on Coupled Processes at Yucca Mountain. 1992. Ground Water at Yucca Mountain: How High Can It Rise? National Academy Press, Washington, DC.

Raloff, J., and I. Peterson. 1987. Trouble with EPA's Radwaste Rules. *Science News* 132(5): 73.

Rawls, J. 1971. *A Theory of Justice*. Harvard University Press, Cambridge, MA.

Roxburgh, I. 1987. *Geology of High-Level Nuclear Waste Disposal*. Chapman and Hall, London.

Sagoff, M. 1988. Property Rights and Environmental Law. *Philosophy and Public Policy* 8(2): 9-12.

Sawyer, G. 1990. Report of the State of Nevada Commission on Nuclear Projects. Nevada Commission on Nuclear Projects, Carson City, NV.

Scheinman, L. 1987. *The International Atomic Energy Agency and World Nuclear Order*. Resources for the Future, Washington, DC.

Shrader-Frechette, K.S. 1983. *Nuclear Power and Public Policy: The Social and Ethical Problems of Fission Technology*. Reidel Publishing Company, Dordrecht, Netherlands.

Shrader-Frechette, K.S. 1985. Nuclear Arms and Nuclear Power: Philosophical Connections. In *Nuclear War: Philosophical Perspectives*, M.A. Fox and L. Groarke, eds. Peter Lang Publishers, New York, pp. 85-100.

Shrader-Frechette, K.S. 1991a. Ethical Dilemmas and Radioactive Waste: A Survey of the Issues. *Environmental Ethics* 13(4): 327-343.

Shrader-Frechette, K.S. 1991b. *Risk and Rationality*. University of California Press, Berkeley, CA.

Shrader-Frechette, K.S. 1993a. *Burying Uncertainty: Risk and the Case Against Geological Disposal of Nuclear Waste*. University of California Press, Berkeley, CA.

Shrader-Frechette, K.S. 1993b. Locke and Limits on Land and Ownership. *Journal of the History of Ideas* 54(2): 201-219.

Sierra Club. 1984. *Low-Level Nuclear Waste: Options for Storage*. Sierra Club Radioactive Waste Campaign, Buffalo, NY.

Slovic, P., J. Flynn, and M. Layman. 1991. Perceived Risk, Trust, and the Politics of Nuclear Waste. *Science* 254: 1603-1607.

Smith, C. et al., 1982. Population Risks from Disposal of High-Level Radioactive Wastes in Geologic Repositories. EPA-520/3-80-006. U.S. Environmental Protection Agency, Washington, DC, pp. 10-51.

Smith, H.D. 1988a. Electrochemical Corrosion-Scoping Experiments: An Evaluation of the Results. Westinghouse Hanford Co., Richland, WA.

Smith, H.D. 1988b. Initial Report on Stress-Corrosion-Cracking Experiments Using Zircaloy-4 Spent Fuel Cladding C-Rings. Westinghouse Hanford Co., Richland, WA.

Stein R., and P. Collyer. 1984. Pilot Research Projects for Underground Disposal of Radioactive Wastes in the United States of America. In *Radioactive Waste Management, Vol. 3. Proceedings of an International Conference Held by the IAEA in Seattle, 16-20 May 1983*. U.S. Department of Energy, Washington, DC.

Tolkien, J.R.R. 1965. *The Fellowship of the Ring*. Ballantine Books, New York.

U.S. Congress. 1987. Nuclear Waste Program. Hearings Before the Committee on Energy and Natural Resources, U.S. Senate, 100th Congress, April 29 and May 7, 1987, Part 3. U.S. Government Printing Office, Washington, DC.

U.S. Congress. 1988. Nuclear Waste Policy Act. Hearing Before the Subcommittee on Energy and the Environment of the Committee on Interior and Insular Affairs, House of Representatives, 100th Congress, First Session, September 18, 1987. U.S. Government Printing Office, Washington, DC.

U.S. Congress. 1990. Waste Isolation Pilot Plant Land Withdrawal. Hearings Before the Committee on Energy and Natural Resources, U.S. Senate, 101st Congress, Second Session, April 3 and 26, 1990. U.S. Government Printing Office, Washington, DC.

Walker, I. 1962. *Geologic and Hydrogeologic Evaluation of a Proposed Site*. State Department of Health, Frankfort, KY.

Watson, R.A. 1969. Explanation and Prediction in Geology. *Journal of Geology* 77: 488-494.

Williams, R. 1980. *A Technique for the Geothermic Modeling of Underground Surfaces*. Sandia National Labs, Albuquerque, NM.

Younker, J.L., S.L. Albrecht, W.J. Arabasz, J.H. Bell, F.W. Cambray, S.W. Carothers, J.I. Drever, J.T. Einaudi, D.E. French, K.V. Hodges, R.H. Jones, D.K. Kreamer, W.G. Pariseau, T.A. Vogel, T. Webb, W.B. Andrews, G.A. Fasano, S.R. Mattson, R.C. Murray, L.B. Ballou, M.A. Revelli, A.R. Ducharme, L.E. Shephard, W.W. Dudley, D.T. Hoxie, R.J. Herbst, E.A. Patera, B.R. Judd, J.A. Docka, L.R. Rickertsen, J.M. Boak, and J.R. Stockey. 1992. Report of the Peer Review Panel on the Early Site Suitability Evaluation of the Potential Repository Site at Yucca Mountain, Nevada. SAIC-91/8001. U.S. Department of Energy, Washington, DC.

Chapter 11
SUMMARY OF THE SCIENTIFIC, ETHICAL, AND PUBLIC POLICY RECOMMENDATIONS FOR SUSTAINABLE DEVELOPMENT

John Lemons[1]
Donald A. Brown[2]

Developing public policy programs to foster sustainable development recommendations in Agenda 21 will require the application of disciplinary expertise from the areas of science, economics, law, and ethics. This means that concerned citizens and policymakers must examine and understand the capabilities of these disciplines as tools to assist in the implementation of the recommendations. In an attempt to promote this understanding, Chapters 2 through 5 of this book examine the general roles of science, ethics, economics, and law in sustainable development decisionmaking by identifying and assessing the capabilities, limitations, strengths, and weaknesses of each discipline's potential contribution to implementation of sustainable development recommendations. Chapters 6 through 10 explore the use and capabilities of the disciplines as applied to specific global sustainable development problems.

Agenda 21 recommends the development of four scientific program areas: (1)strengthening the scientific basis for sustainable management, (2)enhancing scientific understanding, (3)improving long-term scientific assessment, and (4)building scientific capacity and capability. Many other chapters in Agenda 21 contain language calling for an expanded role for science in resolving problems of sustainability and environmental protection, and in our opinion, they seem to rely heavily upon science to yield reasonably certain predictive information for decisionmakers in their implementation of sustainable development recommendations.

The recommendations and language contained in Chapter 35 of Agenda 21 regarding the use of science were one of the least contentious issues at the Earth Summit. Despite the lack of controversy surrounding these recommendations, as we have generally shown in Chapter 2 and more specifically in Chapters 6 through 10 dealing with problems of biodiversity, climate change, water resources, hazardous and toxic wastes, and nuclear waste, an overreliance on the capabilities of science to yield reasonably certain predictions concerning the environmental and human health consequences of many human activities in these respective areas is not warranted due to the existence of pervasive uncertainty. If decisionmakers utilize methods of cost-benefit analysis, then the uncertainties will result in discounting the worth or value of the environmental amenities that cannot be quantified with reasonable certainty. If decisionmakers wait until more certain information is acquired, then

[1]Department of Life Sciences, University of New England, Biddeford, ME 04005, U.S.A.; [2]Bureau of Hazardous Sites and Superfund Enforcement, Pennsylvania Department of Environmental Resources, Commonwealth of Pennsylvania, 400 Market St., Harrisburg, PA 17101-2301, U.S.A.

J. Lemons and D. A. Brown (eds.), Sustainable Development: Science, Ethics, and Public Policy, 275–278.

the status quo behaviors and public policies responsible for unsustainable practices will continue. Consequently, we conclude that the precautionary approach recommended in Agenda 21 should be adopted in national and international legislation and public policy. This means that the burden of proof should be shifted from those who attempt to demonstrate that there will be harm to environmental resources or human health from human activities to those who propose that developmental or technological activities will result in negligible harm. It also means that the role of science in public policy must be understood to be to identify useful indicators of environmental and human health change rather than to make precise predictions. In this way, science can serve more adequately in the formulation of sustainable development policy goals.

In addition, as demonstrated in Chapters 6 through 10, both the methodologies and tools of science, and how scientists deal with uncertainty are value-laden. This needs to be recognized by decisionmakers lest the results of scientific studies appear to be more quantifiable or value-free than is warranted and so that the often hidden ethical implications of the use of the value-laden methods and tools of science are disclosed and therefore subject to evaluation.

Agenda 21 contains many explicit as well as implicit ethical assumptions and implications, many of which are potentially conflicting. Examples of explicit ethical assumptions include a concern for future persons, the eradication of poverty, and recommendations that for the most part are based upon an anthropocentric approach implying that duties or obligations are owed to the environment or to nonhuman species primarily insofar as they might possess value to humans. Chapter 3 discusses the ethics of sustainable development in general terms, while Chapters 6 through 10 provide examples of ethical assumptions and implications used in specific resource problems, including the need to: (1)deal with scientific uncertainty, (2)assign the burden of proof in scientific reasoning, (3)decide what resources will be spent on problem analysis, (4)choose which disciplines will be used in analysis of problems and how to synthesize disparate disciplines in the analysis, (5)use cost-benefit analysis or other methods of valuation in decisionmaking, and (7)make metaphysical assumptions about the nature of reality.

Because most of the language of Agenda 21 is embedded in traditional discourses of Western science, economics, and law, most ethical arguments concerning the implementation of Agenda 21 recommendations are likely to be based upon Western ethical arguments and theories. Examples of such theories include: (1)utilitarianism, (2)theories of rights and duties, (3)theories of justice, and (4)anthropocentric versus biocentric ethics. Agenda 21 does not prescribe the use of any particular ethical argument or theory, although it contains a considerable amount of language highly indicative of an anthropocentric approach. It also contains language based upon utilitarianism (e.g., the use of cost-benefit analysis), while at other times it speaks of the rights of all present and future people to a decent life with at least basic necessities. Because the use of different ethical theories and arguments will lead to different sustainable development policies and consequences for different people and segments of the environment, it would appear that the international community must therefore decide which theories and arguments are most appropriate as a basis for decisionmaking. Such a decision will be difficult given the fact that philosophers have not yet reached consensus on what theories and arguments are most appropriate. Despite this difficulty, the authors of all of the chapters in this volume generally have criticized either or both a purely anthropocentric and utilitarian approach to specific sustainable development problems and have recommended that the intrinsic interests of nonhuman species as well as concepts of justice for present and future people be reflected in sustainable development policies.

One of the most difficult problems confronting decisionmakers will be whether to use traditional market mechanisms or, alternatively, some form of so-called ecological economics as a basis for sustainable development policymaking. In Chapter 4, the rationales for and against traditional economic approaches are discussed briefly. The rationale supporting the use of traditional economic approaches is based upon claims that they: (1)maximize efficiency, (2)promote liberty, (3)provide benefits to those most deserving of them, (4)result in the greatest mutual advantage to citizens, and (5)have been responsible for the unparalleled levels of prosperity in Western democracies. The criticisms against the use of traditional economic approaches are based upon claims that they: (1)fail to account for and mitigate market externalities; (2)treat all environmental entities as economic or instrumental commodities and therefore do not consider the intrinsic interests of nonhuman species; (3)fail to produce certain required or desirable public goods such as wilderness areas or habitats for other species; (4)are based upon utilitarian and cost-benefit justifications for public policy that ignore or do not account sufficiently for the interests of nonhumans, the rights of some present or future people, or problems of inequitable distribution of benefits and costs; and (5)fail to include the loss of natural capital that occurs with development in gross national product calculations. As discussed in Chapter 4, ecological economists have made a number of recommendations to overcome these criticisms in order to promote sustainable development goals. Chapters 5 and 6 discuss some of the implications of using various economic theories and methods in problems of biodiversity and climate change, respectively.

One of the most significant issues to arise from the resource problems examined in this volume is whether and to what extent law designed to foster sustainable development and environmental protection should reflect a precautionary approach; that is, given pervasive scientific uncertainty, should law and administrative discretion in decisionmaking err on the side of protecting the environment and human health in a sustainable manner? As we have seen, traditional scientific standards that minimize type I error and require results at, say, the 95 percent confidence level place a burden of proof upon those seeking to protect the environment and human health that is difficult to achieve. In addition, rules governing the admissibility of scientific evidence do not generally allow the admission of such evidence that establishes a reasonable basis for concern about harm but does not conclusively prove causation. Consequently, most law in the United States is not consistent with the precautionary principles articulated in Agenda 21. In addition, factors influencing administrative decisionmaking even under conditions where discretion exists work against the adoption of the precautionary principle. Importantly, it should be understood that although the question of where the burden of proof should be placed is an ethical question, it typically is decided by public policymakers and the legal community.

Finally, we conclude with a statement that our analysis of Agenda 21 recommendations suggests a need for a greater understanding and integration of the disciplines than has ever been achieved. All disciplines are relevant to the implementation of Agenda 21 recommendations, but the capabilities, limitations, strengths, weaknesses, and ethical implications of their methods and tools often are not understood by public policymakers and decisionmakers. For example, the value-laden aspects of the methods and tools of cost-benefit analysis and their practical and ethical consequences to problems of sustainable development and environmental protection need to be understood and taken into account in order to solve problems of biodiversity, climate change, protection of human health from toxic chemicals, and provision for the just distribution of water resources. If the value-laden methods and tools of science are not understood, then many aspects of the problem of dealing with nuclear waste will appear to be based upon sound science instead of undisclosed values, assumptions, and inferences of scientists. Unless the difficult task of integration is achieved, neither will the recommendations for sustainable development and environmental protection be reflected in law or public policy decisionmaking.

As the chapters in this book have pointed out, linkages among the disciplines exist for each resource problem we have assessed. In addition, linkages exist among the different types of sustainability problems, including those among: (1)one environmental problem and another, such as between biodiversity and climate change; (2)different human activities such as environmental protection and development generally; (3)developed and developing countries; (4)present and future generations; (5)protection of natural resources, other species, and basic human needs; (6)ecology and economics; and (7)social equity and economic efficiency. The success in solving any one type of problem may require that one or more of the other problems be solved simultaneously because of the nature of the linkages that exist among them.

INDEX

Environmental Science and Technology Library

1. A. Caetano, M.N. De Pinho, E. Drioli and H. Muntau (eds.), *Membrane Technology: Applications to Industrial Wastewater Treatment.* 1995 ISBN 0-7923-3209-1
2. Z. Zlatev: *Computer Treatment of Large Air Pollution Models.* 1995 ISBN 0-7923-3328-4
3. J. Lemons and D.A. Brown (eds.): *Sustainable Development: Science, Ethics, and Public Policy.* 1995 ISBN 0-7923-3500-7

KLUWER ACADEMIC PUBLISHERS – DORDRECHT / BOSTON / LONDON

10517623R0

Made in the USA
Lexington, KY
30 July 2011